Andreas Fichtner

Full Seismic Waveform Modelling and Inversion

With contributions by Florian Bleibinhaus
and Yann Capdeville

Springer

Andreas Fichtner
Utrecht University
Department of Earth Sciences
Budapestlaan 4
3584 CE Utrecht
The Netherlands
fichtner@geo.uu.nl

Advances in Geophysical and Environmental Mechanics and Mathematics
ISSN 1866-8348 e-ISSN 1866-8356
ISBN 978-3-642-15806-3 e-ISBN 978-3-642-15807-0
DOI 10.1007/978-3-642-15807-0
Springer Heidelberg Dordrecht London New York

Cover design: deblik, Berlin

Printed on acid-free paper

Springer is part of Springer Science+Business Media (www.springer.com)

To my family and Carolin

Foreword

Our planet is permanently vibrating, excited by oceans, atmosphere, earthquakes, or man-made sources. Luckily, Earth's physical properties are such that these vibrations – elastic waves to be more specific – often propagate to large distances carrying information on the medium they encounter along the way. The problem of making an educated guess at the subsurface structure from observations of ground motions is as old as instrumental seismology itself (so not that old, maybe a century or so). Let us call the problem *seismic tomography* akin to CT scanning in medicine, a field seismologists have always envied. Because of limitations in illuminating the Earth with sufficient coverage we have never obtained the sharp and detailed internal structures so familiar from medical imaging.

Up to now we have mostly cut corners in the way we model and fit our seismic data. We typically reduce long, wiggly seismograms to a few bytes of information (e.g. travel times, phase velocities) and try to explain these data with approximate theories. This has been for a good reason. Our computers were simply not fast and big enough to allow the calculation of complete wave fields through 3D Earth structures. Frequently the data just do not warrant the use of sophisticated physics.

The situation regarding computational power in connection with 3D wave propagation has dramatically changed in the past few years. Even on a global scale the calculation of wave fields across the complete observed frequency range is in sight. On smaller scales (continents, basins, volcanoes, reservoirs) we are already witnessing the emergence of 3D wave propagation as *the* tool for data modelling, inversion and parameter studies.

This was Albert Tarantola's (and others) dream 25 years ago: just simulate the physics correctly and let the data (i.e. the misfit to a theoretical seismogram) decide whether the Earth model is good or not. In his world (the probabilistic approach) this should be done using a Monte Carlo-type approach: calculate zillions of models and use all the results to estimate parameters and uncertainties. Unfortunately, we are not there yet. We still need to resort to linearisations around (hopefully good) starting models and employ adjoint-type techniques to update our Earth models. Fortunately, in many situations, good starting models can be found, making iterative waveform inversion the preferred tool to improve our Earth models using most or all of the observed data.

The book in your hands is the first to provide a broad overview on how to solve the forward problem to calculate accurate seismograms in 3D Earth models, to compare theory with observations and to iteratively update Earth models until a satisfactory fit to observations is achieved. Many questions remain, in particular in connection with properly assessing uncertainties. Yet, the tools described here will be essential in the coming years to move forward in reconstructing the structure of Earth's interior on all scales. In that sense they will have a strong impact on many related fields in Earth sciences such as geodynamics, earthquake physics, exploration geophysics, tectonics, volcanology. The book should be essential reading for everyone interested in getting the best out of seismic observations in terms of Earth's physical properties.

Munich, Germany
October 2010

Heiner Igel

Preface

Full waveform inversion is a novel variant of seismic tomography that is characterised by the numerical solution of the equations of motion, the exploitation of full waveform information and the iterative improvement of the tomographic images that accounts for non-linearity in the relation between model parameters and synthetic data. The numerical solutions ensure the accurate modelling of seismic wave propagation through realistically heterogeneous Earth models, thus making full waveform inversion the tomographic method of choice when the medium properties are highly variable. Thanks to the combination of numerical solutions and adjoint techniques, any type of waveform can be exploited for the benefit of improved tomographic resolution – without the need to identify particular waveforms in terms of classical seismic phases such as P or S.

The purpose of this book is to present the necessary ingredients for a full waveform inversion applied to real data. It is intended to serve as an accessible introduction to the topic for advanced students and professionals. The level is such that it could be used as a source of advanced teaching material for specialised seminars, courses and schools. Throughout the text we assume basic knowledge of elastic wave propagation in a seismological context, as it can be found, for instance, in the works of Båth (1979), Bullen & Bolt (1985), Lay & Wallace (1995), Udías (1999), Kennett (2001) or Aki & Richards (2002).

This book is organised in three parts that represent the major steps to be taken in order to solve a full waveform inversion problem.

Part I is dedicated to the numerical solution of the elastic wave equation. In the course of the past few decades many numerical methods for the discretisation of the wave equation have been developed – each being particularly well suited for specific applications. The choice to cover only finite-difference and spectral-element methods in detail has therefore been particularly difficult. In addition to being most frequently used in full waveform inversion, finite-difference and spectral-element methods also offer the opportunity to introduce many of the concepts that are fundamental in numerical wave propagation. Therefore, I hope that the content of Chaps. 2, 3, and 4 may serve as a useful basis for the understanding of numerical methods that I was not able to cover in a book that is focused on the solution of the inverse problem. Chapter 5 is concerned with the description and the numerical implementation of visco-elastic dissipation. Special attention is given to the concept of memory variables and the construction of Q models with a specified frequency dependence. Efficient absorbing boundaries are of outstanding importance when the computational domain does not comprise the whole Earth. Chapter 6 therefore describes absorbing boundary conditions and absorbing boundary layer approaches such as the Gaussian taper method and perfectly matched layers (PML).

The subject of Part II is the solution of the full waveform tomographic inverse problem. Chapter 7 reviews the foundations of non-linear optimisation, including the concept of local and global minima, uniqueness, convexity, regularisation and a selection of gradient-based minimisation algorithms. The adjoint method, treated in Chaps. 8, 9 and 10, is one of the methodological cornerstones of full waveform inversion because it allows us to efficiently compute the partial derivatives of seismic observables that are needed in gradient-based minimisation schemes. The general operator formulation of the adjoint method, its application to the elastic wave equation, the extension to second derivatives and the derivation of explicit expressions for Fréchet and Hessian kernels are covered with particular detail. Of outstanding importance in full waveform inversion is the choice of suitable misfit functionals that extract as much waveform information as possible while conforming to the restrictions imposed by the data and the physics of wave propagation. Chapter 11 summarises several misfit functionals that have proven effective in applications to real data. Regardless of any technological advances, physical intuition remains the most important ingredient in the solution of any inverse problem. In the case of full waveform inversion, this intuition mostly comes from the interpretation of Fréchet kernels, a collection of which is presented in Chap. 12.

While full waveform inversion is still a comparatively young method, there is already an important conclusion to be drawn: Its application is highly problem-dependent! A numerical method that is efficient for the modelling of body waves on a regional scale may be inefficient for the simulation of global surface wave propagation. Misfit functionals that are well suited to image sharp reflectors may not be able to recover the long-wavelength structure of the Earth. Owing to this extreme problem dependence, Part III presents a collection of case studies where full waveform inversion has been applied successfully. Chapter 13 is concerned with a continental-scale problem, giving special emphasis to the numerical modelling, data selection and misfit quantification. A local-scale problem is described in Chap. 14, written by Florian Bleibinhaus, who used the acoustic wave equation to image the 2D velocity structure near the San Andreas fault. A highly efficient data reduction scheme applied to a synthetic problem at the global scale is the topic of Chap. 15, written by Yann Capdeville.

Writing a book on full seismic waveform modelling and inversion turned out to be more challenging than I expected, because it requires to cover a large variety of different topics without creating a monster that hardly anyone would attempt to read from the first to the last page. I have chosen the content such that it hopefully allows the reader to develop and apply his or her own full waveform modelling and inversion methods. My writing process was marked by many tough decisions concerning interesting topics that I was not able to present at all or not in detail. For instance, the large variety of numerical methods for the solution of the wave equation certainly deserve at least one separate book. The same is true for the ample fields of nonlinear optimisation and absorbing boundary methods that I had to present in a rather condensed form.

Utrecht, The Netherlands *Andreas Fichtner*
August 2010

Acknowledgements

The present book is the result of the continued support by my family, friends and colleagues. It summarises part of the tremendous progress that has been made in full seismic waveform modelling and inversion thanks to the cooperation of several scientific communities, and I am grateful for the opportunity to communicate some of the acquired knowledge to both students and professionals.

The more theoretical chapters on solutions to the seismic forward and inverse problems are complemented by case studies, two of which have been contributed by Florian Bleibinhaus and Yann Capdeville. Their work highlights the potential of full waveform inversion and points towards future research directions.

This book would not exist without the help of Heidi Höfer who invited me to speak in the weekly geoscience seminar at Frankfurt University where she arranged my first meeting with Chris Bendall from Springer Germany. Thank you very much Heidi!

Several chapters are based on my Ph.D. thesis written at the University of Munich and under the supervision of Heiner Igel, Hans-Peter Bunge and Brian Kennett. My supervisors provided me with the ideal balance of freedom and guidance that allowed me to develop and follow my interests.

I am particularly grateful to my colleagues and friends who read different parts of the book, suggesting improvements and finding mistakes that I would never have found myself. I can hardly overstate the influence that Brian Kennett had on the development of this book by forming my way of doing science. He willingly shared his book writing experience and helped me to improve my less than perfect English. Moritz Bernauer went through every single equation with impressive patience, rigour and curiosity. Moritz reminded me to see the various chapters through the eyes of a student. Heiner Igel's profound understanding of numerical wave propagation helped me to shape the first part of this book which is on the solution of the forward problem. The influence of Peter Moczo is particularly pronounced in the chapter on finite-difference modelling which improved substantially thanks to his careful criticism. Dimitri Komatitsch kindly agreed to review the chapter on the spectral-element method and provided numerous helpful suggestions and comments. Furthermore, I am indebted to Christian Böhm, Thomas Bohlen, Josep de la Puente, Gaetano Festa, Martin Käser, Paul Käufl, Carène Larmat, Qinya Liu, Jean-Paul Montagner, Hanneke Paulssen, Daniel Peter, Kabir Roy-Chowdhury,

Tarje Nissen-Meyer, Malcolm Sambridge, Anne Sieminski, Karin Sigloch, Marco Stupazzini, Carl Tape, Benoît Tauzin, Hrvoje Tkalčić, Jeannot Trampert, Jeroen Tromp, Jean-Pierre Vilotte and many other members of the seismic modelling and inversion community who provided invaluable inspiration and support.

Moreover, I would like to thank my teachers Bernhard Forkmann, Wolfgang Sprößig and Kenneth Creager for initiating and supporting my interest in seismology and mathematics.

Much of the research described in this book could have never been accomplished without the support of the Bavarian elite network, the German Academic Exchange Service (DAAD), the German Science Foundation (DFG) and the vivid academic atmosphere created by the European SPICE network.

At Utrecht University I found the ideal environment for writing this book within less than 6 months, and I am truly grateful that Jeannot Trampert gave me the opportunity to spend my postdoc time in his group. I would like to thank my colleagues for countless discussions and for taking the time to answer all my questions.

The collaboration with Chris Bendall and Agatha Oelschlaeger from Springer Germany has always been a pleasure. Thank you for giving me the opportunity to write this book and for patiently answering my e-mails.

Finally, and most importantly, I would like to thank my family and Carolin for their love, understanding and support.

Contents

Chapter 1
Preliminaries

This first chapter is intended to serve as a preparation for the journey through this book. To see full waveform inversion in the context of previous developments in seismic tomography, we start with a brief historical overview that begins in the early days of modern seismology. This will be followed in Sect. 1.2 by a deliberately lax formulation of the full waveform tomographic inverse problem. Section 1.3 is concerned with the terminological confusion that appears to be common to all young branches of science.

1.1 A Brief Historical Overview

The development of full waveform inversion techniques has always been considered a crucial step towards a more detailed understanding of subsurface properties. To see the motivation for and the high expectations in full waveform inversion, we briefly review the history of our seismologically derived knowledge of the Earth's internal structure.

This history probably starts in 1760 when J. Michell (1724–1793) first associated earthquakes with waves that travel through the Earth's crust with a speed of at least 20 miles/min, that is roughly 0.5 km/s (Michell, 1760). Michell's observation that waves propagate through the Earth could be explained with the theory of elasticity that was developed in the eighteenth and nineteenth centuries. A. L. Cauchy (1789–1857), S. D. Poisson (1781–1840), G. G. Stokes (1819–1903) and many others studied the elastic wave equation, which still forms the basis of modern-day seismological applications. P and S waves travelling with different speeds were identified as possible analytical solutions in homogeneous and unbounded media. Solutions for arbitrarily heterogeneous media have not become available until recently, thanks to advances in computational science and numerical mathematics. Between 1852 and 1858, R. Mallet (1810–1881) and his son J. W. Mallet (1832–1912) performed what is likely to be the first active-source experiment by measuring the propagation speed of seismic waves using gun powder explosions. They linked wave speed variations to variations of material properties, thus solving what we would classify today as a seismic inverse problem. In 1889, E. L. A. von Rebeur-Paschwitz (1861–1895) accidentally recorded the first teleseismic earthquake in Potsdam

A. Fichtner, *Full Seismic Waveform Modelling and Inversion*, Advances in Geophysical and Environmental Mechanics and Mathematics,
DOI 10.1007/978-3-642-15807-0_1,

(von Rebeur-Paschwitz, 1889) using a horizontally swinging pendulum that was designed for astronomical measurements. The epicentre was near Tokyo. Teleseismically recorded elastic waves are today's principal source of information on deep Earth structure.

The theoretically predicted P and S waves were first clearly identified by R. D. Oldham (1858–1936) in 1900 (Oldham, 1900). Six years later he discovered the rapid decay of P wave amplitudes at epicentral distances greater than 100°. He correctly inferred the existence of the Earth's outer core (Oldham, 1906), the radius of which was determined by B. Gutenberg (1889–1960) in 1913 (Gutenberg, 1913). K. B. Zoeppritz (1881–1908) compiled travel time tables for waves observed at teleseismic distances (Zoeppritz, 1907), and he translated them into 1D models of the Earth's mantle. Much of his visionary work was published by his colleagues L. Geiger and B. Gutenberg after he died at the age of 26 (Zoeppritz & Geiger, 1909; Zoeppritz et al., 1912). In 1909 A. Mohorovičić (1857–1936) studied regional earthquakes, and he observed two types of P waves (today's Pn and Pg) and two types of S waves (today's Sn and Sg). He explained their travel time curves with a discontinuity at 54 km depth – the crust–mantle discontinuity that now bears his name (Mohorovičić, 1910). H. Jeffreys (1891–1989) combined results from seismology and studies of Earth tides to conclude that 'there seems to be no reason to deny that the Earth's metallic core is truly fluid' (Jeffreys, 1926). Also in 1926, B. Gutenberg provided the first seismological evidence for a low-velocity zone around 100 km depth – the asthenosphere (Gutenberg, 1926). In 1936, I. Lehmann (1888–1993) observed unidentified P waves at large epicentral distances, today's PKIKP, and inferred the existence of the inner core (Lehmann, 1936). Another milestone in the discovery of the Earth's spherical structure was H. Jeffreys' and K. Bullen's (1906–1976) compilation of travel time tables that were used to infer complete radially symmetric Earth models (Jeffreys & Bullen, 1940).

The second half of the twentieth century was marked by both the refinement of spherical Earth models (Dziewonski et al., 1975; Dziewonski & Anderson, 1981; Kennett et al., 1995) and the discovery of 3D heterogeneities through systematic tomographic approaches. Aki et al. (1977) were among the first to use teleseismic data in a linearised tomography for regional 3D structure. The analysis of nearly 700,000 P wave arrival time residuals allowed Dziewonski et al. (1977) to derive an early model of large-scale heterogeneities in the deep mantle. Tomographic methods were also used to determine 3D variations of seismic anisotropy (e.g. Montagner, 1985) and visco-elastic dissipation (e.g. Romanowicz, 1990). Increased data quality and data coverage contributed to the continuously improving resolution of tomographic images that could then be linked to mantle convection (e.g. van der Hilst et al., 1997; Ritsema & van Heijst, 2000), lithospheric deformation (e.g. Debayle & Kennett, 2000; Montagner, 2002), chemical heterogeneities (e.g. Jordan, 1978; van der Hilst & Kárason, 1999) and the tectonic evolution of continents (e.g. Zielhuis & Nolet, 1994; Zielhuis & van der Hilst, 1996).

Most seismological inferences concerning the structure of the Earth – including the existence of the inner core, the asthenosphere and the major seismic discontinuities – are based on the simplifying assumption that seismic waves can be represented by rays. This concept is closely related to geometrical optics. Within the ray-theoretical framework, the arrival times of seismic waves are sensitive to seismic wave speeds only along a curve connecting source and receiver. The intensive use of ray theory (e.g. Červený, 2001) stems from its simplicity, its intuitive interpretation and from the difficulty of finding solutions of the complete elastic wave equation for realistically heterogeneous Earth models.

The limitations of ray theory in the context of seismic tomography have become a major concern during the past two decades (e.g. Williamson, 1991; Williamson & Worthington, 1993; Spetzler et al., 2001). Ray theory is valid when the length scales of 3D heterogeneities are small compared to the dominant wavelength. This condition imposes an upper limit on the resolution of tomographic images derived from ray theory. Efforts to overcome the limitations of ray theory – and thus to further improve the resolution of tomographic images – include finite-frequency tomography (e.g. Yomogida, 1992; Friederich, 1999, 2003; Dahlen et al., 2000; Yoshizawa & Kennett, 2004, 2005; Zhou et al., 2005; Sigloch et al., 2008) and full waveform inversion (e.g. Bamberger et al., 1982; Tarantola, 1988; Ikelle et al., 1988; Pica et al., 1990; Igel et al., 1996; Pratt, 1999; Djikpéssé & Tarantola, 1999; Dessa et al., 2004; Ravaut et al., 2004; Bleibinhaus et al., 2007; Choi et al., 2008; Fichtner et al., 2009, 2010; Tape et al., 2009, 2010). Finite-frequency tomography is a modification of the ray-theoretical seismic inverse problem that correctly accounts for the spatially extended 3D sensitivity distribution of waves with a finite-frequency content. Full waveform inversion goes one step further than finite-frequency tomography by replacing semi-analytical solutions to the wave equation (e.g. ray-theoretical solutions or mode summations) by fully numerical solutions. This ensures that the propagation of seismic waves through heterogeneous media is modelled accurately. Furthermore, numerical solutions provide complete seismic waveforms that can be exploited for the benefit of improved tomographic resolution, without the need to identify specific seismic phases. For an extensive review of full waveform inversion, the reader is referred to Virieux & Operto (2009).

1.2 The Full Waveform Tomographic Inverse Problem – Probabilistic vs. Deterministic

Full waveform inversion is one of the latest contributions to our quest for information on the structure of the Earth. Yet, regardless of any technological advances, the amount of independent information effectively contained in seismic data is naturally limited, thus leading to the inherent non-uniqueness of tomographic models.

The straightforward recognition of non-uniqueness has led to the universally accepted opinion that solving an inverse problem also means to appraise the variety of solutions that are compatible with the data. This goal can be achieved most elegantly within the framework of *probabilistic inverse theory* (e.g. Keilis-Borok & Yanovskaya, 1967; Press, 1968; Tarantola, 2005), where each Earth model $\mathbf{m}$ is assigned a probability $p(\mathbf{m})$ of representing the true Earth. The probability depends on two factors: First, the degree to which $\mathbf{m}$ is consistent with our prior knowledge of the Earth's structure and second, the misfit between the observed data $\mathbf{u}^0$ and the synthetic data $\mathbf{u}(\mathbf{m})$. The probability distribution $p(\mathbf{m})$ is considered *the* solution to the inverse problem.

For nearly all realistic problems, $p(\mathbf{m})$ is not known explicitly. It must therefore be approximated by *sampling*, that is by evaluating p for specific models $\mathbf{m}$ called *samples* (e.g. Sambridge & Mosegaard, 2002). Each evaluation requires the solution of the wave equation, and the number of samples needed to approximate p sufficiently well grows roughly exponentially with the number of model parameters. Thus, despite its unquestionable conceptual beauty, probabilistic inverse theory is not well suited for full waveform inversion where the number of model parameters is on the order of several thousand to several million.

The pragmatic, though less complete, alternative is *deterministic inverse theory*. This is based on the definition of a misfit functional $\chi(\mathbf{m})$ that quantifies the discrepancy between the observed data and the synthetic data corresponding to the model $\mathbf{m}$. Solving the inverse problem in the deterministic framework then means finding an optimal model $\tilde{\mathbf{m}}$ in the sense that $\chi(\tilde{\mathbf{m}})$ is minimal. Methods for non-linear minimisation, treated in Chap. 7, can be used to approximate $\tilde{\mathbf{m}}$ iteratively. The iteration starts from an initial model $\mathbf{m}_0$ that is sufficiently close to the minimum to ensure convergence.

The possibility to constrain a number of model parameters that is much larger than in probabilistic inverse problems comes at the expense of having only limited information on non-uniqueness. Synthetic inversions in the form of checkerboard or spike tests are therefore often used as proxies for the trustworthiness of structural features seen in tomographic images. This is despite well-known deficiencies of synthetic inversions that may not always provide meaningful quantifications of non-uniqueness (Lévêque et al., 1993). Estimating uncertainties and trade-offs in the deterministic framework mostly requires plausible reasoning that is to some degree subjective.

In this book we adopt the deterministic point of view, meaning that we are trying to construct *one* model that explains the data optimally while being physically reasonable. The emphasis is therefore on the reliable estimation of as many model parameters as possible, and less importance is given to the quantification of their uncertainties.

1.3 Terminology: Full Language Confusion

Full waveform inversion is still in its development stage, and just as many other young branches of science it is subject to language confusion. What does 'full' refer to, and what precisely is a 'waveform' ?

The term 'full waveform inversion' was probably coined by Pan et al. (1988) who used complete plane wave seismograms to retrieve the structure of a stratified medium. Earlier articles classify the same approach simply as waveform inversion (e.g. Tarantola, 1984; Gauthier et al., 1986), not giving special emphasis to the potential use of the entire seismograms – from the onset of the P wave to the final recognisable oscillation in the surface wave train. In a very original work, Bamberger et al. (1977) did not explicitly label their new method that would today certainly be named 'full waveform inversion'.

It is in practice rarely efficient to really exploit complete waveform information in the sense that zero misfit implies data=synthetics. Surface waves are commonly ignored in exploration applications, and amplitudes are often disregarded in regional- to global-scale studies. The classification 'full waveform inversion' is therefore rarely fully justified.

'Full waveform tomography' is nearly synonymous with 'full waveform inversion'. It stresses an explicit transmission tomography component in the solution of the inverse problem that was not present in the original full waveform inversions based on the minimisation of the L_2 distance between observed and synthetic waveforms (e.g. Gauthier et al., 1986; Igel et al., 1996). Brenders & Pratt (2007), for instance, used a classical travel time transmission tomography to construct an initial model that was sufficiently close to the optimum to ensure the convergence of the subsequent full waveform inversion. Fichtner et al. (2008, 2009) and van Leeuwen & Mulder (2010) designed measures of full waveform misfit that are particularly well suited to infer the transmission properties of the Earth.

Chen et al. (2007) introduced 'full 3D tomography' where the initial model, the model updates, the sensitivity kernels and the finite-difference wave field modelling are 3D. The adjective 'full' mostly refers to the inversion procedure and to a lesser extent to the exploitation of as much waveform information as possible.

Emphasising the importance of the adjoint method in tomographic inversions based on numerical wave propagation, Tape et al. (2009, 2010) labelled their approach 'adjoint tomography'. This terminology is honest in the sense that it does not suggest the practically almost impossible use of complete seismograms.

At this stage of the development, terminology still seems to be a matter of personal preference. Similar methods are named differently, depending on the subjective importance of one or the other aspect.

In this book we employ the term 'full waveform inversion'. This is intended to indicate the exploitation of as much information *as is physically reasonable*. In our understanding, 'full waveform inversion' does not necessarily mean the exploitation of *all available* information, regardless of the constraints imposed by the data and the physics of the problem. 'Full waveform tomography' is interpreted as a variant of full waveform inversion where a transmission tomography component appears explicitly.

Part I
Numerical Solution of the Elastic Wave Equation

The numerical solution of the equations of motion is one of the defining characteristics of full waveform inversion and an indispensable tool in seismic ground motion studies. Numerical solutions allow us to accurately model the propagation of seismic waves through strongly heterogeneous media, including, for instance, the Earth's lithosphere on a global scale, sedimentary basins and volcanic edifices.

Seismic wave motion in the Earth is governed by the wave equation that relates the displacement field to external forces and to the distributions of density and elastic parameters. Since there are no exact analytical solutions to the wave equation in realistically heterogeneous media, we construct approximate solutions by discretising derivatives. The spatial discretisation leads to a system of ordinary differential equations in time that can be solved numerically either in the frequency domain or by iterative time stepping (Chap. 2).

Several methods have been developed for the numerical solution of the wave equation, each being particularly well suited for specific types of applications. In the context of full waveform inversion, the finite-difference and the spectral-element methods are most frequently used.

In the finite-difference method (Chap. 3), the spatial derivatives are approximated by difference quotients. The approximation leads, as in any other numerical method, to a dispersion error that depends, among other factors, on the number of grid points per wavelength. The feasibility of finite-difference modelling in 3D rests on the definition of a staggered grid where the field variables are evaluated at different grid positions. While increasing the computational efficiency, the staggered grid requires special care in the implementation of anisotropy and the free surface.

The spectral-element method (Chap. 4) is based on the weak or variational formulation of the equations of motion. The computational domain is divided into disjoint elements that are mapped onto a reference element. Within the reference element, the dynamic fields are expressed in terms of Lagrange polynomials, collocated at the Gauss–Lobatto–Legendre points. The integrals of the weak form are then approximated by Gauss–Lobatto–Legendre quadrature, which leads to an algebro-differential equation with a diagonal mass matrix.

The dissipation of seismic waves through the conversion of elastic energy into heat is implemented by a convolution of the time-dependent elastic tensor and the strain rate tensor (Chap. 5). The elastic tensor components are expressed as a superposition of standard linear solids. These are controlled by a set of relaxation parameters that can be chosen such that the quality factor Q follows a pre-defined frequency dependence. The time convolution of the equations of motion can be eliminated with the help of memory variables.

To reduce numerical costs, the computational domain is often restricted to only part of the Earth. This introduces artificial reflections that pollute the solution. Absorbing boundaries (Chap. 6) prevent the occurrence of artificial reflections either by imposing special boundary conditions or by modifying the original wave equation.

Chapter 2
Introduction

In this chapter we present the basic elements for the numerical modelling of seismic wave propagation. Following a summary of notational conventions, we introduce the elastic wave equation in its different formulations (Sect. 2.2). The acoustic wave equation is treated as a special case in Sect. 2.3. While numerical methods differ in the details of the spatio-temporal discretisation, they can still be treated within a unifying framework. The approximation of the spatial derivatives generally leads to a system of ordinary differential equations in time that is commonly referred to as the semi-discrete form of the wave equation. The semi-discrete form can be written in terms of mass and stiffness matrices (Sect. 2.4). Depending on the specifics of an application, the remaining time derivatives can then be approximated using either the Fourier transform or time-stepping algorithms such as the Newmark or leapfrog methods (Sect. 2.5).

2.1 Notational Conventions

Throughout this book we have tried to stay close to notations commonly found in the seismological literature. While the meaning of the different symbols is mostly clear from the context, we start with a small collection of conventions that we shall use consistently.

The Fourier transform of a function f is defined as

$$\tilde{f}(\omega) := F[f](\omega) := \frac{1}{\sqrt{2\pi}} \int_{\mathbb{R}} f(t) \mathrm{e}^{-\mathbf{i}\omega t}\, \mathrm{d}t\,, \quad \mathbf{i} := \sqrt{-1}\,, \tag{2.1}$$

where the symbol := means that the expression to the left is defined by the expression to the right. The boldface $\mathbf{i}$ is intended to distinguish the imaginary unit from the frequently occurring index variable i. The inverse Fourier transform corresponding to (2.1) is

$$f(t) = F^{-1}[\tilde{f}](t) = \frac{1}{\sqrt{2\pi}} \int_{\mathbb{R}} \tilde{f}(\omega) \mathrm{e}^{\mathbf{i}\omega t}\, \mathrm{d}\omega\,. \tag{2.2}$$

A. Fichtner, *Full Seismic Waveform Modelling and Inversion*, Advances in Geophysical and Environmental Mechanics and Mathematics,
DOI 10.1007/978-3-642-15807-0_2,

With the exception of **i**, we use bold-faced symbols for vectors and tensors. The scalar product of two vectors $\mathbf{a}, \mathbf{b} \in \mathbb{R}^n$ is denoted by

$$\mathbf{a} \cdot \mathbf{b} = \sum_{i=1}^{n} a_i b_i \,. \tag{2.3}$$

We more generally use the dot to signify contraction over adjacent indices. Using this notation, the i-component of a matrix–vector product is written as

$$(\mathbf{A} \cdot \mathbf{a})_i = \sum_{i=1}^{n} A_{ij} a_j \,, \quad \mathbf{A} \in \mathbb{R}^{n \times n} \,, \tag{2.4}$$

and the ij-component of a matrix–matrix product is

$$(\mathbf{A} \cdot \mathbf{B})_{ij} = \sum_{k=1}^{n} A_{ik} B_{kj} \,. \tag{2.5}$$

Following this scheme, a double dot denotes a contraction over two adjacent indices, for instance

$$\mathbf{A} : \mathbf{B} = \sum_{i,j=1}^{n} A_{ij} B_{ij} \,, \quad \mathbf{B} \in \mathbb{R}^{n \times n} \tag{2.6}$$

and

$$(\mathbf{B} : \mathbf{C})_{kl} = \sum_{i,j=1}^{n} B_{ij} C_{ijkl} \,, \quad \mathbf{C} \in \mathbb{R}^{n \times n \times n \times n} \,. \tag{2.7}$$

For the real part of a complex-valued variable $z = x + \mathbf{i} y$, we use the fraktur symbol $\mathfrak{Re}$, i.e.

$$\mathfrak{Re}\, z = x = \frac{1}{2}(z + z^*) \,, \tag{2.8}$$

where $z^* = x - \mathbf{i}\, y$ denotes the complex conjugate of z. The symbol $\mathfrak{Im}$ denotes the imaginary part of z:

$$\mathfrak{Im}\, z = y = \frac{1}{2}(z - z^*) \,. \tag{2.9}$$

Of outstanding importance in any deterministic inverse problem is the definition of a misfit functional χ that quantifies the difference between observed and synthetic data. The misfit functional depends on an Earth model $\mathbf{m} \in \mathfrak{M}$, where $\mathfrak{M}$ is the model space. The functional or the Fréchet derivative of χ with respect to $\mathbf{m}$ in a direction $\delta\mathbf{m} \in \mathfrak{M}$ is defined by

$$\nabla_m \chi(\mathbf{m})\, \delta\mathbf{m} := \lim_{\varepsilon \to 0} \frac{1}{\varepsilon} \big[\chi(\mathbf{m} + \varepsilon\, \delta\mathbf{m}) - \chi(\mathbf{m})\big]. \tag{2.10}$$

The derivative $\nabla_m \chi(\mathbf{m})$ is a linear operator acting on the differentiation direction $\delta\mathbf{m}$. In the special case where $\mathbf{m}$ is a vector, the Fréchet derivative $\nabla_m \chi(\mathbf{m})\, \delta\mathbf{m}$ coincides with the directional derivative $\delta\mathbf{m} \cdot \nabla_m \chi(\mathbf{m})$. For convenience, we will mostly use the term 'derivative' instead of 'Fréchet derivative'. The symbol ∇, without subscript, signifies the regular gradient with respect to the position vector $\mathbf{x}$.

2.2 The Elastic Wave Equation

2.2.1 Governing Equations

Full waveform inversion is founded on the solution of the *forward problem*, which consists in the simulation of seismic wave propagation through an Earth model $\mathbf{m}$ and the computation of synthetic seismograms. The propagation of seismic waves in the Earth can be modelled with the *elastic wave equation*

$$\rho(\mathbf{x})\ddot{\mathbf{u}}(\mathbf{x}, t) - \nabla \cdot \boldsymbol{\sigma}(\mathbf{x}, t) = \mathbf{f}(\mathbf{x}, t), \quad \mathbf{x} \in G \subset \mathbb{R}^3, \quad t \in [t_0, t_1] \subset \mathbb{R} \tag{2.11}$$

that relates the displacement field $\mathbf{u}$ in the Earth $G \subset \mathbb{R}^3$ to its mass density ρ, the stress tensor $\boldsymbol{\sigma}$ and an external force density $\mathbf{f}$. A truly marvellous matter of fact! Equation (2.11) is the linearised version of Newton's second law that balances the momentum of particle displacement $\rho(\mathbf{x})\ddot{\mathbf{u}}(\mathbf{x}, t)$, forces resulting from internal stresses $\nabla \cdot \boldsymbol{\sigma}(\mathbf{x}, t)$ and external forces $\mathbf{f}(\mathbf{x}, t)$ that represent the sources of seismic wave motion. For detailed derivations of Eq. (2.11) the reader is referred to Dahlen & Tromp (1998), Kennett (2001) or Aki & Richards (2002). At the surface ∂G of the Earth, the normal components of the stress tensor $\boldsymbol{\sigma}$ vanish, i.e.

$$\boldsymbol{\sigma} \cdot \mathbf{n}|_{\mathbf{x} \in \partial G} = \mathbf{0}, \tag{2.12}$$

where $\mathbf{n}$ is the unit normal on ∂G. Equation (2.12) is the *free surface boundary condition*. Both the displacement field $\mathbf{u}$ and the velocity field $\mathbf{v} = \dot{\mathbf{u}}$ are required to satisfy the *initial condition* of being equal to zero prior to $t = t_0$ when the external force $\mathbf{f}$ starts to act:

$$\mathbf{u}|_{t \le t_0} = \mathbf{v}|_{t \le t_0} = \mathbf{0}. \tag{2.13}$$

For convenience we will mostly choose $t_0 = 0$. To obtain a complete set of equations, the stress tensor $\boldsymbol{\sigma}$ must be related to the displacement field $\mathbf{u}$. For this we assume that $\boldsymbol{\sigma}$ depends linearly on the history of the strain tensor $\boldsymbol{\varepsilon} := \frac{1}{2}(\nabla\mathbf{u} + \nabla\mathbf{u}^T)$:

$$\boldsymbol{\sigma}(\mathbf{x}, t) = \int_{-\infty}^{\infty} \dot{\mathbf{C}}(\mathbf{x}, t - t') : \boldsymbol{\varepsilon}(\mathbf{x}, t')\, dt' . \tag{2.14}$$

Equation (2.14) defines a *linear visco-elastic rheology*. The 4th-order tensor $\mathbf{C}$ is the *elastic tensor*. Since the current stress cannot depend on future strain, we require the elastic tensor to be causal:

$$\mathbf{C}(t)|_{t<t_0} = \mathbf{0} . \tag{2.15}$$

The symmetry of $\boldsymbol{\varepsilon}$, the conservation of angular momentum and the relation of $\mathbf{C}$ to the internal energy (e.g. Aki & Richards, 2002) require that the components of $\mathbf{C}$ satisfy the following symmetry relations:

$$C_{ijkl} = C_{klij} = C_{jikl} . \tag{2.16}$$

The symmetries of the elastic tensor reduce the number of its independent components to 21, and they allow us to write (2.14) directly in terms of the displacement gradient $\nabla \mathbf{u}$:

$$\boldsymbol{\sigma}(\mathbf{x}, t) = \int_{-\infty}^{\infty} \dot{\mathbf{C}}(\mathbf{x}, t - t') : \nabla \mathbf{u}(\mathbf{x}, t')\, dt' . \tag{2.17}$$

The number of non-zero independent elastic tensor components – also referred to as elastic parameters or elastic moduli – determines the *anisotropic properties* of the medium. For instance, a triclinic crystal such as plagioclase requires all 21 independent parameters for its complete description and 3 elastic parameters are needed to describe crystals with cubic symmetry such as garnet (Babuska & Cara, 1991). On a macroscopic scale the Earth can often be described sufficiently well as an isotropic body where the elastic tensor components are linear combinations of only two elastic moduli: the *Lamé parameters* λ and μ:

$$C_{ijkl} = \lambda\, \delta_{ij}\delta_{kl} + \mu\, \delta_{ik}\delta_{jl} + \mu\, \delta_{il}\delta_{jk} . \tag{2.18}$$

The parameter μ relates strain to shear stresses and is therefore called *shear modulus*. Since λ has no intuitive physical meaning, it is commonly replaced by the *bulk modulus* $\kappa = \lambda + \frac{2}{3}$ that relates strain to the scalar pressure, defined as $p =: -\kappa\, \nabla \cdot \mathbf{u}$.

The time dependence of the elastic tensor is responsible for visco-elastic dissipation, that is the process of transforming elastic energy into heat. In the case of a non-dissipative medium, the time dependence of $\mathbf{C}$ takes the form of a unit step or Heaviside function $H(t)$:

$$\mathbf{C}(\mathbf{x}, t) = \mathbf{C}(\mathbf{x})\, H(t) . \tag{2.19}$$

The constitutive relation (2.17) then takes the form

$$\boldsymbol{\sigma}(\mathbf{x}, t) = \mathbf{C}(\mathbf{x}) : \nabla \mathbf{u}(\mathbf{x}, t) \,. \tag{2.20}$$

For simplicity we will assume a non-dissipative medium throughout most of this book. This is also justified because dissipation can be incorporated easily with the help of memory variables, as we shall see in Chap. 5. Memory variables allow us to circumvent the numerically inconvenient convolution in Eq. (2.17).

2.2.2 Formulations of the Elastic Wave Equation

Equations (2.11) and (2.20) constitute the *displacement–stress formulation* of the elastic wave equation in the absence of dissipation. Together with the initial and boundary conditions they uniquely specify the displacement field $\mathbf{u}(\mathbf{x}, t)$.
Different but fully equivalent formulations are possible and sometimes required by a specific numerical method. We may, for instance, eliminate the stress tensor $\boldsymbol{\sigma}$ by combining Eqs.(2.11) and (2.20). This results in the *displacement formulation* of the elastic wave equation:

$$\rho(\mathbf{x})\, \ddot{\mathbf{u}}(\mathbf{x}, t) - \nabla \cdot [\mathbf{C}(\mathbf{x}) : \nabla \mathbf{u}(\mathbf{x}, t)] = \mathbf{f}(\mathbf{x}, t) \,. \tag{2.21}$$

Of particular relevance in numerical modelling is the *velocity–stress formulation* where the wave equation is written as a first-order system in both time and space. We find the velocity–stress formulation by simply differentiating Eq. (2.20) with respect to time and then substituting $\mathbf{v}$ for $\dot{\mathbf{u}}$:

$$\rho(\mathbf{x})\, \dot{\mathbf{v}}(\mathbf{x}, t) - \nabla \cdot \boldsymbol{\sigma}(\mathbf{x}, t) = \mathbf{f}(\mathbf{x}, t) \,, \tag{2.22a}$$

$$\dot{\boldsymbol{\sigma}}(\mathbf{x}, t) - \mathbf{C}(\mathbf{x}) : \nabla \mathbf{v}(\mathbf{x}, t) = \mathbf{0} \,. \tag{2.22b}$$

The elastic wave equation in its different formulations has been studied extensively. Analytical solutions exist for numerous classes of models including, for instance, the isotropic and homogeneous half space (e.g. Lamb, 1904; de Hoop, 1958; Aki & Richards, 2002), stratified media (e.g. Kennett, 1979, 1980, 1981) and spherically symmetric globes with radial anisotropy (e.g. Takeuchi & Saito, 1978; Woodhouse, 1988; Friederich & Dalkolmo, 1995). Ray theory (e.g. Červený, 2001) and perturbation methods (e.g. Woodhouse & Dahlen, 1978; Maupin, 2001) can be used to approximate solutions to the elastic wave equation for mildly heterogeneous media. In full waveform inversion, the focus is on strongly heterogeneous regions of the Earth where analytical or perturbation methods are not applicable. This is the domain of numerical methods that are the subject of the first part of this book.

2.3 The Acoustic Wave Equation

The equations of motion simplify significantly in the fluid regions of the Earth (e.g. the outer core, oceans, and melt bodies), where the shear modulus μ is effectively zero. Inserting $\mu = 0$ into the isotropic constitutive relation (2.18), yields

$$\sigma_{ij} = \kappa\, \delta_{ij}\, \nabla \cdot \mathbf{u} = -p\, \delta_{ij}\,, \tag{2.23}$$

where we introduced the scalar pressure $p := -\kappa\, \nabla \cdot \mathbf{u}$. With the help of Eq. (2.23) the momentum balance law (2.11) reduces to

$$\rho\, \ddot{\mathbf{u}} + \nabla\, p = \mathbf{f}. \tag{2.24}$$

Dividing (2.24) by the density ρ and taking the divergence gives

$$\nabla \cdot \ddot{\mathbf{u}} + \nabla \cdot (\rho^{-1} \nabla\, p) = \nabla \cdot (\rho^{-1}\, \mathbf{f})\,. \tag{2.25}$$

Using the definition of the pressure p, we can eliminate the displacement field $\mathbf{u}$ from Eq. (2.25):

$$\kappa^{-1}\, \ddot{p} - \nabla \cdot (\rho^{-1}\, \nabla\, p) = -\nabla \cdot (\rho^{-1}\, \mathbf{f})\,. \tag{2.26}$$

When density varies much more slowly than the pressure field p and the source $\mathbf{f}$, we can simplify (2.26) to a scalar partial differential equation, known as the *acoustic wave equation*:

$$\ddot{p} - v_a^2\, \Delta p = -v_a^2\, \nabla \cdot \mathbf{f}\,, \tag{2.27}$$

with the *acoustic wave speed* $v_a := \sqrt{\frac{\kappa}{\rho}}$. It follows from (2.27) that wave motion in fluid media can be fully described with a single scalar field (the pressure, p), the properties of which depend only on the source term and the spatial distribution of the acoustic wave speed.

While being strictly valid only in fluid and gaseous media, the acoustic wave equation is frequently used in active-source full waveform inversion (see for example Chap. 14), because its numerical solution is computationally inexpensive compared to the solution of the elastic wave equation. The consequences of this *acoustic approximation* include the restriction to isotropic source radiation patterns and the absence of Rayleigh waves and P-to-S conversions. The acoustic approximation is, nevertheless, justifiable when the data analysis is restricted to the first-arriving P waves and when the seismic sources radiate little S wave energy (e.g. explosions).

With the exception of Chap. 14, the focus of this book will be on elastic wave propagation. The transition to the acoustic case is mostly straightforward.

2.4 Discretisation in Space

Analytical solutions to the elastic wave equation exist only for comparatively simple models that often do not reflect the structural complexities of the Earth. This deficiency motivates the development of numerical methods for the simulation of seismic wave propagation through almost arbitrarily heterogeneous Earth models. While being different in the technical details, all numerical methods have one point in common: the discrete spatial approximation of the continuously defined wave field $\mathbf{u}(\mathbf{x}, t)$. This means that $\mathbf{u}(\mathbf{x}, t)$ is approximated by a finite number of time-dependent coefficients $\bar{u}_1(t), \ldots, \bar{u}_N(t)$ that can be summarised in an N-dimensional vector $\bar{\mathbf{u}}(t)$. Depending on the specifics of the numerical method used, the coefficients $\bar{u}_i(t)$ $(i = 1, \ldots, N)$ may represent, for instance, discrete values of $\mathbf{u}(\mathbf{x}, t)$ sampled at a finite number of points, or polynomial coefficients when $\mathbf{u}(\mathbf{x}, t)$ is approximated by a polynomial.

Following spatial discretisation, the displacement formulation of the elastic wave equation, given in (2.21), turns into an algebro-differential equation that can always be written in the following canonical form:

$$\mathbf{M} \cdot \ddot{\bar{\mathbf{u}}}(t) + \mathbf{K} \cdot \bar{\mathbf{u}}(t) = \bar{\mathbf{f}}(t)\,. \tag{2.28}$$

The matrices $\mathbf{M}$ and $\mathbf{K}$ are referred to as the *mass matrix* and the *stiffness matrix*, respectively. The vector $\bar{\mathbf{f}}$ represents a discrete version of the force density $\mathbf{f}$. Both $\mathbf{M}$ and $\mathbf{K}$ tend to be sparse. In practice, the mass and the stiffness matrices are rarely computed explicitly because only the vector–matrix products are needed in actual computations. They are, nevertheless, useful tools in theoretical developments, as we will soon discover.

Depending on the numerical method, it may be more advantageous to discretise the displacement–stress formulation Eqs. (2.11) and (2.20) or the velocity-stress formulation (Eq. 2.22) of the elastic wave equation. The corresponding space-discrete systems are then

$$\mathbf{M} \cdot \ddot{\bar{\mathbf{u}}}(t) + \mathbf{K}_1 \cdot \bar{\mathbf{s}}(t) = \bar{\mathbf{f}}(t)\,, \tag{2.29a}$$

$$\bar{\mathbf{s}}(t) = \mathbf{K}_2 \cdot \bar{\mathbf{u}}(t) \tag{2.29b}$$

for the displacement–stress formulation and

$$\mathbf{M} \cdot \dot{\bar{\mathbf{v}}}(t) + \mathbf{K}_1 \cdot \bar{\mathbf{s}}(t) = \bar{\mathbf{f}}(t)\,, \tag{2.30a}$$

$$\dot{\bar{\mathbf{s}}}(t) - \mathbf{K}_2 \cdot \bar{\mathbf{v}}(t) = \mathbf{0} \tag{2.30b}$$

for the velocity–stress formulation. The matrices $\mathbf{K}_1$ and $\mathbf{K}_2$ are stiffness matrices in a broader sense and $\bar{\mathbf{s}}$ represents a discrete approximation to the stress tensor $\boldsymbol{\sigma}$. The transition from the continuous wave equation in its various formulations to one of the space-discrete systems (2.28), (2.29) or (2.30) reduces the forward problem to the solution of a large algebraic system and an ordinary differential equation in time.

2.5 Discretisation in Time or Frequency

The ordinary differential equations in time that arise from the spatial discretisation of the equations of motion can be solved either in the time domain or in the frequency domain. Both approaches have advantages and disadvantages that must be weighted depending on the particular application and the available computational resources.

2.5.1 *Time-Domain Modelling*

The time-domain modelling of wave propagation is mostly based on the replacement of the time derivatives in the space-discrete equations of motion by suitable finite-difference approximations. These allow us to advance the wave field in discrete time steps Δt. The choice of a particular finite-difference scheme depends on the formulation of the wave equation.

Throughout the following paragraphs we assume that the mass matrix $\mathbf{M}$ can be inverted, noting, however, that the inversion of $\mathbf{M}$ may present a formidable numerical challenge.

2.5.1.1 Displacement and Displacement–Stress Formulation

The space-discrete version of the displacement formulation (2.28) involves the second time derivative of the discrete displacement field $\ddot{\bar{\mathbf{u}}}$, which is explicitly given by

$$\ddot{\bar{\mathbf{u}}}(t) = \mathbf{M}^{-1} \cdot \left[\bar{\mathbf{f}}(t) - \mathbf{K} \cdot \bar{\mathbf{u}}(t)\right]. \tag{2.31}$$

Approximating $\ddot{\bar{\mathbf{u}}}(t)$ by the second-order finite difference

$$\ddot{\bar{\mathbf{u}}}(t) \approx \frac{1}{\Delta t^2} \left[\bar{\mathbf{u}}(t + \Delta t) - 2\bar{\mathbf{u}}(t) + \bar{\mathbf{u}}(t - \Delta t)\right] \tag{2.32}$$

leads to an explicit time-stepping scheme that allows us to compute the displacement at time $t + \Delta t$ from the displacement at times t and $t - \Delta$:

$$\bar{\mathbf{u}}(t + \Delta t) = 2\bar{\mathbf{u}}(t) - \bar{\mathbf{u}}(t - \Delta t) + \Delta t^2 \mathbf{M}^{-1} \cdot \left[\bar{\mathbf{f}}(t) - \mathbf{K} \cdot \bar{\mathbf{u}}(t)\right]. \tag{2.33}$$

For notational clarity we replaced $\approx$ by $=$ in Eq. (2.33), keeping in mind that this is an approximation.

A frequently used alternative to (2.33) is the *Newmark scheme* (Newmark, 1959; Chaljub et al., 2007), defined by

$$\bar{\mathbf{u}}(t + \Delta t) = \bar{\mathbf{u}}(t) + \Delta t\, \bar{\mathbf{v}}(t) + \Delta t^2 \left[\left(\frac{1}{2} - \beta\right) \ddot{\bar{\mathbf{u}}}(t) + \beta\, \ddot{\bar{\mathbf{u}}}(t + \Delta t)\right], \tag{2.34a}$$

$$\bar{\mathbf{v}}(t + \Delta t) = \bar{\mathbf{v}}(t) + \Delta t\left[(1 - \gamma)\, \ddot{\bar{\mathbf{u}}}(t) + \gamma\, \ddot{\bar{\mathbf{u}}}(t + \Delta t)\right], \tag{2.34b}$$

with the parameters $\gamma \in [0, 1]$ and $\beta \in [0, \frac{1}{2}]$. Second-order accuracy instead of first-order accuracy is achieved if and only if $\gamma = \frac{1}{2}$. In the special case $\gamma = \frac{1}{2}$ and $\beta = 0$, we obtain an explicit central-difference scheme:

$$\bar{\mathbf{u}}(t + \Delta t) = \bar{\mathbf{u}}(t) + \Delta t \, \bar{\mathbf{v}}(t) + \frac{1}{2} \Delta t^2 \, \ddot{\bar{\mathbf{u}}}(t) \,, \tag{2.35a}$$

$$\bar{\mathbf{v}}(t + \Delta t) = \bar{\mathbf{v}}(t) + \frac{1}{2} \Delta t \left[\ddot{\bar{\mathbf{u}}}(t) + \ddot{\bar{\mathbf{u}}}(t + \Delta t) \right] . \tag{2.35b}$$

The order of operations in the iterative advancement of $\bar{\mathbf{u}}$ is then as follows. (1) Compute $\bar{\mathbf{u}}$ at time $t+\Delta t$ from the discrete field variables at time t, using Eq. (2.35a). (2) With the help of Eq. (2.31), compute the acceleration $\ddot{\bar{\mathbf{u}}}(t + \Delta t)$ from $\bar{\mathbf{u}}(t + \Delta t)$. (3) Advance the velocity $\bar{\mathbf{v}}$ from time t to $t + \Delta t$, using Eq. (2.35b). (4) Go back to (1) and repeat as often as needed. While being more complicated than (2.33), the Newmark scheme has the advantageous property of conserving linear and angular momentum (e.g. Kane et al., 2003).

We note that the time discretisation schemes for the displacement formulation are immediately applicable to the displacement–stress formulation from Eq. (2.29).

2.5.1.2 Velocity–Stress Formulation

The most commonly used time discretisation scheme for the velocity–stress formulation (2.30) is the *leapfrog method*. This is based on alternating updates of the discrete velocity and stress fields. Starting from $\bar{\mathbf{s}}(t - \Delta t/2)$ and $\bar{\mathbf{v}}(t)$, we obtain $\bar{\mathbf{s}}(t + \Delta t/2)$ via a second-order finite-difference approximation of Eq. (2.30b):

$$\bar{\mathbf{s}}(t + \Delta t/2) = \bar{\mathbf{s}}(t - \Delta t/2) + \Delta t \, \mathbf{K}_2 \cdot \bar{\mathbf{v}}(t) \,. \tag{2.36a}$$

With the help of the discrete stress field $\bar{\mathbf{s}}(t+\Delta t/2)$ we can then advance the velocity field from time t to time $t+\Delta t$, using the same second-order approximation applied to Eq. (2.30a):

$$\bar{\mathbf{v}}(t + \Delta t) = \bar{\mathbf{v}}(t) + \Delta t \, \mathbf{M}^{-1} \cdot [\bar{\mathbf{f}}(t + \Delta t/2) - \mathbf{K}_1 \cdot \bar{\mathbf{s}}(t + \Delta t/2)] \,. \tag{2.36b}$$

Again, for notational convenience, we replaced $\approx$ by $=$ in Eq. (2.36a) and (2.36b).

2.5.1.3 Stability

All of the above time-stepping algorithms are explicit in the sense that the dynamic fields at time $t + \Delta t$ depend only on the dynamic fields at times *prior* to $t + \Delta t$. This implies that the algorithms are only conditionally stable. The stability criterion, named *CFL condition* in honour of R. Courant, K. Friedrichs and H. Lewy (R. Courant, K. Friedrichs and H. Lewy, Courant et al. 1928), typically takes the form

$$\Delta t \leq \text{const.} \frac{\min h}{\max v}, \tag{2.37}$$

where h is the width of the numerical grid and v is the propagation speed of the fastest wave, that is the P wave. The constant on the right-hand side of Eq. (2.37) depends on the methods used for the space and time discretisation. Its order of magnitude is 1. The CFL condition limits the maximum possible time increment and therefore the efficiency of any explicit time stepping. In Sect. 3.1 we derive the CFL condition for a finite-difference approximation of the 1D scalar wave equation.

2.5.1.4 Alternative Methods

The Newmark and the leapfrog schemes are by far the most frequently used methods for the time discretisation of the space-discrete equations of motion. This may appear surprising given the availability of numerous alternatives such as predictor–corrector variants of implicit multi-step methods or higher order Runge–Kutta methods (e.g. Quarteroni et al., 2000).

The restriction to conditionally stable methods of comparatively low order can be explained not only by the larger computational requirements of implicit and higher order methods but also by the good performance of the explicit second-order schemes. Empirical studies show that the numerical error is often dominated by the inaccuracies of the *spatial* discretisation. This is true even when the number of time steps is large, that is on the order of several thousands. Whether the conventional Newmark and leapfrog schemes are sufficient for future applications, such as global-scale wave propagation at periods around 2 s, is still an open question.

2.5.2 *Frequency-Domain Modelling*

Frequency-domain modelling is based on the Fourier-transformed version of the space-discrete displacement formulation (2.28):

$$-\omega^2 \mathbf{M} \cdot \bar{\mathbf{u}}(\omega) + \mathbf{K} \cdot \bar{\mathbf{u}}(\omega) = \bar{\mathbf{f}}(\omega). \tag{2.38}$$

Upon defining the *impedance matrix* **L**

$$\mathbf{L}(\omega) := -\omega^2 \mathbf{M} + \mathbf{K}, \tag{2.39}$$

Eq. (2.38) simplifies to the linear system

$$\mathbf{L}(\omega) \cdot \bar{\mathbf{u}}(\omega) = \bar{\mathbf{f}}(\omega). \tag{2.40}$$

The solution of (2.40) is the discrete wave field $\bar{\mathbf{u}}$ at one frequency ω.

It is particularly attractive to solve the system (2.40) using direct matrix factorisation methods, such as LU decomposition (e.g. Press et al., 2007). Once the matrix factors are known, they can be reused to solve the forward problem for any new source $\bar{\mathbf{f}}$ at very low computational cost (e.g. Pratt et al., 1998; Pratt, 1999). However, in realistic 3D applications, the memory requirements of direct methods can become prohibitive so that iterative solvers must be used (e.g. Quarteroni et al., 2000).

Frequency-domain modelling is the method of choice when solutions for a few dominant or well-chosen frequencies are needed or when the problem is 2D thus permitting the application of direct linear system solvers.

One of the outstanding advantages of this approach is the easy implementation of visco-elastic dissipation. This is because the convolution in Eq. (2.14) translates to a simple product in the frequency domain. Additional memory variables as in time-domain modelling are therefore not needed (see Chap. 5). In Chap. 14 we present an application of frequency-domain modelling in the context of full waveform tomography with active-source data.

2.6 Summary of Numerical Methods

In the course of the past decades a large number of numerical methods for the solution of the seismic wave equation have been developed – often for the purposes of ground motion prediction or waveform inversion. Each method comes with advantages and disadvantages that need to be weighted carefully in the light of a specific application.

The most significant distinction between different approaches concerns the spatial discretisation, that is, the transformation of the exact spatial derivatives in the wave equation into an algebraic system. The following is a list of some methods that try to highlight their major similarities and differences.

Finite-difference methods: Numerical modelling in nearly all physical sciences started with the finite-difference method, and seismology is no exception. Early applications can be found in Alterman & Karal (1968), Boore (1970, 1972), Alford et al. (1974) and Kelly et al. (1976). To illustrate the finite-difference concept, we consider a generic function $f(x)$ that represents the dynamic fields that appear in the wave equation (e.g. stress, strain, and displacement). The fundamental idea is to consider only a finite number of evenly spaced grid points x_i (i=1, ..., N) and to replace the derivative $\partial_x f(x_i)$ at grid point x_i by a *finite-difference approximation* that involves f evaluated at neighbouring grid points. The best-known example is the second-order central finite-difference approximation

$$\partial_x f(x_i) = \frac{1}{2\,\Delta x}\,[f(x_i + \Delta x) - f(x_i - \Delta x)] + \mathcal{O}(\Delta x^2)\,, \tag{2.41}$$

where Δx is the grid spacing. The feasibility of finite-difference modelling in three dimensions rests on the definition of a *staggered grid* where the dynamic fields are defined at different grid positions (e.g. Madariaga, 1976; Virieux, 1984, 1986, Igel et al., 1995). This is in contrast to the *conventional grid* where all field variables are defined at coincident grid positions. The staggered grid results in a reduced average grid spacing that greatly reduces the *numerical dispersion*, i.e. the artificial dispersion introduced by the discretisation of the original equations of motion.

The popularity of finite-difference modelling is largely due to the comparatively low computational costs and the accuracy especially in the modelling of body wave propagation. In Chap. 3 we treat the finite-difference method in detail.

Optimal operators: The discretisation of the equations of motion introduces numerical errors that are particularly prominent near the eigenfrequencies of the elastic medium. This observation led Geller & Takeuchi (1995) to the construction of optimal operators that are designed to minimise the discretisation error first of all in the vicinity of the eigenfrequencies.

The very general criterion for an operator to be optimal is that the inner product of an exact eigenfunction with the net error of the discrete equations of motion at the corresponding eigenfrequency is approximately 0. This criterion is independent of the actual space discretisation scheme. In the particular framework of finite-difference methods, optimality means that the lowest order errors of the time and the space discretisations cancel, thus leading to highly accurate numerical schemes (Geller & Takeuchi, 1998; Takeuchi & Geller, 2000).

Since time-domain optimal operators are inherently implicit, predictor–corrector algorithms must be used to avoid the solution of large algebraic systems in the time stepping. The increased solution accuracy clearly compensates the additional computational costs of the predictor–corrector scheme (Mizutani et al., 2000; Kristek & Moczo, 2006).

Pseudospectral methods: Like finite-difference methods, pseudospectral methods directly discretise the spatial derivatives in the equations of motion (e.g. Kosloff & Baysal, 1982; Furumura et al., 1998). The discretisation proceeds in three steps. Firstly, the wave field sampled at a finite number of grid points is transformed to the wave number domain using the fast Fourier transform. The transformed wave field is then multiplied by $\mathbf{i}k$, where k is the wave number. This multiplication corresponds to a space derivative. Finally, using the inverse fast Fourier transform, the wave field is transformed back to the space domain. Since the derivative is exact up to the Nyquist wave number, as few as two grid points per wavelength are theoretically sufficient for the spatial sampling of the wave field.

Kosloff et al. (1990) proposed a variant of the Fourier pseudospectral method where a Chebyshev transform is used in the vertical direction in order to account more accurately for the free surface condition. An extension to the 3D case with surface topography can be found in Tessmer & Kosloff (1994). The application of the Chebyshev pseudospectral method to wave propagation on the scale of the mantle is presented in Igel (1999).

Pseudospectral methods outperform finite-difference methods regarding the very small amount of numerical dispersion. However, due to the global nature of the derivative approximation, they are restricted to comparatively smooth media (Mizutani et al., 2000). Furthermore, pseudospectral methods face issues of parallelisation because global memory access is required for the computation of the Fourier and Chebyshev transforms.

Finite-element methods: Finite-element methods are based on the *weak* or *the variational form* of the wave equation that we describe in Sect. 4.1.1. The computational domain is decomposed into disjoint subdomains, called the elements. Within each element the dynamic fields are approximated by polynomials of low order (e.g. piecewise linear functions Bao et al., 1998), and continuity between the elements is imposed explicitly. The elastic wave equation then reduces to a space-discrete system for the polynomial coefficients.

Despite its capability to correctly account for irregular geometries and the free surface, applications of the pure finite-element method to elastic wave propagation are comparatively rare (e.g. Lysmer & Drake, 1972; Toshinawa & Ohmachi, 1992; Bao et al., 1998). This is mostly due to the comparatively large numerical dispersion that results from the low-order polynomial approximations. Moreover, the mass matrix in finite-element methods is not diagonal, which renders its inversion computationally expensive.

While the pure finite-element method does not appear to be well suited for wave propagation, hybrid schemes have been used very successfully. Moczo et al. (1997, 2007), for instance, combined the finite-element and finite-difference methods for the simulation of wave propagation along irregular surface topography. They discretised the equations of motion in the interior of the computational domain using finite differences. A rim of finite elements was then used to mesh the topography.

Spectral-element methods: Spectral-element methods are half way between finite-element and pseudospectral methods, combining the advantages of both approaches while avoiding many of their drawbacks. Like in finite-element methods, the computational domain is subdivided into non-overlapping elements that can be adapted to irregular geometries. Inside the elements a high-order spectral approximation is used for the dynamic fields. The spectral-element method as originally developed for fluid mechanics (Patera, 1984) and seismic wave modelling (Priolo et al., 1994; Seriani et al., 1995; Seriani, 1998) indeed uses Chebyshev polynomials as basis functions, thus establishing a direct link to the Chebyshev pseudospectral methods.

In a widely used spectral-element variant the Chebyshev polynomials are replaced by Lagrange polynomials collocated at the Gauss–Lobatto–Legendre (GLL) points. The combination with GLL quadrature leads to a diagonal mass matrix that can be trivially inverted. In Chap. 4 and Appendix A we provide a detailed introduction to the spectral-element method. Applications on a variety of scales can be found in Faccioli et al. (1997), Komatitsch (1997), Komatitsch & Vilotte (1998), Komatitsch & Tromp (2002), Chaljub et al. (2003), Chaljub & Valette (2004) and Nissen-Meyer et al. (2007, 2008)

Direct solution method: The direct solution method was introduced in a series of papers by Geller & Ohminato (1994) and Cummins et al. (1994a,b). As finite-element and spectral-element methods, it is founded on the weak form of the equations of motion. What distinguishes the direct solution method is the choice of basis functions: linear splines in the radial direction and spherical harmonics in the horizontal directions. The space-discrete equations are then solved in the frequency domain.

Takeuchi (2000) applied the optimal operator formalism of Geller & Takeuchi (1995) to the direct solution method and investigated its applicability to waveform inversion. The method was extended to radially anisotropic media by Kawai et al. (2006) and then applied to waveform inversion for localised heterogeneities by Konishi et al. (2009) and Kawai & Geller (2010).

Discontinuous Galerkin methods: Discontinuous Galerkin methods for seismic wave propagation have been developed only recently (e.g. Dumbser & Käser, 2006; Käser et al., 2007; de la Puente et al., 2007, 2008). They represent a class of finite-element methods where neighbouring elements are linked by numerical fluxes and not by continuity constraints. This allows for solutions that are discontinuous across element boundaries. Discontinuous Galerkin methods are therefore particularly well suited for the modelling of earthquake rupture processes (de la Puente et al., 2009). An application of a discontinuous Galerkin method in the context of full waveform inversion can be found in Brossier et al. (2009).

In Chaps. 3 and 4 two of the above-mentioned discretisation approaches are described in more detail: the finite-element and the spectral-element method.

Choosing two out of many possible methods was a difficult but inevitable decision. Since this book is intended to cover both the forward and the inverse problems, I gave preference to the methods that are currently employed most frequently in the context of waveform tomography. The finite-difference and spectral-element methods also offer the opportunity to introduce a large number of concepts that are of general importance in numerical seismology.

A comprehensive treatise on all the different approaches to the numerical solution of the seismic wave equation is still to be written, and I offer my apologies to colleagues whose work I was not able to present here.

Chapter 3
Finite-Difference Methods

The finite-difference method can be considered *the* classical and most frequently applied method for the numerical simulation of seismic wave propagation. It is based on the approximation of an exact derivative $\partial_x f(x_i)$ at a grid position x_i in terms of the function f evaluated at a finite number of neighbouring grid points.

Early implementations of the finite-difference method in a seismological context used a *conventional grid* where all field variables (e.g. displacement, stress, and strain) are defined at the same grid positions. Examples may be found in Alterman & Karal (1968), Boore (1970, 1972), Alford et al. (1974) and Kelly et al. (1976).

The breakthrough in finite-difference modelling was the application of the *staggered-grid* approach (Madariaga, 1976; Virieux, 1984, 1986). In the staggered grid, field variables are defined at different grid positions, which reduces the effective grid spacing compared to the conventional grid. Further developments in finite-difference wave propagation focused, for instance, on the modelling of the free surface (e.g. Graves, 1996; Kristek et al., 2002), the incorporation of general anisotropy (Igel et al., 1995) and the correct implementation of material discontinuities (e.g. Moczo et al., 2002). Applications of finite-difference modelling include studies of seismic ground motion in densely populated areas (e.g. Frankel & Vidale, 1992; Wang et al., 2008; Moczo et al., 2007; see Fig. 3.1), the simulation of wave propagation through random media (e.g. Frankel & Clayton, 1984, 1986) and full waveform inversion (e.g. Igel et al., 1996; Pratt, 1999; Dessa et al., 2004; Bleibinhaus et al., 2007). For an excellent review of finite-difference methods for wave propagation, the reader is referred to Moczo et al. (2007).

To introduce the basic finite-difference concepts, we start in Sect. 3.1 with the 1D scalar wave equation. This simple example also allows us to study the stability and the numerical dispersion of the discretised equations. Based on the staggered-grid approach we then make the transition to the 3D elastic case. We pay special attention to the implementation of anisotropy, the free surface and arbitrarily located point sources.

A. Fichtner, *Full Seismic Waveform Modelling and Inversion*, Advances in Geophysical and Environmental Mechanics and Mathematics,
DOI 10.1007/978-3-642-15807-0_3,

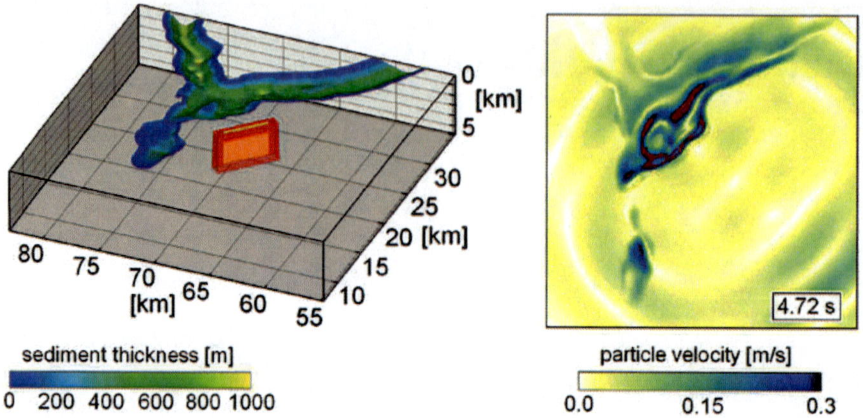

Fig. 3.1 *Left*: Geometrical configuration of the Grenoble (France) sedimentary basin. The location of a fault zone is marked by a red box. *Right*: Snapshot of the absolute value of the horizontal-component particle velocity, 4.72 s after the rupture initiation. The simulation is based on a fourth-order finite-difference discretisation of the elastic wave equation, as described in Sect. 3.2.1. The horizontal displacement velocity is largest above the densely populated sedimentary basin. (Figure modified from Moczo et al. (2007), with permission from the authors)

3.1 Basic Concepts in One Dimension

The 1D scalar wave equation is particularly well suited not only for an introduction to the finite-difference method itself but also for the illustration of fundamental concepts in numerical analysis, including stability and grid dispersion. We start our development with the description of two methods for the construction of finite-difference approximations. These are then used to replace the exact derivatives in the wave equation. The result of this procedure is an iterative scheme that allows us to advance a discrete representation of the wave field in time. Both numerical experiments and a rigorous analysis reveal the properties of the iterative scheme in general and its stability requirements in particular.

3.1.1 Finite-Difference Approximations

3.1.1.1 Truncated Taylor Expansions

The most straightforward procedure for the construction of finite-difference approximations to the derivative $\partial_x f(x)$ of a generic function f is based on truncated Taylor expansions. To find an approximation that is correct to second order in the grid spacing Δx, we choose the ansatz

$$\partial_x f(x) \approx g\,[f(x+\Delta x) - f(x-\Delta x)] \tag{3.1}$$

with a scalar coefficient g that we seek to determine such that the right-hand side of (3.1) converges to $\partial_x f(x)$ as $\Delta x \to 0$. Expanding $f(x+\Delta x)$ and $f(x-\Delta x)$ into Taylor series centred around x, yields

$$g\,[f(x+\Delta x) - f(x-\Delta x)] = g\left[2\,\partial_x f(x)\,\Delta x + \frac{2}{3!}\,\partial_x^3 f(x)\,\Delta x^3 + \dots\right]. \tag{3.2}$$

Choosing

$$g = \frac{1}{2\,\Delta x} \tag{3.3}$$

results in the well-known second-order finite-difference stencil

$$\partial_x f(x) = \frac{1}{2\,\Delta x}\,[f(x+\Delta x) - f(x-\Delta x)] + \mathcal{O}(\Delta x^2)\,. \tag{3.4}$$

To derive approximations that are of arbitrary order in Δx, we generalise Eq. (3.1) such that it involves f evaluated at $2N$ grid points $x_n = x \pm n\,\Delta x$, with $n = 1, \dots, N$:

$$\partial_x f(x) \approx \sum_{n=1}^{N} g_n\,[f(x+n\,\Delta x) - f(x-n\,\Delta x)]\,. \tag{3.5}$$

To determine the finite-difference coefficients g_n, we replace $f(x \pm n\Delta x)$ by a Taylor series

$$\sum_{n=1}^{N} g_n\,[f(x+n\,\Delta x) - f(x-n\,\Delta x)]$$
$$= \sum_{n=1}^{N} g_n\left[\sum_{k=0}^{\infty}\frac{1}{k!}\partial_x^k f(x)(n\Delta x)^k - \sum_{n=0}^{\infty}\frac{1}{k!}\partial_x^k f(x)(-n\Delta x)^k\right]. \tag{3.6}$$

The terms involving even powers of Δx cancel so that we are left with

$$\sum_{n=1}^{N} g_n\,[f(x+n\,\Delta x) - f(x-n\,\Delta x)] = \sum_{n=1}^{N} g_n\left[\sum_{k=0}^{\infty}\frac{2}{(2k+1)!}\,\partial_x^{2k+1} f(x)(n\Delta x)^{2k+1}\right]. \tag{3.7}$$

Isolating the summands corresponding to $k=0$ yields

$$\sum_{n=1}^{N} g_n\,[f(x+n\,\Delta x) - f(x-n\,\Delta x)] = 2\Delta x\,\partial_x f(x)\sum_{n=1}^{N} n g_n$$
$$+ \sum_{k=1}^{\infty}\frac{2}{(2k+1)!}\partial_x^{2k+1} f(x)\sum_{n=1}^{N} g_n\,(n\Delta x)^{2k+1}\,. \tag{3.8}$$

To ensure that the right-hand side of (3.8) converges to $\partial_x f(x)$ as $\Delta x \to 0$, we impose the condition

$$1 = 2\Delta x \sum_{n=1}^{N} n g_n \,. \tag{3.9}$$

Equation (3.9) leaves us with $N-1$ degrees of freedom for the coefficients g_n that we may now use to eliminate the monomials proportional to $\Delta x^3, \ldots, \Delta x^{2N-1}$ from (3.8). For this we furthermore require that the finite-difference coefficients satisfy the following system of $N-1$ linear equations:

$$0 = \sum_{n=1}^{N} g_n (n\Delta x)^{2k+1}\,, \quad k = 1, \ldots, N-1\,. \tag{3.10}$$

The resulting finite-difference approximation is then of order $2N$ in Δx:

$$\sum_{n=1}^{N} g_n \left[f(x + n\,\Delta x) - f(x - n\,\Delta x)\right] = \partial_x f(x) + \mathcal{O}(\Delta x^{2N})\,. \tag{3.11}$$

By far the most frequently used finite-difference stencils are those of order 4–8. The associated coefficients are the solutions of the linear system given by Eqs. (3.9) and (3.10):

$$\begin{aligned}
&\text{order 4:} \quad g_1 = \frac{2}{3\Delta x}\,, \quad g_2 = -\frac{1}{12\Delta x}\,,\\
&\text{order 6:} \quad g_1 = \frac{3}{4\Delta x}\,, \quad g_2 = -\frac{3}{20\Delta x}\,, \quad g_3 = \frac{1}{60\Delta x}\,,\\
&\text{order 8:} \quad g_1 = \frac{4}{5\Delta x}\,, \quad g_2 = -\frac{1}{5\Delta x}\,, \quad g_3 = \frac{4}{105\Delta x}\,, \quad g_4 = -\frac{1}{280\Delta x}\,.
\end{aligned} \tag{3.12}$$

By construction, higher order approximations converge faster to the exact derivative as $\Delta x \to 0$. Also for (unavoidably) finite grid spacings Δx higher order approximations generally yield more accurate solutions.

The finite-difference approximations considered so far are based on grid points that are evenly spaced and symmetric with respect to the grid point where the derivative is approximated. While this approach appears most natural, there are useful variations. In the staggered-grid discretisation, for instance (see Sect. 3.2.1), the first derivative $\partial_x f(x)$ is approximated correct to fourth-order *between* the grid points and not *at* the grid points:

$$\begin{aligned}
\partial_x f(x) = &\frac{9}{8\,\Delta x}\left[f(x + \Delta x/2) - f(x - \Delta x/2)\right]\\
&- \frac{1}{24\,\Delta x}\left[f(x + 3\Delta x/2) - f(x - 3\Delta x/2)\right] + \mathcal{O}(\Delta x^4)\,.
\end{aligned} \tag{3.13}$$

The *effective grid spacing*, that is the distance between the grid points and the position where the derivative is approximated, is therefore $\pm\Delta x/2$ and $\pm 2\Delta x/3$.
As we will see in Sect. 3.2.3.3, several asymmetric and one-sided approximations play a crucial role in the implementation of the free surface in finite-difference methods.

3.1.1.2 Sampling of Band-Limited Derivative Operators

An alternative way of constructing finite-difference schemes is based on the sampling of a band-limited version of the derivative operator (e.g. Mora, 1986; Igel et al., 1995). For this we note that the exact differentiation of $f(x)$ in the space domain corresponds to a multiplication of its Fourier transform $\tilde{f}(k)$ by $\mathbf{i}k$, where k denotes the spatial wave number.

The discretisation implies that any numerical derivative can be accurate only within a limited wave number range, $[-k_{\max}, k_{\max}]$. The *Nyquist wave number* $k_{\max} = \frac{\pi}{\Delta x}$ is the maximum wave number that can be represented when the grid spacing is Δx. It follows that the wave number domain derivative operator in the range $[-k_{\max}, k_{\max}]$ is given by

$$\tilde{g}(k) = \mathbf{i}k\,[H(k + k_{\max}) - H(k - k_{\max})]\,, \tag{3.14}$$

with H being the Heaviside function. The multiplication $\tilde{g}(k)\tilde{f}(k)$ represents a band-limited differentiation that corresponds to a convolution in the space domain:

$$\widetilde{\partial_x f}(k) \approx \tilde{g}(k)\tilde{f}(k) \quad\Longleftrightarrow\quad \partial_x f(x) \approx \frac{1}{\sqrt{2\pi}} \int_{-\infty}^{\infty} g(y) f(x-y)\,dy\,. \tag{3.15}$$

The space-domain version of the approximate derivative operator $\tilde{g}$ is proportional to the derivative of the sinc function:

$$g(x) = \frac{2k_{\max}}{\sqrt{2\pi}} \frac{\mathrm{d}}{\mathrm{d}x} \operatorname{sinc}(k_{\max}x) = \frac{2}{\sqrt{2\pi}\,x^2} [k_{\max}x\,\cos(k_{\max}x) - \sin(k_{\max}x)]\,. \tag{3.16}$$

To approximate the convolution in (3.15) we replace the integral by a Riemann sum that only involves f and g sampled at a finite number of grid points:

$$y_n = n\,\Delta x\,, \quad n = -N, \dots, 0, \dots, N\,. \tag{3.17}$$

The continuous integral then turns into a discrete convolution sum:

$$\partial_x f(x) \approx \frac{1}{\sqrt{2\pi}} \sum_{n=-N}^{N} g(y_n)\, f(x - y_n)\,\Delta x = \sum_{n=-N}^{N} g_n f(x - n\Delta x)\,. \tag{3.18}$$

The finite-difference coefficients g_n are given by:

$$g_n = \begin{cases} 0\,, & \text{for} \quad n = 0\,, \\ \frac{(-1)^n}{n\Delta x}\,, & \text{for} \quad n \neq 0\,. \end{cases} \tag{3.19}$$

The discrete convolution in Eq. (3.18) constitutes a finite-difference approximation of the derivative $\partial_x f(x)$. The incontestable beauty of this approach lies in its ability to effortlessly generate finite-difference coefficients for any distribution of grid points.

Our detour via the wave number domain suggests to quantify the accuracy of the discrete convolution differentiator in terms of its spectrum. Firstly, we note that the coefficients (3.19) are anti-symmetric with respect to the index n. The spectrum of g_n is therefore purely imaginary. As illustrated in the left panel of Fig. 3.2, the spectrum of the approximate differentiator approaches the spectrum of the exact differentiation $\mathbf{i}k$ as the number of grid points N that contribute to the discrete convolution increases. However, even for impractically large N, we observe strong oscillations that indicate an insufficiently good approximation of the exact derivative.

The extent of this failure becomes most apparent when the discrete convolution from Eqs. (3.18) and (3.19) is compared to the differentiators derived from truncated Taylor expansions. As shown in the right panel of Fig. 3.2, the Taylor method yields finite-difference coefficients that approximate the exact derivative very well for wave numbers smaller than $\approx 50\,\%$ of the Nyquist wave number.

This result does not mean that the sampling of the band-limited derivative operator (3.16) fails per se, because the result depends strongly on the location of the grid points. Sampling (3.16) at the grid points $n\Delta x$ leads, as we have seen, to an

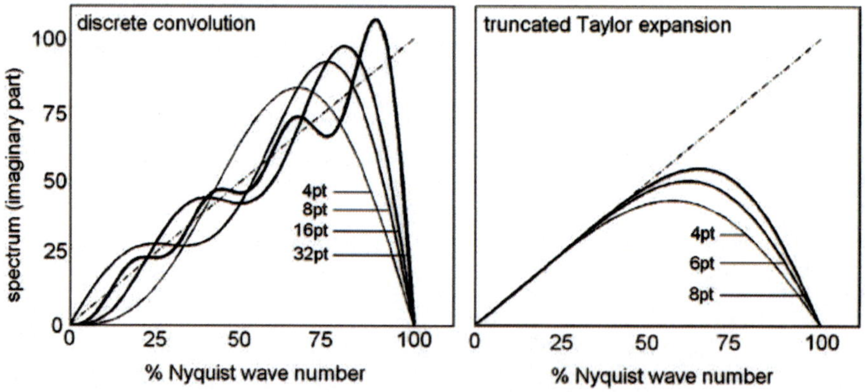

Fig. 3.2 *Left*: Imaginary part of the wave number spectrum corresponding to the discrete convolution differentiator defined by Eqs. (3.18) and (3.19) for variable numbers *grid points*. (4 points $\leftrightarrow$ $N = 2$, 8 points $\leftrightarrow$ $N = 4, \ldots$.) The *dashed line* represents the exact differentiation operator $\mathbf{i}k$. *Right*: The same as on the *left* but for the finite-difference coefficients from Eq. (3.12) that we obtained from the truncation of Taylor expansions

inaccurate approximation of the continuous convolution by the discrete Riemann sum (3.18). However, in the *staggered-grid approach* described in Sect. 3.2.1, the derivative of f is evaluated halfway between the grid points. This means that we approximate $\partial_x f(x)$ in terms of f given at the grid positions $x - (n + 1/2)\,\Delta x$ for $n = -N, \ldots, 0, \ldots, N-1$. The discrete convolution then takes the form

$$\partial_x f(x) \approx \frac{1}{\sqrt{2\pi}} \sum_{n=-N}^{N-1} g(y_n)\, f(x-y_n)\, \Delta x = \sum_{n=-N}^{N-1} g_n f[x-(n+1/2)\Delta x]\,, \quad (3.20)$$

and the finite-difference coefficients g_n are defined through

$$g_n = \frac{(-1)^{n+1}}{\pi\,(n+1/2)^2 \Delta x}\,. \quad (3.21)$$

The corresponding spectrum for the four-point and eight-point convolutions is shown in the left panel of Fig. 3.3. The coefficients from Eq. (3.21) clearly outperform the coefficients in (3.12) that we derived from the Taylor expansion. In fact, the eight-point operator is hardly distinguishable from the exact derivative operator, even for wave numbers close to the Nyquist wave number. Examples where the remarkable properties of the discrete convolution defined in (3.20) and (3.21) have been used for 3D elastic wave propagation on a staggered grid can, for instance, be found in Igel et al. (1995).

An intuitive explanation for the large differences in accuracy of the discrete convolutions is provided in the right panel of Fig. 3.3, where we compare the sampling of the derivative of the sinc function (3.16) on the regular grid, $x_n = n\Delta x$, and the staggered grid, $x_n = (n + 1/2)\,\Delta x$. The finite-difference coefficients for the staggered grid decay as n^{-2}, meaning that only coefficients with small n effectively contribute to the discrete convolution. On the regular grid, however, the coefficients

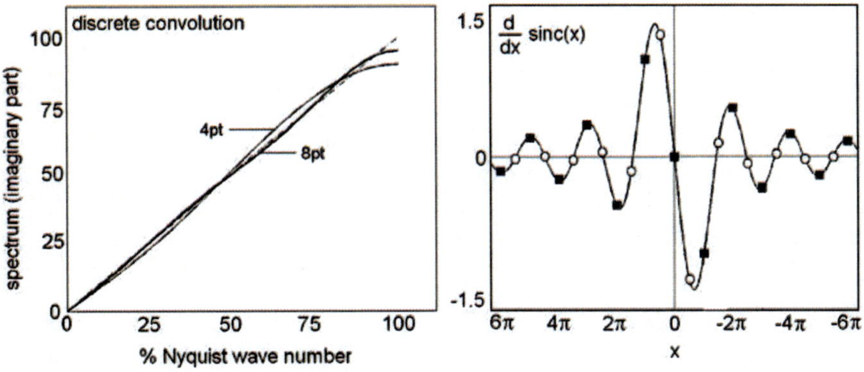

Fig. 3.3 *Left*: Imaginary part of the wave number spectrum corresponding to the four-point and eight-point discrete convolution differentiator for the *staggered grid*. *Right*: The derivative of the sinc function and the sampling points in the *regular grid* (■) and in the *staggered grid* (○)

decay slowly as n^{-1}. Many more coefficients are therefore needed for an accurate approximation of the continuous convolution by the discrete Riemann sum.

It is, in principle, possible to approximate the exact derivative arbitrarily well with finite-difference operators that involve a large number of grid points. In practice, one hardly uses more than eight points in order to balance accuracy and computational costs.

3.1.2 Discretisation of the 1D Wave Equation

To introduce the basic concepts of the finite-difference method in the context of wave propagation, we consider the 1D wave equation

$$\rho(x)\,\ddot{u}(x,t) - \partial_x[\mu(x)\,\partial_x u(x,t)] = 0 \tag{3.22}$$

with the line density ρ and the elastic parameter μ. In the interest of simplicity, we disregard external forces, and we impose the non-zero initial condition

$$u(x,t)|_{t=0} = u_0(x)\,, \quad \dot{u}(x,t)|_{t=0} = 0 \tag{3.23}$$

with the initial displacement field u_0. The computational domain is the interval $[-L, L]$, and as boundary conditions we require zero displacement at $x = \pm L$:

$$u(x,t)|_{x=-L} = u(x,t)|_{x=L} = 0\,. \tag{3.24}$$

Other boundary conditions are possible, but their implementation usually requires additional work that we avoid in this introductory example. The discretisation process starts with the definition of $2N + 1$ grid points x_i that are evenly distributed over the interval $[-L, L]$:

$$x_i = i\,\Delta x\,, \quad i = -N, \ldots, 0, \ldots, N\,. \tag{3.25}$$

The grid spacing Δx is determined by the requirement $N\Delta x = L$. We now approximate the first derivative $\partial_x u$ evaluated at the grid point x_i by the wave field evaluated at neighbouring grid points. Choosing, for instance, the second-order finite-difference approximation from Eq. (3.4), we find

$$\partial_x u(x_i,t) \approx \frac{1}{2\Delta x}\,[u(x_i+\Delta x,t) - u(x_i-\Delta x,t)] = \frac{1}{2\Delta x}\,[\bar{u}^{i+1}(t) - \bar{u}^{i-1}(t)]\,. \tag{3.26}$$

The time-dependent scalars $\bar{u}^{i\pm1}$ are the wave field, sampled at the grid positions $x_i \pm \Delta x$. Repeating this procedure for the second spatial derivative in Eq. (3.22), yields

$$\partial_x[\mu(x)\,\partial_x u(x,t)] \approx \frac{1}{4\Delta x^2}\,\{\mu^{i+1}[\bar{u}^{i+2}(t)-\bar{u}^{i}(t)]-\mu^{i-1}[\bar{u}^{i}(t)-\bar{u}^{i-2}(t)]\}\,, \tag{3.27}$$

where we introduced the variable $\mu^i := \mu(x_i)$, which is the elastic parameter μ evaluated at the grid position x_i. The boundary conditions (3.24) are imposed explicitly by setting

$$\bar{u}^{-N-2} = \bar{u}^{-N-1} = \bar{u}^{-N} = \bar{u}^{N} = \bar{u}^{N+1} = \bar{u}^{N+2} = 0\,. \tag{3.28}$$

We can now assemble the semi-discrete version of the wave equation (3.22):

$$\rho^i\ddot{\bar{u}}^i(t) - \frac{1}{4\Delta x^2}\,\{\mu^{i+1}[\bar{u}^{i+2}(t)-\bar{u}^{i}(t)]-\mu^{i-1}[\bar{u}^{i}(t)-\bar{u}^{i-2}(t)]\} = 0\,, \tag{3.29}$$

with $\rho^i := \rho(x_i)$. It is important to note that Eq. (3.29) is meaningful only under the assumption that both ρ and μ are continuous with variations that can be represented reasonably well with a finite grid spacing. In the presence of material discontinuities, averaging schemes must be applied to ensure accurate numerical solutions (e.g. Moczo et al., 2002; Kristek & Moczo, 2006).

Formally, we can write Eq. (3.29) in terms of a *mass matrix* **M** and a *stiffness matrix* **K** (see Chap. 2):

$$\mathbf{M}\cdot\ddot{\bar{\mathbf{u}}}(t) + \mathbf{K}\cdot\bar{\mathbf{u}}(t) = \mathbf{0}\,. \tag{3.30}$$

The vector $\bar{\mathbf{u}}$ is composed of the $2N+1$ coefficients $\bar{u}^i$. Since **M** is diagonal, it can be trivially inverted. In practice, the mass and the stiffness matrices are never computed explicitly because only the vector–matrix products $\mathbf{M}\cdot\ddot{\bar{\mathbf{u}}}$ and $\mathbf{K}\cdot\bar{\mathbf{u}}$ are needed. We will nevertheless use **M** and **K** for notational convenience.

It now remains to discretise the second-time derivative in (3.30). In order to construct an explicit scheme, we use the second-order finite-difference approximation

$$\ddot{\bar{\mathbf{u}}}(t) \approx \frac{1}{\Delta t^2}\,[\bar{\mathbf{u}}(t+\Delta t) - 2\bar{\mathbf{u}}(t) + \bar{\mathbf{u}}(t-\Delta t)]\,, \tag{3.31}$$

with a suitably chosen time increment Δt which we will study later in more detail. Inserting (3.31) into (3.30) then yields a fully discrete version of the scalar wave equation (3.22):

$$\bar{\mathbf{u}}(t+\Delta t) = 2\bar{\mathbf{u}}(t) - \bar{\mathbf{u}}(t-\Delta t) - \Delta t^2\,\mathbf{M}^{-1}\cdot\mathbf{K}\cdot\bar{\mathbf{u}}(t)\,. \tag{3.32}$$

Equation (3.32) suggests the following recipe for the iterative finite-difference solution of the 1D wave equation. Starting with $\bar{\mathbf{u}}(0)$ and $\bar{\mathbf{u}}(-\Delta t)$ determined by the initial condition (3.23), we compute $\bar{\mathbf{u}}(\Delta t)$. Then with the help of $\bar{\mathbf{u}}(0)$ and $\bar{\mathbf{u}}(\Delta t)$ we find $\bar{\mathbf{u}}(2\Delta t)$. This is repeated as long as required.

The strategy that we followed in the derivation of (3.32) is very general. It can, in particular, be used together with higher order finite-difference operators. The achieved accuracy will generally depend not only on the time increment, the grid spacing, and the properties of the finite-difference approximations but also on the material parameters.

To test the performance of the algorithm, we consider a homogeneous medium where the exact analytical solution $u(x,t)$ is well-known:

$$u(x,t) = \frac{1}{2}\left[u_0(x - vt) + u_0(x + vt)\right]. \tag{3.33}$$

The wave field consists of two wave packages propagating in opposite directions with the velocity $v = \sqrt{\mu/\rho}$. The solution (3.33) is valid as long as the wave field does not reach the boundaries at $x = \pm L$. For our numerical experiment we choose $v = 5\,\text{km/s}$, $\Delta x = 1\,\text{km}$ and $\Delta t = 0.2\,\text{s}$. The initial wave field is the derivative of a Gaussian with variance σ:

$$u_0(x) = \frac{\mathrm{d}}{\mathrm{d}x}\,\mathrm{e}^{-x^2/\sigma^2}. \tag{3.34}$$

Choosing $\sigma = 15\,\text{km}$ results in a dominant wavelength of approximately 30 km. Figure 3.4 shows the results of the numerical simulation and compares them to the analytical solution. For propagation distances up to 200 km, roughly equivalent to 7 wavelengths, the numerical solution is hardly distinguishable from the analytical reference. However, as the simulation proceeds, the numerical errors increase steadily. Clearly, after 450 s (2,250 time steps) the numerical solution becomes almost useless. The snapshot at 900 s (4,500 time steps) reveals that the numerical solution tends to disperse, meaning that higher frequency components appear to propagate slower than lower frequency components. This is despite the non-

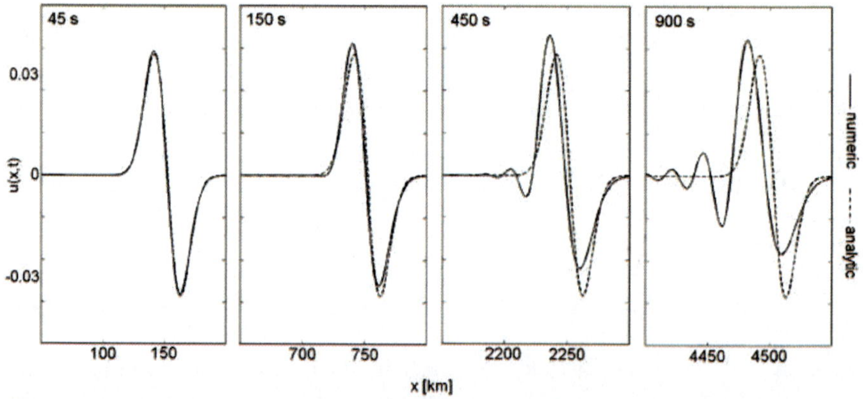

Fig. 3.4 Snapshots of the wave package travelling in positive x-direction for different times. The analytical solution is shown for reference in the form of the *dashed curves*. The numerical error, which is the discrepancy between the exact analytical solution and the numerical solution, increases steadily with increasing propagation distance. Details of the simulation parameters are provided in the text

dispersive character of the analytical solution where all frequencies travel at the same wave speed v. We refer to this undesirable phenomenon as *numerical dispersion* or *grid dispersion*.

To better quantify the discrepancy between analytical and numerical solutions, we use the following measure of the numerical error:

$$E(t) := \frac{\sqrt{\sum_{i=-N}^{N}[\bar{u}^i(t) - u(i\Delta x, t)]^2}}{\sqrt{\sum_{i=-N}^{N} u^2(i\Delta x, t)}}. \tag{3.35}$$

More elaborate measures of misfit can be defined using, for instance, wavelet transforms (Kristekova et al., 2006, 2009), but $E(t)$ as defined in (3.35) is sufficient for our purposes. There are three major factors that control the numerical error: the dominant wavelength relative to the grid spacing Δx, the order of the finite-difference operator and the number of iterations or time steps.

The left part of Fig. 3.5 illustrates the role played by the finite-difference order and the dominant wavelength. Shown is the error $E(t)$ after 1,000 iterations, that is for $t = 200$ s. Clearly, the error drops rapidly as the dominant wavelength increases. To achieve an error below 1 %, the dominant wavelength should not be less than 20 km for the fourth- and eighth-order operators, or 40 km for the second-order operator. Increasing orders generally leads to more accurate results, but the effect is largest when going from order 2 to 4.

The right panel of Fig. 3.5 shows the dependence of the error $E(t)$ on the number of iterations. It demonstrates that numerical inaccuracies are cumulative, meaning that they increase monotonically – and in this case almost exactly linearly – with the propagation distance of the waves.

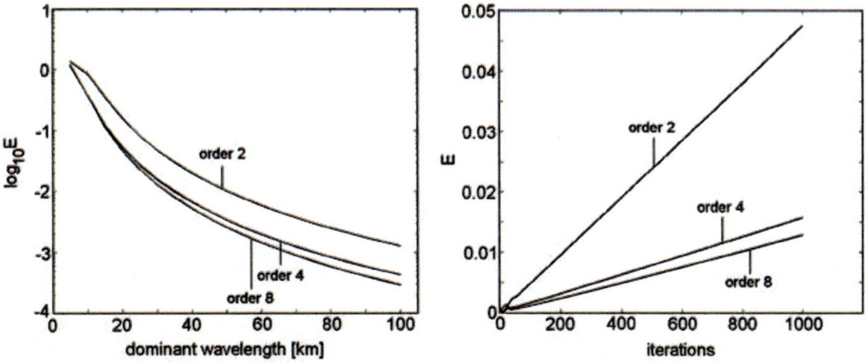

Fig. 3.5 *Left*: Numerical error after 1,000 iterations as a function of the dominant wavelength and the finite-difference order. *Right*: Numerical error as a function of the number of iterations and the finite-difference order

3.1.3 von Neumann Analysis: Stability and Numerical Dispersion

Each method used to approximate the solution of a differential equation requires a detailed analysis in order to learn about errors and stability without relying on specific numerical experiments that may not be representative. *von Neumann analysis* – first applied by Crank & Nicolson (1947) and Charney et al. (1950) – is a powerful tool for the evaluation of finite-difference approximations to hyperbolic partial differential equations. It allows us to study stability and the numerical dispersion that we already observed qualitatively in the previous paragraph.

For our analysis we assume a homogeneous medium defined on the interval $[-\pi, \pi]$, where the wave field $u(x,t)$ can be represented in terms of a Fourier series

$$u(x,t) = \sum_{k=-\infty}^{\infty} \psi_k(t)\, \mathrm{e}^{\mathrm{i}kx} \tag{3.36}$$

with time-dependent Fourier coefficients $\psi_k(t)$. The corresponding representation for the discrete approximation to the wave field is

$$\bar{u}^n(t) = \sum_{k=-N/2}^{N/2} \bar{\psi}_k(t)\, \mathrm{e}^{\mathrm{i}kn\,\Delta x}\,. \tag{3.37}$$

Note that the spatial discretisation with grid spacing $\Delta x = 2\pi/N$ restricts the wave number range to plus/minus the Nyquist wave number $\pi/\Delta x$ that is to $k \in [-\pi/\Delta x, \pi/\Delta x] = [-N/2, N/2]$. Introducing (3.37) into the semi-discrete wave equation (3.29) gives

$$\begin{aligned}\ddot{\bar{u}}^n(t) &= \frac{v^2}{4\,\Delta x^2} \sum_{k=-N/2}^{N/2} \left[\mathrm{e}^{2\mathrm{i}k\,\Delta x} - 2 + \mathrm{e}^{-2\mathrm{i}k\,\Delta x}\right] \bar{\psi}_k(t)\, \mathrm{e}^{\mathrm{i}kn\,\Delta x} \\ &= -\frac{v^2}{\Delta x^2} \sum_{k=-N/2}^{N/2} \sin^2(k\,\Delta x)\, \bar{\psi}_k(t)\, \mathrm{e}^{\mathrm{i}kn\,\Delta x}\,. \end{aligned} \tag{3.38}$$

Replacing $\ddot{\bar{u}}^n(t)$ by the second-order finite-difference approximation (3.31) and substituting the Fourier series (3.37) yields an equation for the coefficients $\bar{\psi}_k$:

$$\bar{\psi}_k(t+\Delta t) - 2\bar{\psi}_k(t) + \bar{\psi}_k(t-\Delta t) = -\frac{v^2 \Delta t^2}{\Delta x^2} \sin^2(k\,\Delta x)\, \bar{\psi}_k(t)\,. \tag{3.39}$$

To eliminate the explicit dependence on $\bar{\psi}_k$ at time $t - \Delta t$, we define the auxiliary variables

$$\bar{\phi}_k(t+\Delta t) := \bar{\psi}_k(t+\Delta t) - \bar{\psi}_k(t)\,. \tag{3.40}$$

With the help of $\bar{\phi}_k$ we can write Eq. (3.39) in the form of a linear system of equations that depends only on coefficients at times t and $t + \Delta t$:

$$\begin{aligned}\bar{\phi}_k(t+\Delta t) - \bar{\phi}_k(t) &= -\frac{v^2 \Delta t^2}{\Delta x^2} \sin^2(k\,\Delta x)\,\bar{\psi}_k(t)\,,\\ \bar{\psi}_k(t+\Delta t) - \bar{\psi}_k(t) &= \bar{\phi}_k(t+\Delta t)\,, \end{aligned} \tag{3.41}$$

or using matrix notation:

$$\begin{pmatrix}\bar{\phi}_k(t+\Delta t)\\ \bar{\psi}_k(t+\Delta t)\end{pmatrix} = \mathbf{A}_k \cdot \begin{pmatrix}\bar{\phi}_k(t)\\ \bar{\psi}_k(t)\end{pmatrix}. \tag{3.42}$$

The 2×2 matrix $\mathbf{A}_k$ is given by

$$\mathbf{A}_k = \begin{pmatrix} 1 & -\frac{v^2\Delta t^2}{\Delta x^2}\sin^2(k\,\Delta x)\\ 1 & 1-\frac{v^2\Delta t^2}{\Delta x^2}\sin^2(k\,\Delta x)\end{pmatrix}. \tag{3.43}$$

It follows that we can advance the Fourier coefficients $\bar{\psi}_k$ in time by the repeated application of $\mathbf{A}_k$ to the initial values of the coefficients:

$$\begin{pmatrix}\bar{\phi}_k(j\,\Delta t)\\ \bar{\psi}_k(j\,\Delta t)\end{pmatrix} = \mathbf{A}_k^j \cdot \begin{pmatrix}\bar{\phi}_k(0)\\ \bar{\psi}_k(0)\end{pmatrix}. \tag{3.44}$$

The following stability and dispersion analysis is founded on the iteration defined by Eq. (3.44).

3.1.3.1 Numerical Stability

All relevant properties of the algorithm, and its stability in particular, are now encapsulated in the eigenvalues λ_k of the matrix $\mathbf{A}_k$. In fact, for the iteration to be stable, the absolute values of the eigenvalues $|\lambda_k|$ must be smaller than or equal to 1. Otherwise, that is, for $|\lambda_k| > 1$, the absolute values of $\bar{\psi}_k$ and $\bar{\phi}_k$ grow indefinitely. To facilitate the eigenvalue analysis, we define

$$\gamma_k := \sin^2(k\,\Delta x) \tag{3.45}$$

and the *Courant number*

$$c := \frac{v\,\Delta t}{\Delta x}. \tag{3.46}$$

In terms of γ_k and c, the eigenvalues λ_k of $\mathbf{A}_k$ are given by

$$\lambda_k = 1 - \frac{1}{2}c^2\gamma_k^2 \pm \sqrt{\frac{1}{4}c^4\gamma_k^4 - c^2\gamma_k^2}\,. \tag{3.47}$$

For $c^2\gamma_k^2 > 4$ the square root is always real, and we can have $|\lambda_k| > 1$. For the algorithm to be stable we therefore require at least $c^2\gamma_k^2 \leq 4$. In this case, the square root becomes imaginary, and the absolute value of λ_k is then

$$|\lambda_k| = \sqrt{\lambda_k \lambda_k^*} = 1\,, \tag{3.48}$$

which implies stability. Since γ_k ranges between 0 and 1, it follows that we need to impose

$$\frac{v\,\Delta t}{\Delta x} = c \leq 2 \tag{3.49}$$

in order to ensure the stability of the iterative solution to the 1D wave equation discretised by second-order finite-difference approximations, as explained in the previous paragraph. The stability condition (3.49) is commonly referred to as the *CFL condition*, named after R. Courant, K. Friedrichs and H. Lewy (Courant et al. 1928). For our example from Fig. 3.4, with $\Delta x = 1\,\text{km}$ and $v = 5\,\text{km/s}$, the CFL condition imposes a maximum time increment of 0.4 s. Smaller time increments are predicted to yield stable solutions, whereas larger time increments are expected to lead to solutions that grow indefinitely as the iteration proceeds. Figure 3.6 proves that the prediction of the CFL condition is remarkably accurate. The solution is stable for $\Delta x = 0.39\,\text{s}$, but it 'explodes' for $\Delta t = 0.41\,\text{s}$.

Our analysis was based on a specific finite-difference discretisation of the 1D wave equation with constant parameters. More generally, CFL conditions of the form

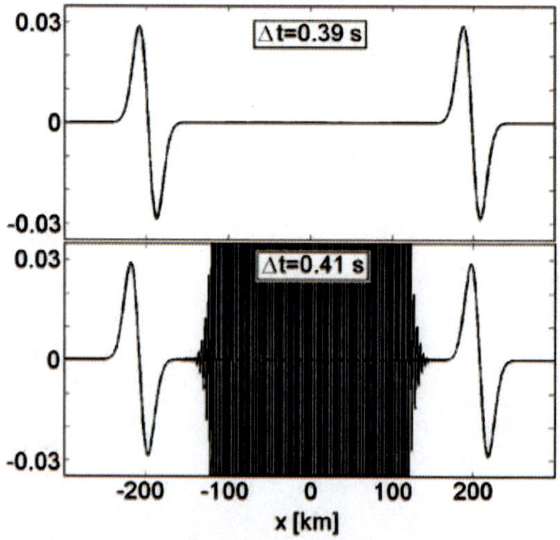

Fig. 3.6 Numerical (in)stability. *Top*: The time increment $\Delta t = 0.39\,\text{s}$ is slightly below the limit of 0.4 s imposed by the CFL condition. The solution is stable. *Bottom*: The time increment $\Delta t = 0.41$ is slightly above the limit of 0.4, leading to instabilities. After few iterations the numerical solution 'explodes'

$$\Delta t \leq \text{const.} \frac{\min h}{\max v} \tag{3.50}$$

with a grid spacing parameter h and a wave speed v are valid for all discretised wave equations where the time-stepping is explicit. The constant on the right-hand side depends on the methods used for the space and time discretisation.

The CFL condition strongly limits the efficiency of numerical methods because it imposes an upper bound for the time increment Δt. Reducing the minimum grid spacing by a factor m in order to achieve more accurate solutions automatically implies a reduction of the time increment from Δt to $\Delta t/m$.

It is, in principle, possible to circumvent the CFL condition using implicit time-stepping schemes. However, their numerical costs usually compensate the benefit of using a larger time increment.

3.1.3.2 Numerical Dispersion

To quantify the numerical dispersion that we observed in Fig. 3.4, we assume, without loss of generality, that the initial state $(\bar{\phi}_k(0), \bar{\psi}_k(0))$ from Eq. (3.44) is an eigenvector of the matrix $\mathbf{A}_k$. The iteration that advances the Fourier coefficients in time then simplifies to

$$\begin{pmatrix} \bar{\phi}_k(j\Delta t) \\ \bar{\psi}_k(j\Delta t) \end{pmatrix} = \lambda_k^j \begin{pmatrix} \bar{\phi}_k(0) \\ \bar{\psi}_k(0) \end{pmatrix} . \tag{3.51}$$

Since λ_k is complex valued within the stable range of time increments Δt, it can be rewritten in Eulerian form

$$\lambda_k = |\lambda_k| \, e^{i\alpha_k} , \tag{3.52}$$

where the phase α_k is generally non-zero. To attach physical meaning to the phase, we write α_k in terms of a numerical wave speed $\bar{v}_k$:

$$\alpha_k =: \Delta t \, k \, \bar{v}_k . \tag{3.53}$$

Introducing Eqs. (3.51), (3.52) and (3.53) into the Fourier series representation of the finite-difference coefficients $\bar{u}^n$ (Eq. 3.37) gives

$$\bar{u}^n(j\Delta t) = \sum_{k=-N/2}^{N/2} \bar{\psi}_k(0) \, |\lambda_k|^j \, e^{ik(j\Delta t \, \bar{v}_k + n\Delta x)} . \tag{3.54}$$

Equation (3.54) reveals that $\bar{v}_k$ indeed plays the role of a wave speed. If the numerical solution were exact, $\bar{v}_k$ would be equal to $v = \sqrt{\mu/\rho}$ and independent of the wave number k. As a result of the discretisation, however, $\bar{v}_k$ is not generally equal to v. Moreover, $\bar{v}_k$ depends on k, meaning that the numerical solution is dispersive, in

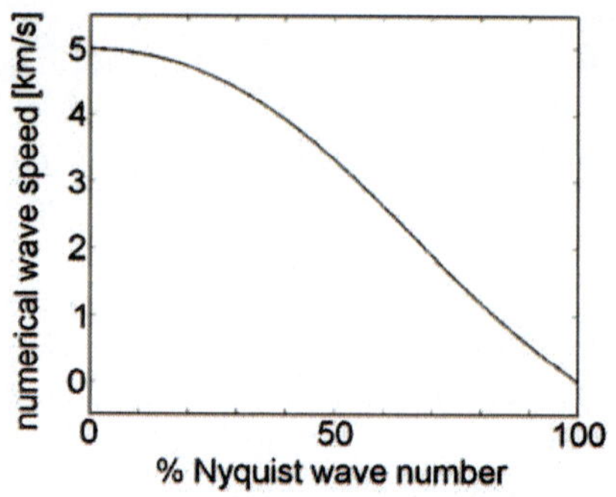

Fig. 3.7 Numerical dispersion. Shown is the numerical wave speed $\bar{v}_k$ as a function of the wave number, k. The wave speed of the exact solution is $v = 5\,\text{km/s}$, independent of k. Due to numerical dispersion, $\bar{v}_k$ decreases with increasing wave number. Shorter wavelength components therefore travel at reduced speed. Wavelengths corresponding to the Nyquist wave number do not propagate at all. Only components with a wave number close to zero travel with the correct speed

contrast to the exact solution. Figure 3.7 illustrates the effect of numerical dispersion for the setup that we already used for the example in Fig. 3.4. There we observed qualitatively that shorter wavelength components travel at a lower speed than longer wavelength components, and that the numerical solution in general seems to be slower than the analytical solution. This is confirmed by our qualitative dispersion analysis. Only components with a wave number close to zero propagate at the correct speed, $v = 5\,\text{km/s}$. As the wave numbers approach the Nyquist wave number, the propagation speed tends to zero.

Numerical dispersion is, of course, not a special property of our particular example. All numerical solutions are dispersive to some degree, regardless of the method used. In many cases, numerical dispersion analysis is not as straightforward as in the case of low-order finite-difference methods. Its effect should therefore, if possible, be assessed by comparison with analytical solutions.

3.2 Extension to the 3D Cartesian Case

The application of the finite-difference method to the 3D elastic wave equation is conceptually similar to the 1D scalar case that we studied in the previous section. Exact derivatives are replaced by finite-difference approximations, and this leads to a discrete scheme that can be advanced in time iteratively.

Yet, the simple scheme applied in one dimension proves inefficient in three dimensions because of the large number of grid points per wavelength needed to achieve accurate solutions. Fortunately, a modification of the spatial discretisation – described in Sect. 3.2.1 and referred to as the *staggered grid* – allows us to effectively reduce the grid spacing without increasing the number of discrete field variables. In Sect. 3.2.2 we explore how the staggered-grid approach can be extended to the anisotropic case with the help of interpolation operators.

All finite-difference operators are non-local in the sense that the approximation of $\partial_x f$ at a grid point x_i requires f at neighbouring grid points. Non-locality has interesting consequences for the implementation of the free surface. To compute the strain tensor and the stress divergence at and immediately below the surface, part of the seismic wave field needs to be known *above* the surface, at least when standard finite-difference operators are used. Since the implementation of the free surface is far from trivial, it receives special attention in Sect. 3.2.3.

Finally, we elaborate on the numerical implementation of arbitrarily located point sources (Sect. 3.4) and discuss the accuracy and efficiency of 3D finite-difference modelling (Sect. 3.5).

3.2.1 The Staggered Grid

To translate the finite-difference method to the 3D cartesian case, we consider the velocity–stress formulation of the elastic wave equation in the half-space $G = \mathbb{R}^2 \times (-\infty, z_0]$

$$\rho(\mathbf{x})\,\dot{\mathbf{v}}(\mathbf{x},t) - \nabla\cdot\boldsymbol{\sigma}(\mathbf{x},t) = \mathbf{f}(\mathbf{x},t)\,, \tag{3.55}$$

$$\dot{\boldsymbol{\sigma}}(\mathbf{x},t) = \mathbf{C}(\mathbf{x}) : \dot{\boldsymbol{\varepsilon}}(\mathbf{x},t) \tag{3.56}$$

subject to the free surface boundary condition

$$\mathbf{e}_z \cdot \boldsymbol{\sigma}|_{z=z_0} = \mathbf{0} \tag{3.57}$$

and the initial conditions

$$\mathbf{u}|_{t\le t_0} = \mathbf{v}|_{t\le t_0} = \mathbf{0}\,. \tag{3.58}$$

The symbol $\mathbf{e}_z$ denotes the unit vector in vertical direction. For the moment we disregard dissipation because its description (see Chap. 5) is largely independent of the spatial discretisation of the equations of motion. The first step towards the finite-difference approximation of Eqs. (3.55) and (3.56) is to write the stress divergence $\nabla \cdot \boldsymbol{\sigma}$ and the strain rate tensor $\dot{\boldsymbol{\varepsilon}}$ in explicit form:

$$\begin{aligned}
(\nabla\cdot\boldsymbol{\sigma})_x &= \partial_x\sigma_{xx} + \partial_y\sigma_{xy} + \partial_z\sigma_{xz}\,,\\
(\nabla\cdot\boldsymbol{\sigma})_y &= \partial_x\sigma_{yx} + \partial_y\sigma_{yy} + \partial_z\sigma_{yz}\,,\\
(\nabla\cdot\boldsymbol{\sigma})_z &= \partial_x\sigma_{zx} + \partial_y\sigma_{zy} + \partial_z\sigma_{zz}\,,
\end{aligned} \tag{3.59}$$

$$\begin{aligned}
&\dot{\varepsilon}_{xx} = \partial_x v_x\,, \quad \dot{\varepsilon}_{yy} = \partial_y v_y\,, \quad \dot{\varepsilon}_{zz} = \partial_z v_z\,,\\
&\dot{\varepsilon}_{xy} = \frac{1}{2}(\partial_x v_y + \partial_y v_x)\,, \quad \dot{\varepsilon}_{xz} = \frac{1}{2}(\partial_x v_z + \partial_z v_x)\,, \quad \dot{\varepsilon}_{zy} = \frac{1}{2}(\partial_z v_y + \partial_y v_z)\,.
\end{aligned} \tag{3.60}$$

Our goal is to replace the derivatives in the above equations by finite-difference approximations. The practicality of this approach in three dimensions rests on the definition of a *staggered grid* (e.g. Virieux, 1984, 1986; Levander, 1988; Igel et al., 1995; Graves, 1996). To illustrate the staggered-grid concept we start by defining the discrete versions of the velocity components v_x, v_y and v_z on three different grids:

$$\bar{v}_x^{i,j,k} := v_x(x_i + \Delta x/2, y_j, z_k)\,, \tag{3.61}$$

$$\bar{v}_y^{i,j,k} := v_y(x_i, y_j + \Delta y/2, z_k)\,, \tag{3.62}$$

$$\bar{v}_z^{i,j,k} := v_z(x_i, y_j, z_k + \Delta z/2)\,, \tag{3.63}$$

where $\mathbf{x}_{ijk} = (x_i, y_j, z_k)$ is a generic grid point. Based on the above discretisations we can approximate the derivatives of the velocity field, as they appear in Eq. (3.60). For instance, the fourth-order approximation of $\partial_y v_x(x_i + \Delta x/2, y_j + \Delta y/2, z_k)$ in terms of v_x evaluated at the positions $(x_i + \Delta x/2, y_j + n\Delta y, z_k)$ with $n = -1, 0, 1, 2$, is given by (see Eq. 3.13)

$$\begin{aligned}
&\partial_y v_x(x_i + \Delta x/2, y_j + \Delta y/2, z_k) \\
&\quad\approx \frac{9}{8\,\Delta y}\left[v_x(x_i + \Delta x/2, y_j + \Delta y, z_k) - v_x(x_i + \Delta x/2, y_j, z_k)\right] \\
&\quad- \frac{1}{24\,\Delta y}\left[v_x(x_i + \Delta x/2, y_j + 2\Delta y, z_k) - v_x(x_i + \Delta x/2, y_j - \Delta y, z_k)\right] \\
&\quad= \frac{9}{8\,\Delta y}\left[\bar{v}_x^{i,j+1,k} - \bar{v}_x^{i,j,k}\right] - \frac{1}{24\,\Delta y}\left[\bar{v}_x^{i,j+2,k} - \bar{v}_x^{i,j-1,k}\right].
\end{aligned} \tag{3.64}$$

The corresponding approximation for $\partial_x v_y$ is

$$\begin{aligned}
&\partial_x v_y(x_i + \Delta x/2, y_j + \Delta y/2, z_k) \\
&\quad\approx \frac{9}{8\,\Delta x}\left[\bar{v}_y^{i+1,j,k} - \bar{v}_y^{i,j,k}\right] - \frac{1}{24\,\Delta x}\left[\bar{v}_y^{i+2,j,k} - \bar{v}_y^{i-1,j,k}\right].
\end{aligned} \tag{3.65}$$

At this point we note that the approximations of $\partial_y v_x$ and $\partial_x v_y$ are available at the same grid point, namely at $(x_i + \Delta x/2, y_j + \Delta y/2, z_k)$. We may thus combine (3.64) and (3.65) into the fourth-order approximation of the strain rate component $\dot{\varepsilon}_{xy}$:

$$\begin{aligned}
\dot{\varepsilon}_{xy}^{i,j,k} &= \dot{\varepsilon}_{xy}(x_i + \Delta x/2, y_j + \Delta y/2, z_k) \\
&= \frac{1}{2}\left[\partial_y v_x(x_i + \Delta x/2, y_j + \Delta y/2, z_k) + \partial_x v_y(x_i + \Delta x/2, y_j + \Delta y/2, z_k)\right] \\
&\approx \frac{9}{16\,\Delta x}\left[\bar{v}_x^{i,j+1,k} - \bar{v}_x^{i,j,k} + \bar{v}_y^{i+1,j,k} - \bar{v}_y^{i,j,k}\right] \\
&- \frac{1}{48\,\Delta y}\left[\bar{v}_x^{i,j+2,k} - \bar{v}_x^{i,j-1,k} + \bar{v}_y^{i+2,j,k} - \bar{v}_y^{i-1,j,k}\right].
\end{aligned} \tag{3.66}$$

Following this example we find the grid points where the finite-difference approximations of the remaining strain rate components are defined:

$$
\begin{aligned}
\dot{\bar{\varepsilon}}_{xx}^{i,j,k} &= \dot{\varepsilon}_{xx}(x_i, y_j, z_k)\,, \quad \dot{\bar{\varepsilon}}_{yy}^{i,j,k} = \dot{\varepsilon}_{yy}(x_i, y_j, z_k)\,, \quad \dot{\bar{\varepsilon}}_{zz}^{i,j,k} = \dot{\varepsilon}_{zz}(x_i, y_j, z_k)\,,\\
\dot{\bar{\varepsilon}}_{xy}^{i,j,k} &= \dot{\varepsilon}_{xy}(x_i + \Delta x/2, y_j + \Delta y/2, z_k)\,,\\
\dot{\bar{\varepsilon}}_{xz}^{i,j,k} &= \dot{\varepsilon}_{xz}(x_i + \Delta x/2, y_j, z_k + \Delta z/2)\,,\\
\dot{\bar{\varepsilon}}_{yz}^{i,j,k} &= \dot{\varepsilon}_{yz}(x_i, y_j + \Delta y/2, z_k + \Delta z/2)\,.
\end{aligned}
\tag{3.67}
$$

Assuming an isotropic medium, the stress and strain rate tensors are connected via the constitutive relation

$$
\sigma_{ij} = \sum_{k,l=1}^{3} (\lambda\,\delta_{ij}\delta_{kl} + \mu\,\delta_{ik}\delta_{jl} + \mu\,\delta_{il}\delta_{jk})\,\varepsilon_{kl}\,,
\tag{3.68}
$$

where λ and μ are the Lamé parameters. Using (3.68) we can compute the components of the discrete stress rate tensor:

$$
\begin{aligned}
\dot{\bar{\sigma}}_{xx}^{ij} &= (\lambda + 2\mu)\,\dot{\bar{\varepsilon}}_{xx}^{ij} + \lambda\,(\dot{\bar{\varepsilon}}_{yy}^{ij} + \dot{\bar{\varepsilon}}_{zz}^{ij})\,,\\
\dot{\bar{\sigma}}_{yy}^{ij} &= (\lambda + 2\mu)\,\dot{\bar{\varepsilon}}_{yy}^{ij} + \lambda\,(\dot{\bar{\varepsilon}}_{xx}^{ij} + \dot{\bar{\varepsilon}}_{zz}^{ij})\,,\\
\dot{\bar{\sigma}}_{zz}^{ij} &= (\lambda + 2\mu)\,\dot{\bar{\varepsilon}}_{zz}^{ij} + \lambda\,(\dot{\bar{\varepsilon}}_{xx}^{ij} + \dot{\bar{\varepsilon}}_{yy}^{ij})\,,\\
\dot{\bar{\sigma}}_{xy}^{ij} &= 2\mu\,\dot{\bar{\varepsilon}}_{xy}^{ij}\,, \quad \dot{\bar{\sigma}}_{xz}^{ij} = 2\mu\,\dot{\bar{\varepsilon}}_{xz}^{ij}\,, \quad \dot{\bar{\sigma}}_{yz}^{ij} = 2\mu\,\dot{\bar{\varepsilon}}_{yz}^{ij}\,.
\end{aligned}
\tag{3.69}
$$

Equations (3.69) reveal that the discrete stress rate components $\dot{\bar{\sigma}}_{mn}^{ij}$ are located at the same grid positions as the discrete strain rate components $\dot{\bar{\varepsilon}}_{mn}^{ij}$. The collocation of stress and strain components holds only in isotropic media, as we will see in Sect. (3.2.3). Note also that the elastic parameters λ and μ are needed at different grid positions. We did not translate this requirement into the notation in order to keep the treatment readable. Based on (3.69) we can advance the discrete stress field $\bar{\sigma}$ in time. It remains to discretise the stress divergence. For the x-component of $\nabla \cdot \boldsymbol{\sigma}$ we find the following fourth-order approximation:

$$
\begin{aligned}
&[\nabla \cdot \boldsymbol{\sigma}(x_i + \Delta x/2, y_j, z_k)]_x = \rho\,\dot{v}_x(x_i + \Delta x/2, y_j, z_k) - f_x(x_i + \Delta x/2, y_j, z_k)\\
&\approx \frac{9}{8\Delta x}\left[\sigma_{xx}(x_i + \Delta x, y_j, z_k) - \sigma_{xx}(x_i, y_j, z_k)\right]\\
&- \frac{1}{24\Delta x}\left[\sigma_{xx}(x_i + 2\Delta x, y_j, z_k) - \sigma_{xx}(x_i - \Delta x, y_j, z_k)\right]\\
&+ \frac{9}{8\Delta y}\left[\sigma_{xy}(x_i + \Delta x/2, y_j + \Delta y/2, z_k) - \sigma_{xy}(x_i + \Delta x/2, y_j - \Delta y/2, z_k)\right]\\
&- \frac{1}{24\Delta y}\left[\sigma_{xy}(x_i + \Delta x/2, y_j + 3\Delta y/2, z_k) - \sigma_{xy}(x_i + \Delta x/2, y_j - 3\Delta y/2, z_k)\right]
\end{aligned}
$$

$$
+ \frac{9}{8\Delta z}\left[\sigma_{xz}(x_i+\Delta x/2, y_j, z_k+\Delta z/2) - \sigma_{xz}(x_i+\Delta x/2, y_j, z_k-\Delta z/2)\right]
- \frac{1}{24\Delta z}\left[\sigma_{xz}(x_i+\Delta x/2, y_j, z_k+3\Delta z/2) - \sigma_{xz}(x_i+\Delta x/2, y_j, z_k-3\Delta z/2)\right]. \tag{3.70}
$$

Written in terms of the discrete field variables, Eq. (3.70) becomes

$$
\begin{aligned}
\rho\,\dot{\bar{v}}_x^{i,j,k} - f_x^{i,j,k} &= \frac{9}{8\Delta x}\left[\bar{\sigma}_{xx}^{i+1,j,k} - \bar{\sigma}_{xx}^{i,j,k}\right] - \frac{1}{24\Delta x}\left[\bar{\sigma}_{xx}^{i+2,j,k} - \sigma_{xx}^{i-1,j,k}\right] \\
&+ \frac{9}{8\Delta y}\left[\bar{\sigma}_{xy}^{i,j,k} - \bar{\sigma}_{xy}^{i,j-1,k}\right] - \frac{1}{24\Delta y}\left[\bar{\sigma}_{xy}^{i,j+1,k} - \sigma_{xy}^{i,j-2,k}\right] \\
&+ \frac{9}{8\Delta z}\left[\bar{\sigma}_{xz}^{i,j,k} - \bar{\sigma}_{xz}^{i,j,k-1}\right] - \frac{1}{24\Delta z}\left[\bar{\sigma}_{xz}^{i,j,k+1} - \sigma_{xz}^{i,j,k-2}\right].
\end{aligned} \tag{3.71}
$$

Following this scheme, we find that the grid positions of $\bar{v}_x^{i,j,k}$, $\bar{v}_y^{i,j,k}$ and $\bar{v}_z^{i,j,k}$ coincide with those that we originally defined in Eqs. (3.61), (3.62) and (3.63). The staggered grid is therefore in itself consistent. Figure 3.8 summarises the positions of the different discrete quantities.

The above development suggests a simple recipe for iteratively advancing the discrete wave field with the help of the leapfrog time discretisation (see Sect. 2.5.1). Starting from the zero initial conditions

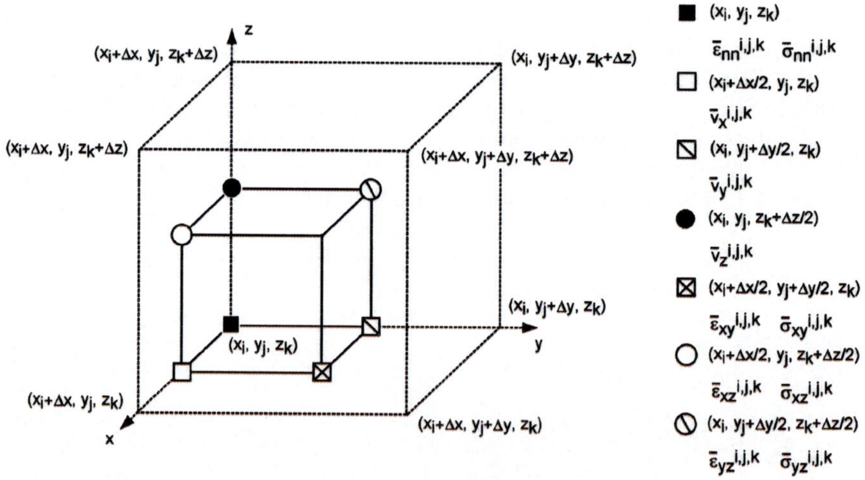

Fig. 3.8 Illustration of the *staggered-grid* concept. The diagonal elements of the discrete stress and strain tensors $\bar{\varepsilon}_{nn}^{i,j,k}$ and $\bar{\sigma}_{nn}^{i,j,k}$ are located at the generic *grid point* $\mathbf{x}_{ijk} = (x_i, y_j, z_k)$. The remaining quantities are displaced from $\mathbf{x}_{ijk}$ by half a grid spacing.

1. compute the strain rate tensor $\dot{\bar{\boldsymbol{\varepsilon}}}(t)$ from the velocity $\bar{\mathbf{v}}(t)$,
2. using the constitutive relation, compute the strain rate $\dot{\bar{\boldsymbol{\sigma}}}(t)$ from $\dot{\bar{\boldsymbol{\varepsilon}}}(t)$,
3. using $\dot{\bar{\boldsymbol{\sigma}}}(t)$, compute $\bar{\boldsymbol{\sigma}}(t + \Delta t/2)$ via a second-order finite-difference approximation,
4. from $\bar{\boldsymbol{\sigma}}(t + \Delta t/2)$, compute the discrete version of $\nabla \cdot \bar{\boldsymbol{\sigma}}(t + \Delta t/2)$,
5. using $\nabla \cdot \bar{\boldsymbol{\sigma}}(t + \Delta t/2)$ and the discrete momentum balance, compute $\bar{\mathbf{v}}(t + \Delta t)$ with the help of the second-order finite-difference approximation, and then

repeat as often as required. The beauty of the staggered grid is that it allows us to decrease the effective grid spacing without increasing the number of discrete field variables. In the conventional grid used, for instance, in our 1D example (Sect. 3.1), the effective grid spacing in x-direction is a multiple of $\pm\Delta x$. This is reduced to $\pm\Delta x/2$ and $\pm 2\Delta x/3$. The result is a much lower numerical dispersion that comes without any increase in computational costs.

3.2.2 Anisotropy and Interpolation

A defining property of the staggered grid is that the diagonal and off-diagonal elements of $\dot{\bar{\boldsymbol{\sigma}}}$ and $\dot{\bar{\boldsymbol{\varepsilon}}}$ are defined at different grid positions. In the isotropic case, the diagonal elements of $\dot{\bar{\boldsymbol{\sigma}}}$ are related only to the diagonal elements of $\dot{\bar{\boldsymbol{\varepsilon}}}$, and each off-diagonal element of $\dot{\bar{\boldsymbol{\sigma}}}$ is linked to the same off-diagonal element of $\dot{\bar{\boldsymbol{\varepsilon}}}$. In the presence of anisotropy, however, mixing between diagonal and off-diagonal elements occurs because the stress rate component $\dot{\bar{\sigma}}_{ij}$ is determined by a linear combination of strain rate components $\dot{\bar{\varepsilon}}_{kl}$:

$$\dot{\bar{\sigma}}_{ij} = \sum_{kl=1}^{3} C_{ijkl} \dot{\bar{\varepsilon}}_{kl} , \tag{3.72}$$

where C_{ijkl} are the components of the elastic tensor. Anisotropy therefore breaks the self-consistency of the staggered grid, and interpolation becomes necessary in order to evaluate the Hooke sum (3.72).

To interpolate, for instance, $\dot{\varepsilon}_{xx}(x_i, y_j, z_k)$ to a location halfway between the grid points in x-direction, i.e. to $\dot{\varepsilon}_{xx}(x_i + \Delta x/2, y_j, z_k)$, we require a discrete interpolation operator. To derive discrete interpolation operators we proceed as in Sect. 3.1.1.2 where we constructed finite-difference approximations by sampling band-limited versions of the exact derivative operator. Following Igel et al. (1995), we consider the generic function $f(x)$ that we wish to interpolate to $f(x + \Delta x/2)$. The interpolation can be expressed in terms of an inverse Fourier transform:

$$f(x + \Delta x/2) = \frac{1}{\sqrt{2\pi}} \int_{-\infty}^{\infty} \mathrm{e}^{\mathrm{i}k\,\Delta x/2} \tilde{f}(k)\, \mathrm{e}^{\mathrm{i}kx}\, \mathrm{d}k . \tag{3.73}$$

It follows that $\mathrm{e}^{\mathrm{i}k\,\Delta x/2}$ is the exact interpolation operator in the wave number domain. The spatial discretisation restricts the wave number range to $[-k_{\max}, k_{\max}]$, where $k_{\max} = \pi/\Delta x$ is, again, the Nyquist wave number. The optimal band-limited version of the interpolation operator in the wave number domain is therefore given by

$$\tilde{h}(k) = \mathrm{e}^{\mathrm{i}k\,\Delta x/2}\,[H(k+k_{\max}) - H(k-k_{\max})]\,. \tag{3.74}$$

The space-domain representation of (3.74) is

$$h(x) = \frac{2k_{\max}}{\sqrt{2\pi}}\,\mathrm{sinc}[k_{\max}(x+\Delta x/2)]\,. \tag{3.75}$$

We may now approximate the interpolation by a convolution of $f(x)$ with $h(x)$:

$$f(x+\Delta x/2) \approx \frac{1}{\sqrt{2\pi}} \int\limits_{-\infty}^{\infty} h(y) f(x-y)\,\mathrm{d}y\,. \tag{3.76}$$

Replacing the continuous convolution in (3.76) by a finite Riemann sum involving the grid positions $x \pm n\Delta x$ results in the following discrete interpolation operator:

$$f(x+\Delta x/2) \approx \sum_{n=-N}^{N} h_n f(x-n\Delta x)\,, \tag{3.77}$$

where the coefficients h_n are defined by

$$h_n := \frac{\Delta x}{\sqrt{2\pi}}\,h(n\Delta x) = \frac{(-1)^n}{\pi\,(n+1/2)}\,. \tag{3.78}$$

The discrete convolutional interpolation defined by Eqs. (3.77) and (3.78) can now be used to interpolate the components of the strain rate tensor to the positions where they are needed for the evaluation of the Hooke sum.
Following the example from Sect. 3.1.1.2, we assess the quality of the interpolation by comparing the wave number spectra of the exact interpolation operator $\mathrm{e}^{\mathrm{i}k\Delta x/2}$, and the approximate interpolation operators defined by Eqs. (3.77) and (3.78). This comparison is shown in the left part of Fig. 3.9 for discrete convolutions with two, four and eight points. The amplitude spectra of the two-point and four-point operators clearly indicate insufficient accuracy by being far from the amplitude spectrum of the exact interpolation, which is equal to 1. The eight-point convolution provides a better approximation of the exact interpolation operator, but its spectrum shows an undesirable oscillatory behaviour. As illustrated in the right part of Fig. 3.9, the oscillations can be suppressed by tapering the coefficients h_n with a Gaussian $\mathrm{e}^{-\alpha n^2}$, where the parameter α is around 0.07.

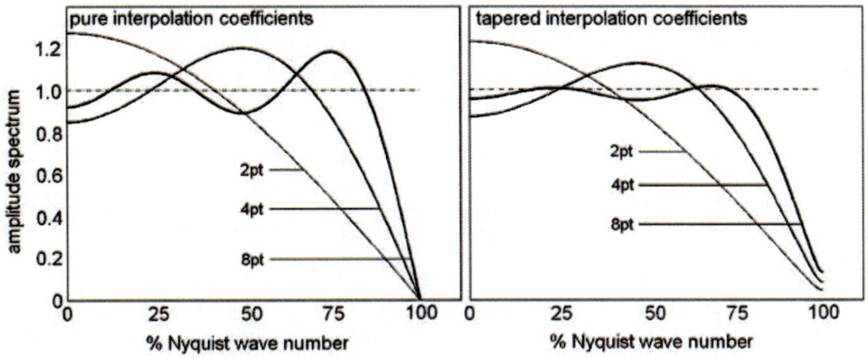

Fig. 3.9 *Left*: Amplitude spectra of the discrete interpolation operators defined in Eqs. (3.77) and (3.78), for convolutions involving two, four and eight points. The amplitude spectrum of the exact interpolation operator is shown in the form of a *dash-dotted line*. *Right*: The same as on the *left* but for coefficients tapered by the Gaussian $e^{-0.07\,n^2}$. The tapering suppresses the undesirable oscillations of the eight-point operator

As noted by Igel et al. (1995), the interpolation introduces an additional numerical error. The error is, however, small when the anisotropy is as weak as in the Earth seen at macroscopic length scales.

3.2.3 Implementation of the Free Surface

We have so far limited ourselves to the spatial discretisation in the interior of the computational domain, disregarding its boundaries. The numerical treatment of the free surface deserves special attention because it does not follow naturally from the staggered-grid scheme. In this sense, finite-difference methods are not holistic as finite-element methods where the free surface condition is implicitly satisfied.
Implementing the free surface accurately is difficult due to the non-local nature of the finite-difference approximations. It implies that velocities above the free surface are required to compute the discrete strain rate tensor at or immediately below the surface, as illustrated in Fig. 3.10. In fact, it follows directly from Eqs. (3.66) and (3.67) that $\bar{v}_x^{i,j,-1}$, $\bar{v}_y^{i,j,-1}$, $\bar{v}_z^{i,j,-1}$ and $\bar{v}_z^{i,j,-2}$ are needed to compute the fourth-order finite-difference approximations $\dot{\bar{\varepsilon}}_{xz}^{i,j,0}$, $\dot{\bar{\varepsilon}}_{yz}^{i,j,0}$ and $\dot{\bar{\varepsilon}}_{zz}^{i,j,0}$. Moreover, the discrete stress tensor $\bar{\sigma}$ above the surface is needed for the discrete momentum balance in Eq. (3.71).

The non-trivial nature of the free surface in finite-difference modelling has sparked the development of an impressive variety of numerical techniques. Today, the most widely used methods for the implementation of the free surface in finite-difference modelling fall into three categories: *vacuum formulations*, *image methods* and *interior methods*.

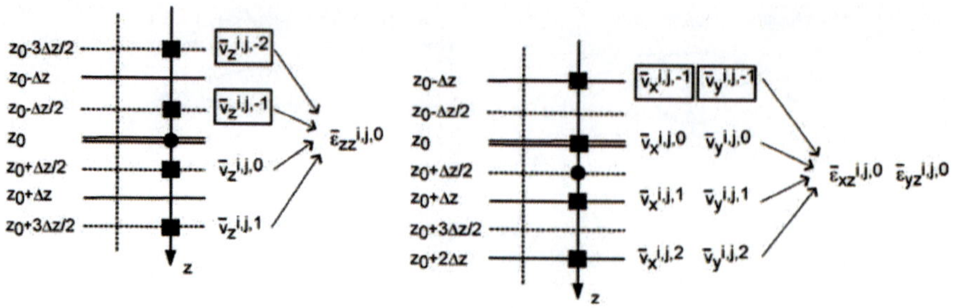

Fig. 3.10 Illustration of the spatial discretisation in the vicinity of the free surface at $z = z_0$, indicated by the *horizontal double line*. *Left*: The four-point finite-difference stencil implies that the velocity components $\bar{v}_z^{i,j,-1}$ and $\bar{v}_z^{i,j,-2}$ above the surface are needed to compute the strain rate component $\dot{\bar{\varepsilon}}_{zz}^{i,j,0}$. *Right*: The computation of $\dot{\bar{\varepsilon}}_{xz}^{i,j,0}$ and $\dot{\bar{\varepsilon}}_{yz}^{i,j,0}$ also requires knowledge of two velocity components above the surface, namely $\bar{v}_x^{i,j,-1}$ and $\bar{v}_y^{i,j,-1}$

3.2.3.1 Vacuum Formulation

A straightforward approach to the implementation of the free surface is to set the elastic parameters above the free surface to zero and density close to zero, to avoid division by zero. Since this is intended to approximate a vacuum, the method is commonly referred to as vacuum formulation. The free surface boundary condition is thus not treated explicitly. It is, instead, assumed to be implicitly fulfilled through the distribution of elastic parameters on the grid. Applications of the vacuum formulation can be found, for instance, in Zahradník & Urban (1984), Zahradník et al. (1993), Ohminato & Chouet (1997) and Bohlen & Saenger (2006).

The vacuum formulation is attractive because of its trivial implementation and the possibility to model irregular topography. Despite obvious advantages, the vacuum formulation has received criticism. Based on numerical experiments, Graves (1996) concluded that the method is stable only for two-point operators that are not sufficiently accurate for realistic applications. Modifications of the method led to stable fourth-order schemes, however, at the expense of decreased solution quality. Bohlen & Saenger (2006) investigated the ability of the vacuum formulation to model irregular topography and found that more than 60 grid points per minimum wavelength are required in a second-order scheme to obtain acceptable results when the surface is not planar.

3.2.3.2 Image Methods

Approximations to the discrete velocity components above the surface can be deduced from the boundary condition (3.57), the component-wise version of which is

$$\sigma_{xz}|_{z=z_0} = \sigma_{yz}|_{z=z_0} = \sigma_{zz}|_{z=z_0} = 0\,. \qquad (3.79)$$

Only $\sigma_{zz}|_{z=z_0} = 0$ can be translated directly into a discrete condition

$$\sigma_{zz}(x_i, y_i, z_0) = \bar{\sigma}_{zz}^{i,j,0} = 0\,. \tag{3.80a}$$

The discrete stress components $\bar{\sigma}_{xz}$ and $\bar{\sigma}_{yz}$ are not defined at $z = z_0$. We therefore approximate $\bar{\sigma}_{xz}|_{z=z_0} = 0$ and $\bar{\sigma}_{yz}|_{z=z_0} = 0$ by the following anti-symmetry conditions or *stress imaging conditions* (e.g. Levander, 1988; Graves, 1996):

$$\begin{aligned}
\sigma_{xz}(x_i + \Delta x/2, y_j, z_0 + \Delta z/2) &= \bar{\sigma}_{xz}^{i,j,0} = -\bar{\sigma}_{xz}^{i,j,-1} \\
&= -\sigma_{xz}(x_i + \Delta x/2, y_j, z_0 - \Delta z/2)\,, \qquad (3.80b)\\
\sigma_{yz}(x_i, y_j + \Delta y/2, z_0 + \Delta z/2) &= \bar{\sigma}_{yz}^{i,j,0} = -\bar{\sigma}_{yz}^{i,j,-1} \\
&= -\sigma_{yz}(x_i, y_j + \Delta y/2, z_0 - \Delta z/2)\,. \qquad (3.80c)
\end{aligned}$$

The discrete conditions (3.80a), (3.80b) and (3.80c) provide three linear equations for the four unknowns $\bar{v}_x^{i,j,-1}$, $\bar{v}_y^{i,j,-1}$, $\bar{v}_z^{i,j,-1}$ and $\bar{v}_z^{i,j,-2}$. Thus, to uniquely determine the required velocity components, we need to impose an additional anti-symmetry relation that approximates $\sigma_{zz}|_{z=z_0} = 0$ in terms of $\bar{\sigma}_{zz}^{i,j,1}$ and $\bar{\sigma}_{zz}^{i,j,-1}$:

$$\sigma_{zz}(x_i, y_j, z_0 + \Delta z) = \bar{\sigma}_{zz}^{i,j,1} = -\bar{\sigma}_{zz}^{i,j,-1} = -\sigma_{zz}(x_i, y_j, z_0 - \Delta z)\,. \tag{3.80d}$$

Assuming an isotropic medium, we can compute $\bar{v}_z^{i,j,-1}$ from Eq. (3.80a):

$$0 = \dot{\bar{\sigma}}_{zz}^{i,j,0} = (\lambda + 2\mu)\,\dot{\bar{\varepsilon}}_{zz}^{i,j,0} + \lambda\left(\dot{\bar{\varepsilon}}_{xx}^{i,j,0} + \dot{\bar{\varepsilon}}_{yy}^{i,j,0}\right). \tag{3.81}$$

Substituting for $\dot{\bar{\varepsilon}}_{zz}^{i,j,0}$ the second-order finite-difference approximation

$$\dot{\bar{\varepsilon}}_{zz}^{i,j,0} = \frac{1}{\Delta z}\left(\bar{v}_z^{i,j,0} - \bar{v}_z^{i,j,-1}\right) \tag{3.82}$$

yields

$$\bar{v}_z^{i,j,-1} = \bar{v}_z^{i,j,0} + \frac{\lambda\,\Delta z}{\lambda + 2\mu}\left(\dot{\bar{\varepsilon}}_{xx}^{i,j,0} + \dot{\bar{\varepsilon}}_{yy}^{i,j,0}\right). \tag{3.83}$$

With the help of $\bar{v}_z^{i,j,-1}$ we can approximate $\bar{v}_x^{i,j,-1}$ and $\bar{v}_y^{i,j,-1}$ from Eqs. (3.80b) and (3.80c): Inserting the isotropic constitutive relation into (3.80b) gives

$$\dot{\bar{\varepsilon}}_{xz}^{i,j,0} = -\dot{\bar{\varepsilon}}_{xz}^{i,j,-1}\,. \tag{3.84}$$

Using second-order approximations of the spatial derivatives $\partial_x v_z$ and $\partial_z v_x$ transforms (3.84) into

$$\frac{1}{\Delta z}\left(\bar{v}_x^{i,j,1}-\bar{v}_x^{i,j,0}\right)+\frac{1}{\Delta x}\left(\bar{v}_z^{i+1,j,0}-\bar{v}_z^{i,j,0}\right)$$
$$=-\frac{1}{\Delta z}\left(\bar{v}_x^{i,j,0}-\bar{v}_x^{i,j,-1}\right)-\frac{1}{\Delta x}\left(\bar{v}_z^{i+1,j,-1}-\bar{v}_z^{i,j,-1}\right). \qquad (3.85)$$

From (3.85) we can eliminate $\bar{v}_x^{i,j,-1}$:

$$\bar{v}_x^{i,j,-1}=\bar{v}_x^{i,j,1}+\frac{\Delta z}{\Delta x}\left(\bar{v}_z^{i+1,j,0}-\bar{v}_z^{i,j,0}+\bar{v}_z^{i+1,j,-1}-\bar{v}_z^{i,j,-1}\right). \qquad (3.86)$$

The analogous equation for $\bar{v}_y^{i,j,-1}$ is then

$$\bar{v}_y^{i,j,-1}=\bar{v}_y^{i,j,1}+\frac{\Delta z}{\Delta y}\left(\bar{v}_z^{i,j+1,0}-\bar{v}_z^{i,j,0}+\bar{v}_z^{i,j+1,-1}-\bar{v}_z^{i,j,-1}\right). \qquad (3.87)$$

Finally, we can compute $\bar{v}_z^{i,j,-2}$ from (3.80d):

$$\dot{\bar{\sigma}}_{zz}^{i,j,1}=(\lambda+2\mu)\,\dot{\bar{\varepsilon}}_{zz}^{i,j,1}+\lambda\left(\dot{\bar{\varepsilon}}_{xx}^{i,j,1}+\dot{\bar{\varepsilon}}_{yy}^{i,j,1}\right)$$
$$=-(\lambda+2\mu)\,\dot{\bar{\varepsilon}}_{zz}^{i,j,-1}-\lambda\left(\dot{\bar{\varepsilon}}_{xx}^{i,j,-1}+\dot{\bar{\varepsilon}}_{yy}^{i,j,-1}\right)=-\dot{\bar{\sigma}}_{zz}^{i,j,-1}. \qquad (3.88)$$

Since $\dot{\bar{\varepsilon}}_{xx}^{i,j,-1}$ and $\dot{\bar{\varepsilon}}_{yy}^{i,j,-1}$ can be approximated from $\bar{v}_x^{i,j,-1}$ and $\bar{v}_y^{i,j,-1}$, it suffices to replace $\dot{\bar{\varepsilon}}_{zz}^{i,j,-1}$ by the second-order finite difference

$$\dot{\bar{\varepsilon}}_{zz}^{i,j,-1}=\frac{1}{\Delta z}\left(\bar{v}_z^{i,j,-1}-\bar{v}_z^{i,j,-2}\right). \qquad (3.89)$$

The resulting formula for $\bar{v}_z^{i,j,-2}$ is then

$$\bar{v}_z^{i,j,-2}=\bar{v}_z^{i,j,-1}+\Delta z\,\dot{\bar{\varepsilon}}_{zz}^{i,j,1}+\frac{\lambda\,\Delta z}{\lambda+2\mu}\left(\dot{\bar{\varepsilon}}_{xx}^{i,j,1}+\dot{\bar{\varepsilon}}_{yy}^{i,j,1}+\dot{\bar{\varepsilon}}_{xx}^{i,j,-1}+\dot{\bar{\varepsilon}}_{yy}^{i,j,-1}\right). \qquad (3.90)$$

In contrast to the finite-difference scheme in the interior of the medium, the image method implementation of the free surface is correct only to second order. To obtain accurate solutions despite the reduced order, the grid needs to be refined, which automatically increases the computational costs.

The location of the free surface in the plane that coincides with the diagonal elements of the stress and strain tensors is not as unique as it may initially seem. As an alternative to the so-called *H formulation* described above, one may adopt the *W formulation* where the free surface is located half a grid spacing deeper, i.e. coincident with the shear-stress components (e.g. Gottschämmer & Olsen, 2001; Kristek et al., 2002).

Ohminato & Chouet (1997) made constructive use of the ambiguity in the free surface location. They implemented the W formulation along an irregular surface that was approximated by a staircase and set the elastic parameters outside the medium to zero, as in the vacuum method. The W formulation ensures that the boundary condition can be applied directly to two of the shear stresses on the boundary. Only the normal stress then needs to be approximated by second-order finite differences.

Robertsson (1996) proposed a very pragmatic variant of the image method. Instead of explicitly computing the velocity components above the surface, they are collectively set to zero. Only the discrete stresses are imaged as in Eqs. (3.80b), (3.80c) and (3.80d). Numerical experiments indicated good agreement with analytical solutions, and the method proved to be easily extendable to staircase topographies.

The staircase approximation of an irregular but smoothly varying topography (e.g. Robertsson, 1996; Ohminato & Chouet, 1997) results in artificial diffraction that can pollute the numerical solution (Muir et al., 1992; Moczo et al., 1997). Hestholm (1999) therefore proposed to rotate an arbitrarily oriented surface element into a local coordinate system where the surface normal is aligned with the z-axis. The boundary condition is then imposed in the rotated system and transformed back to the physical coordinates.

3.2.3.3 Interior Methods

Imaging methods have been widely used for the modelling of seismic wave propagation near the free surface both with and without topographic variations. Nevertheless, the introduction of image stresses above the surface is not a natural ingredient of finite-difference methods, and it constitutes only a second-order approximation of the exact free surface boundary condition. Recently developed interior methods circumvent the need for velocities and stresses above the surface with the help of either hybrid approaches or adjusted finite-difference approximations.

Moczo et al. (1997) proposed a hybrid method that combines a finite-difference discretisation in the interior of the medium with a finite-element approximation along the surface. The finite-element method – just as the spectral-element method treated in Chap. 4 – automatically accounts for the free surface condition and it allows for a smoothly varying irregular topography that does not suffer from the staircase problem.

An elegant solution that remains within the finite-difference framework was proposed by Kristek et al. (2002). The authors replaced the symmetric finite-difference approximations that we used for the interior of the medium by the following set of fourth-order asymmetric or *adjusted* approximations:

$$\partial_z f(z_0) = \frac{1}{\Delta z}\left[-\frac{352}{105}f(z_0) + \frac{35}{8}f(z_0+\Delta z/2) - \frac{35}{24}f(z_0+3\Delta z/2) + \frac{21}{40}f(z_0+5\Delta z/2) - \frac{5}{56}f(z_0+7\Delta z/2)\right] + \mathcal{O}(\Delta z^4)\,, \tag{3.91a}$$

$$\partial_z f(z_0 + \Delta z/2) = \frac{1}{\Delta z}\left[-\frac{11}{12}f(z_0) + \frac{17}{24}f(z_0 + \Delta z) + \frac{3}{8}f(z_0 + 5\Delta z/2) - \frac{5}{24}f(z_0 + 7\Delta z/2) + \frac{1}{24}f(z_0 + 9\Delta z/2)\right] + \mathcal{O}(\Delta z^4), \tag{3.91b}$$

$$\partial_z f(z_0 + \Delta z) = \frac{1}{\Delta z}\left[-\frac{\Delta z}{22}\partial_z f(z_0) - \frac{577}{528}f(z_0 + \Delta z/2) + \frac{201}{176}f(z_0 + 3\Delta z/2) - \frac{9}{176}f(z_0 + 5\Delta z/2) + \frac{1}{528}f(z_0 + 7\Delta z/2)\right] + \mathcal{O}(\Delta z^4), \tag{3.91c}$$

$$\partial_z f(z + \Delta z) = \frac{1}{\Delta z}\left[\frac{16}{105}f(z_0) - \frac{31}{24}f(z_0 + \Delta z/2) + \frac{29}{24}f(z_0 + 3\Delta z/2) - \frac{3}{40}f(z_0 + 5\Delta z/2) + \frac{1}{168}f(z_0 + 7\Delta z/2)\right] + \mathcal{O}(\Delta z^4). \tag{3.91d}$$

The adjusted finite-difference approximations from Eqs. (3.91) allow us to compute the strain rate tensor and the stress divergence immediately at and below the surface without having to introduce artificial velocities and stresses above the surface. In this sense, adjusted approximations appear to result in the most natural finite difference implementation of a planar free surface. Following a series of benchmark tests, Kristek et al. (2002) and Moczo et al. (2004) concluded that adjusted finite-differences near the surface lead to accurate numerical solutions that outperform the classical H and W formulations of the stress imaging method.

3.3 The 3D Spherical Case

Finite-difference wave propagation in Cartesian coordinates is a powerful tool in local- and regional-scale studies, where the curvature of the Earth's surface can be neglected. However, for epicentral distances exceeding several hundred kilometres, the spherical shape of the globe must be taken into account. For this we consider a spherical section where the position vector $\mathbf{x}$ is parameterised in terms of radius r, longitude ϕ and colatitude θ (see Fig. 3.11):

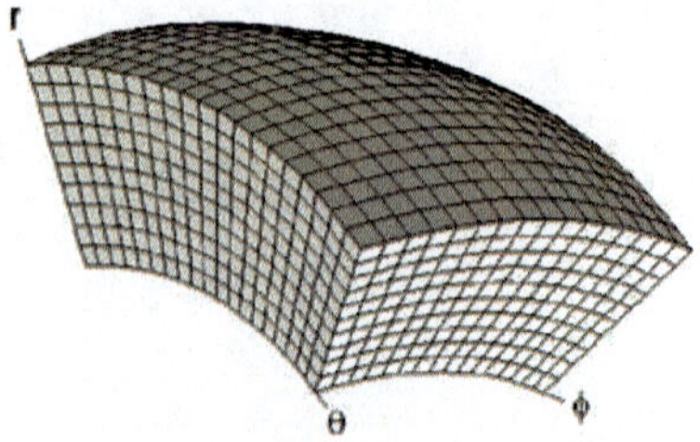

Fig. 3.11 Illustration of the spherical section, parameterised in terms of radius r, longitude ϕ and colatitude θ

$$\mathbf{x} = (x, y, z) = (r \cos\phi \sin\theta, r \sin\phi \sin\theta, r \cos\theta). \tag{3.92}$$

Following Nissen-Meyer (2001) and Igel et al. (2002), we again work with the velocity–stress formulation of the elastic wave equation, given in Eqs. (3.55) and (3.56). In the spherical coordinate system, the components of the stress divergence $\nabla \cdot \boldsymbol{\sigma}$ and the strain rate tensor $\dot{\boldsymbol{\varepsilon}}$ are given by

$$(\nabla \cdot \boldsymbol{\sigma})_r = \partial_r \sigma_{rr} + \frac{1}{r \sin\theta} \partial_\phi \sigma_{\phi r} + \frac{1}{r} \partial_\theta \sigma_{\theta r} + \frac{1}{r} \underbrace{(2\sigma_{rr} + \sigma_{\theta r} \cot\theta - \sigma_{\phi\phi} - \sigma_{\theta\theta})}_{\text{interpolation}}, \tag{3.93a}$$

$$(\nabla \cdot \boldsymbol{\sigma})_\phi = \partial_r \sigma_{\phi r} + \frac{1}{r \sin\theta} \partial_\phi \sigma_{\phi\phi} + \frac{1}{r} \partial_\theta \sigma_{\theta\phi} + \frac{1}{r} \underbrace{(3\sigma_{r\phi} + 2\sigma_{\theta\phi} \cot\theta)}_{\text{interpolation}}, \tag{3.93b}$$

$$(\nabla \cdot \boldsymbol{\sigma})_\theta = \partial_r \sigma_{r\theta} + \frac{1}{r \sin\theta} \partial_\phi \sigma_{\phi\theta} + \frac{1}{r} \partial_\theta \sigma_{\theta\theta} + \frac{1}{r} \underbrace{(3\sigma_{r\theta} + \sigma_{\theta\theta} \cot\theta - \sigma_{\phi\phi} \cot\theta)}_{\text{interpolation}} \tag{3.93c}$$

and

$$\dot{\varepsilon}_{rr} = \partial_r v_r\,, \tag{3.94a}$$

$$\dot{\varepsilon}_{r\phi} = \frac{1}{2} \left(\partial_r v_\phi + \frac{1}{r \sin\theta} \partial_\phi v_\mathrm{r} - \frac{1}{r} \underbrace{v_\phi}_{\text{interp.}} \right), \tag{3.94b}$$

$$\dot{\varepsilon}_{r\theta} = \frac{1}{2} \left(\partial_r v_\theta + \frac{1}{r} \partial_\theta v_\mathrm{r} - \frac{1}{r} \underbrace{v_\theta}_{\text{interp.}} \right), \tag{3.94c}$$

$$\dot{\varepsilon}_{\phi\phi} = \frac{1}{r} \left(\frac{1}{\sin\theta} \partial_\phi v_\phi + \underbrace{v_\mathrm{r} + v_\theta \cot\theta}_{\text{interpolation}} \right), \tag{3.94d}$$

$$\dot{\varepsilon}_{\phi\theta} = \frac{1}{2r} \left(\frac{1}{\sin\theta} \partial_\phi v_\theta - \underbrace{v_\phi \cot\theta + \partial_\theta v_\phi}_{\text{interpolation}} \right), \tag{3.94e}$$

$$\dot{\varepsilon}_{\theta\theta} = \frac{1}{r} \partial_\theta v_\theta + \frac{1}{r} \underbrace{v_\mathrm{r}}_{\text{interp.}}. \tag{3.94f}$$

To translate the staggered-grid approach from the Cartesian to the spherical case, we let θ, ϕ and r play the roles of x, y and z, respectively. The grid spacings in the spherical coordinate directions are $\Delta\theta$, $\Delta\phi$ and Δr. An additional complication is the appearance of non-differentiated terms in Eqs. (3.93) and (3.94). The strain rate component $\dot{\varepsilon}_{\theta\theta}$, for instance, is the sum of $\frac{1}{r}\partial_\theta v_\theta$ and $\frac{1}{r}v_r$. While $\partial_\theta v_\theta$ is given at the grid positions (θ_i, ϕ_j, r_k), the radial velocity v_r is defined at $(\theta_i, \phi_j, r_k + \Delta r/2)$. To make the non-differentiated terms available at the correct positions, interpolation becomes necessary – just as in the case of general anisotropy (see Sect. 3.2.2).

Figure 3.12 shows a wave field snapshot in a spherical section for a 400-km-deep source. Clearly visible are surface waves in the early stage of their development, the direct S wave, as well as the surface-reflected phases sS and sP. Radial-component velocity seismograms for a shallow source (30 km) and at 60° epicentral distance are displayed in Fig. 3.13. The arrivals of the most prominent body wave phases agree well with the ray-theoretical predictions.

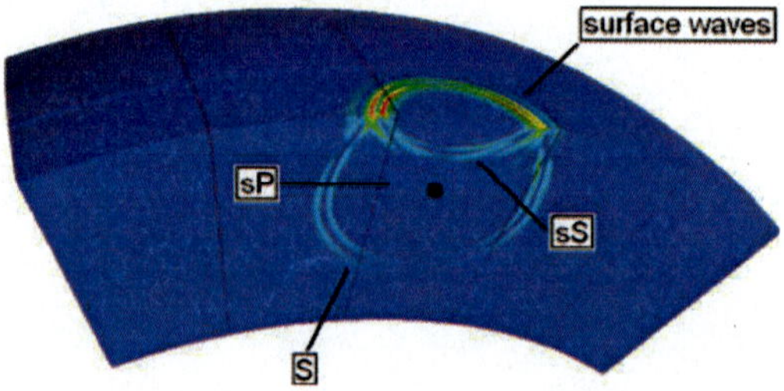

Fig. 3.12 Snapshot of the velocity field amplitude $|\bar{\mathbf{v}}|$ in the radially symmetric model PREM (Dziewonski & Anderson), 3.5 min. after the source initiation. A *filled black circle* (•) marks the hypocentre location at 400 km depth. The image is dominated by surface waves which decay rapidly with depth. Also, the direct S wave and its surface reflections sS and sP are clearly visible. The amplitude of the direct P wave is too small to be seen

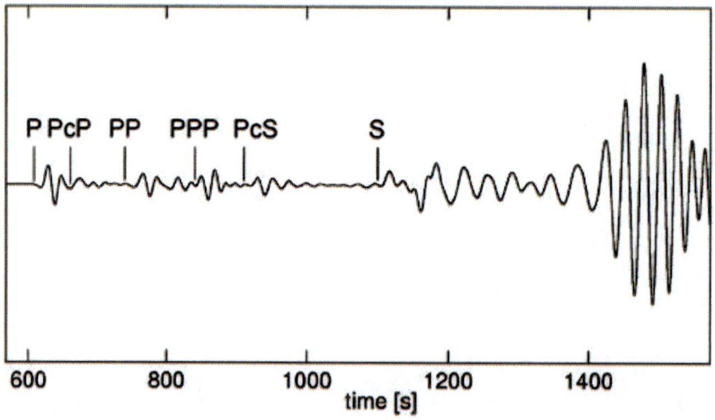

Fig. 3.13 Synthetic radial-component velocity seismogram at an epicentral distance of 60° and a source at 30 km depth. The dominant period is 15 s. The arrivals of the body wave phases P, PP, PPP, PcP, PcS and S are clearly visible. Their travel times agree well with the ray-theoretical predictions, indicated by vertical lines

The finite-difference modelling in spherical coordinates is particularly efficient for the simulation of body wave propagation on continental scales, that is for epicentral distances of up to several tens of degrees. Limiting factors are the constant angular increments $\Delta\theta$ and $\Delta\phi$ and the singularities in Eqs. (3.93) and (3.94) for $r = 0$. The singularity forces us to exclude the centre of the Earth. Moreover, as the spherical section extends further towards the poles and the centre, the grid spacing tends to zero. To comply with the CFL stability condition, the time increment Δt must decrease as the grid spacing decreases, and this can lead to an inefficient algorithm when the spherical section is too large.

3.4 Point Source Implementation

Point-localised sources of the form

$$\mathbf{f}(\mathbf{x}, t) = \mathbf{s}(t)\, \delta(\mathbf{x} - \mathbf{x}^{\mathrm{s}}) , \tag{3.95}$$

with the vectorial source time function $\mathbf{s}$ and the source location $\mathbf{x}^{\mathrm{s}}$, are of outstanding importance in computational seismology. They are used as standard sources in benchmarks and for the approximate representation of physical sources.

The implementation of point sources deserves special attention in any numerical method. This is because the location of a point source usually does not coincide with a grid node. Positioning the source at the nearest node can lead to unacceptably large errors. These can be avoided by the implementation of a band-limited version of the delta function $\delta(\mathbf{x} - \mathbf{x}^{\mathrm{s}})$. To illustrate the concept, we consider the 1D case. The extension to three dimensions is then straightforward.

In the wave number domain, the 1D delta function $\delta(x - x^{\mathrm{s}})$ is given by

$$\tilde{\delta}(k) = \frac{1}{\sqrt{2\pi}}\, \mathrm{e}^{-\mathrm{i}kx^{\mathrm{s}}} . \tag{3.96}$$

As a result of the spatial discretisation, the wave number range is limited to $[-k_{\max}, k_{\max}]$, where $k_{\max} = \pi/\Delta x$ is again the Nyquist wave number. It follows that the optimal band-limited version of $\tilde{\delta}$ is

$$\tilde{d}(k) = \frac{1}{\sqrt{2\pi}} [H(k + k_{\max}) - H(k - k_{\max})]\, \mathrm{e}^{-\mathrm{i}kx^{\mathrm{s}}} , \tag{3.97}$$

or

$$d(x) = \frac{k_{\max}}{\pi} \operatorname{sinc}[k_{\max}\,(x - x^{s})] , \tag{3.98}$$

in the space domain. A discrete band-limited delta function can then be found by sampling $d(x)$ at the grid points $x_n = n\Delta x$. In the special case where the source location x^{s} coincides with a grid point, i.e. $x^{\mathrm{s}} = s\Delta x$ for some integer s, we find

$$d(n\Delta x) = \frac{1}{\Delta x}\,\delta_{ns}\,. \tag{3.99}$$

This means that the discrete source is non-zero only at the grid point with index s. For any source location that does not coincide with a grid point, $d(n\Delta x)$ is generally non-zero so that the band-limited source is distributed through all of space, as illustrated in the left part of Fig. 3.14.

To restrict the active numerical source region to a small volume surrounding the true source position, $d(n\Delta x)$ needs to be windowed such that its spectrum is disturbed as little as possible within the wave number range of interest, $[-k_{\max}, k_{\max}]$.

Possible window functions are Gaussians or, as suggested by Hicks (2002), *Kaiser windows* defined by

$$K(x) = \begin{cases} \frac{I_0(\beta\sqrt{1-(x/r)^2})}{I_0(\beta)} & , \quad \text{for } x \in [-r, r] \\ 0 & , \quad \text{otherwise}\,, \end{cases} \tag{3.100}$$

where I_0 denotes the zero-order Bessel function of the first kind. The shape of the Kaiser window is controlled by the half-width r and the parameter b. Two examples are shown in the right part of Fig. 3.14. In general, there is a trade-off between the width of the windows and the extent to which the spectrum of the discrete source (3.98) is disturbed. Small windows lead to small source volumes but inaccurate source spectra. Large windows preserve the source spectrum but require a large source volume. A rule of thumb is that the source volume should extend approximately four grid points in each coordinate direction, but the optimal value needs to be found on a case-by-case basis.

The above method can be easily generalised to moment tensor sources

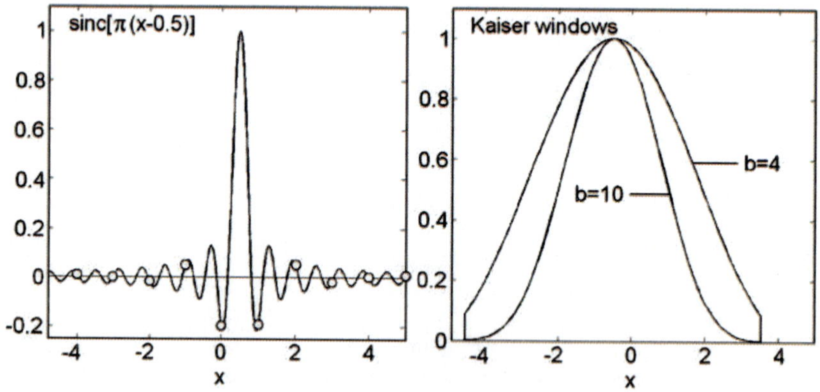

Fig. 3.14 *Left*: Band-limited point source distribution as defined in Eq. (3.98) for $k_{\max} = \pi$ and $x^s = 0.5$. *Circles* (∘) indicate the discrete sampling locations at $x_n = n\Delta x$. *Right*: Kaiser windows for $b = 4$ and 10, and a half-width $r = 4$

$$\mathbf{f}(\mathbf{x}, t) = -\nabla \cdot [\mathbf{M}(t)\, \delta(\mathbf{x} - \mathbf{x}^{\mathrm{s}})] \,, \tag{3.101}$$

where $\mathbf{M}$ denotes the moment tensor. It suffices to replace the band-limited version of $\delta(x - x^{\mathrm{s}})$, given in Eq. (3.98), by its derivative:

$$\frac{\mathrm{d}}{\mathrm{d}x}\delta(x - x^{\mathrm{s}}) \approx \frac{\mathrm{d}}{\mathrm{d}x}\frac{k_{\max}}{\pi}\, \mathrm{sinc}[k_{\max}\,(x - x^{\mathrm{s}})] \,. \tag{3.102}$$

In the 3D case, the band-limited and windowed sources need to be implemented for each of the force directions or moment tensor components.

3.5 Accuracy and Efficiency

The efficiency of numerical modelling schemes is particularly important because their applicability is usually limited by the available computational resources. To be efficient, a numerical method should require as few grid points as possible, to achieve accurate solutions for the shortest spatial wavelength. The term *accurate* is naturally problem specific and so is the necessary number of grid points per minimum wavelength $n_{\min}$. Among other factors, $n_{\min}$ depends on the spatio-temporal discretisation, the types of waves to be modelled (body or surface waves), the presence of irregular topography, the distance travelled by the waves (see Fig. 3.5) and the data to which the synthetics are to be compared. The following paragraphs provide a collection of recommendations for $n_{\min}$ in a variety of scenarios.

For plane body waves propagating through the fourth-order staggered grid introduced in Sect. 3.2.1, Robertsson et al. (1994) and Bohlen (2002) found that $n_{\min}$ should be around 4–8 in order to reduce the dispersion error below 2–5%. Moczo et al. (2000) conclude that the group velocity error for S waves is 2.5% for $n_{\min} = 6$. Those values must be considered in the context of the heterogeneous Earth where velocity variations on the order of 1% can be relevant. Thus, for applications in body wave delay time tomography, $n_{\min} \approx 20$ is certainly recommendable.

In a systematic comparison of several finite-difference and finite-element schemes, Moczo et al. (2010) analysed the dependence of the numerical error on the P wave speed to S wave speed ratio $v_{\mathrm{P}}/v_{\mathrm{S}}$. While staggered-grid approaches appear to be rather insensitive to $v_{\mathrm{P}}/v_{\mathrm{S}}$, the conventional grid produces excessive numerical errors when $v_{\mathrm{P}}/v_{\mathrm{S}}$ is larger than about 5, even when $n_{\min}$ is on the order of 10. This renders the conventional grid impractical for the simulation of wave propagation through sedimentary basins, where $v_{\mathrm{P}}/v_{\mathrm{S}}$ can reach values of 5 and larger.

Numerous techniques aiming at the reduction of $n_{\min}$ can be found in the finite-difference literature. The most intuitive approach is to increase the length of the finite-difference operator. Based on a qualitative error analysis, Dablain (1986) suggested that three grid points per minimum wavelength are sufficient when a 10th-order finite-difference approximation is used for the 2D scalar wave equation. Holberg (1987) optimised the finite-difference coefficients such that the group velocity error is minimal. For a pre-defined maximum group velocity error, he found $n_{\min}$

to decrease roughly exponentially with increasing operator length. In the case of a 30-point operator, $n_{\text{min}} \approx 2.5$ is theoretically sufficient to reduce the group velocity error below 3%. Furumura & Chen (2004) and Kennett & Furumura (2008) used a 16th-order finite-difference approximation to propagate waves in a 2D model across several thousand wavelengths.

In principle, the operator length can be increased indefinitely so that the finite-difference method approaches the pseudo-spectral method where the discrete derivative operator is global and where n_{min} is close to its theoretical minimum of 2 (see Sect. 2.6).

An interesting alternative to the increasing operator length was presented by Cole (1994, 1997) who optimised a two-point finite-difference operator such that it is nearly exact, at least within a narrow wave number range. Jo et al. (1996) and Šteckl & Pratt (1998) used rotated finite-difference stencils to substantially reduce the numerical dispersion without increasing the operator length.

The high-accuracy variants of the finite-difference method mentioned above have been developed for the interior of the computational domain, where they have proven efficient. However, their consistent application along the free surface is difficult. Furthermore, high-order schemes face issues of parallelisation because the amount of data communicated between processors increases rapidly with increasing operator length.

The finite-difference modelling of surface wave propagation can be challenging because the discretisation scheme for the interior of the computational domain is not directly applicable along the free surface. In a series of numerical experiments, Kristek et al. (2002) compared the performance of the fourth-order staggered grid (Sect. 3.2.1) combined with different implementations of a planar free surface. For a Rayleigh wave propagating across 15 dominant wavelengths, a value of $n_{\text{min}} = 10$ was needed to reduce the average envelope misfit to below 40% when the stress imaging technique (Levander, 1988; Graves, 1996; Sect. 3.2.3.2) was used. The large numerical errors are expected because the stress imaging implements the free surface boundary condition only correct to second order. The implementation of the free surface based on the adjusted finite-difference approximations from Eqs. (3.91) (Kristek et al., 2002) reduce the errors by a factor of 2 while using only six grid points per minimum wavelength. To ensure that the energy misfit

$$E := \frac{\int_T [\dot{\tilde{\mathbf{u}}}(\mathbf{x}^r, t) - \dot{\mathbf{u}}(\mathbf{x}^r, t)]^2 \, \mathrm{d}t}{\int_T \dot{\mathbf{u}}^2(\mathbf{x}^r, t) \, \mathrm{d}t} \tag{3.103}$$

is below 1% for propagation distances of around 10 wavelengths, Igel et al. (2002) suggest $n_{\text{min}} \approx 20$.

The required number of grid points per wavelength needed in the presence of irregular topography depends strongly on the topography itself. Based on a visual comparison and assuming that $n_{\text{min}} = 5$ is sufficient in the interior of the fourth-order staggered grid, Robertsson (1996) proposed to sample the wave field with at least $n_{\text{min}} = 15$. In a more quantitative analysis, Bohlen & Saenger (2006) recommend to use around 30–60 grid points per minimum wavelength to ensure acceptable solutions in the presence of a dipping interface.

The important conclusion to be drawn is that the numerical setup in general, and the number of grid points per wavelength in particular, is highly problem dependent. The only rule is that the numerical grid should be chosen such that the resulting synthetics can be compared to the data in a meaningful way. This means that the numerical error should be much smaller than the typical differences that we expect between data and synthetics due to undiscovered Earth structure.

Chapter 4
Spectral-Element Methods

The spectral-element method is a high-order numerical method that allows us to solve the seismic wave equation in 3D heterogeneous Earth models. The method enables adaptation of the mesh to the irregular surface topography and to the variable wavelengths inside the Earth. Moreover, the spectral-element method yields accurate solutions for surface waves without increasing the number of grid points per wavelength, therefore overcoming some of the most severe deficiencies of the finite-difference method.

Originally developed in fluid dynamics (Patera, 1984; Maday & Patera, 1989), the spectral-element method was first applied to the elastic wave equation and in a seismological context by Seriani et al. (1995), Faccioli et al. (1997) and Komatitsch (1997). Numerical solutions with high accuracy have been obtained in a large number of studies (e.g. Komatitsch & Vilotte, 1998; Seriani, 1998; Komatitsch et al., 2004; Fichtner et al., 2009). Applications to global wave propagation in the presence of self-gravitation, rotation and fluid regions can be found in Komatitsch & Tromp (2002), Chaljub et al. (2003) and Chaljub & Valette (2004). For excellent reviews of the spectral-element method the reader is referred to Komatitsch et al. (2005) and Chaljub et al. (2007).

We start our development with an illustration of the basic spectral-element concepts in 1D. The necessary mathematical tools, including Lagrange interpolation and Gauss–Lobatto–Legendre quadrature, can be found in Appendix A. The extension to the 3D elastic case, treated in Sect. 4.2, is straightforward but requires a few words on grid generation. The implementation of point sources in the spectral-element method is the subject of Sect. 4.2.

4.1 Basic Concepts in One Dimension

We follow the classical approach and introduce the basic concepts of the spectral-element method with an example in one dimension. For this we first introduce the *weak form* of the equations of motion and then continue with a description of the *Galerkin method*.

A. Fichtner, *Full Seismic Waveform Modelling and Inversion*, Advances in Geophysical and Environmental Mechanics and Mathematics,
DOI 10.1007/978-3-642-15807-0_4, © Springer-Verlag Berlin Heidelberg 2011

4.1.1 Weak Solution of the Wave Equation

We consider the 1D scalar wave equation

$$\rho(x)\,\ddot{u}(x,t) - \partial_x[\mu(x)\,\partial_x u(x,t)] = f(x,t)\,, \tag{4.1}$$

with the space variable $x \in G = [0, L]$ and time $t \in [0, T]$. The displacement field u is subject to the Neumann boundary conditions

$$\partial_x u(x,t)|_{x=0} = \partial_x u(x,t)|_{x=L} = 0 \tag{4.2}$$

and the initial conditions

$$u|_{t=0} = \dot{u}|_{t=0} = 0\,. \tag{4.3}$$

Equation (4.1) together with the boundary and initial conditions (4.2) and (4.3) is referred to as the *strong form* of the wave equation. To derive the associated *weak* or *variational* form, we multiply (4.1) by an arbitrary, time-independent test function, $w : G \to \mathbb{R}$, and integrate over space:

$$\int_G \rho\, w\, \ddot{u}\, \mathrm{d}x - \int_G w\, \partial_x\,(\mu\, \partial_x u)\; \mathrm{d}x = \int_G w\, f\, \mathrm{d}x\,. \tag{4.4}$$

Integrating the second term on the left-hand side by parts and inserting the boundary condition (4.2) gives

$$\int_G \rho\, w\, \ddot{u}\, \mathrm{d}x + \int_G \mu\, \partial_x w\, \partial_x u\, \mathrm{d}x = \int_G w\, f\, \mathrm{d}x\,. \tag{4.5}$$

Solving the weak form of the wave equation now means finding a wave field u such that it satisfies (4.5) *for any* suitable test function w and subjecting to the initial conditions

$$\int_G \rho\, w\, u|_{t=0}\, \mathrm{d}x = \int_G \rho\, w\, \dot{u}|_{t=0}\, \mathrm{d}x = 0\,. \tag{4.6}$$

The weak form of the wave equation has an important advantageous property from a numerical point of view. The boundary condition (4.2), corresponding to the free surface condition in 3D, is implicitly satisfied. It therefore need not be treated explicitly as in finite-difference methods where the accurate implementation of the free surface can be a tedious task (see Sect. 3.2.3).

4.1.2 Spatial Discretisation and the Galerkin Method

Analytical solutions of both the strong and the weak forms of the wave equation often do not exist when the mass density ρ and the elastic parameter μ are spatially variable. In the *Galerkin method* we approximate the exact solution $u(x,t)$ by a finite superposition of n basis functions ψ_i $(i = 1, \dots, n)$ that depend only on space and not on time. We denote this approximation by $\bar{u}(x,t)$:

$$u(x,t) \approx \bar{u}(x,t) = \sum_{i=1}^{n} u_i(t)\psi_i(x)\,, \tag{4.7}$$

where $u_i(t)$ are the time-dependent expansion coefficients. The quality of this approximation depends on the choice of the basis functions ψ_i, the source term f and on the medium properties ρ and μ. Instead of trying to solve the exact weak formulation, we limit ourselves to the requirement that $\bar{u}$ solves the approximate weak form

$$\int_G \rho\, \psi_i\, \ddot{\bar{u}}\, \mathrm{d}x + \int_G \mu\, \partial_x \psi_i\, \partial_x \bar{u}\, \mathrm{d}x = \int_G \psi_i\, f\, \mathrm{d}x\,, \tag{4.8}$$

for all basis functions ψ_i, with $i = 1, \dots, n$, and subject to the initial conditions

$$\int_G \rho\, \psi_i\, \bar{u}|_{t=0}\, \mathrm{d}x = \int_G \rho\, \psi_i\, \dot{\bar{u}}|_{t=0}\, \mathrm{d}x = 0\,. \tag{4.9}$$

The basis functions ψ_i are thus used in the approximation of the wave field and as test functions in the weak formulation. Equation (4.7) together with the approximate weak formulation transforms the exact weak formulation into the following set of linear equations for the coefficients $u_i(t)$:

$$\sum_{i=1}^{n} \left[\ddot{u}_i(t) \int\limits_G \rho(x)\, \psi_j(x)\, \psi_i(x)\, \mathrm{d}x \right] + \sum_{i=1}^{n} \left[u_i(t) \int\limits_G \mu(x)\, \partial_x \psi_j(x)\, \partial_x \psi_i(x)\, \mathrm{d}x \right]$$
$$= \int\limits_G \psi_j(x)\, f(x,t)\, \mathrm{d}x\,, \tag{4.10}$$

for all $j = 1, \dots, n$. Equation (4.10) is an algebro-differential equation that can conveniently be written in matrix notation:

$$\mathbf{M} \cdot \ddot{\mathbf{u}}(t) + \mathbf{K} \cdot \mathbf{u}(t) = \mathbf{f}(t)\,, \tag{4.11}$$

with the *mass matrix*

$$M_{ji} = \int\limits_G \rho(x)\, \psi_j(x)\, \psi_i(x)\, \mathrm{d}x\,, \tag{4.12}$$

the *stiffness matrix*

$$K_{ji} = \int_G \mu(x)\, \partial_x \psi_j(x)\, \partial_x \psi_i(x)\, \mathrm{d}x\,, \tag{4.13}$$

and the right-hand side

$$f_j(t) = \int_G \psi_j(x)\, f(x,t)\, \mathrm{d}x\,. \tag{4.14}$$

The vector $\mathbf{u}$ – not to be confused with the vectorial displacement field in the complete elastic wave equation – comprises the expansion coefficients u_i. The process of transforming the differential equation (4.1) plus the approximation (4.7) into the algebro-differential equation (4.11) is known as *Galerkin projection*. Once more we emphasise that the free surface boundary condition (4.2) is naturally contained in the stiffness matrix (Eq. 4.13) and does not require any additional work – as is needed in finite-difference methods. What distinguishes the spectral-element method among other numerical methods is the choice of the basis functions ψ_i and the integration scheme used to solve the integrals that appear in the mass and stiffness matrices.

In the next step we decompose the domain G into n_e disjoint subdomains G_e, called the *elements*. Equation (4.10) then transforms to

$$\sum_{i=1}^{n}\left[\ddot{u}_i(t) \sum_{e=1}^{n_e} \int_{G_e} \rho(x)\, \psi_j(x)\, \psi_i(x)\, \mathrm{d}x\right] + \sum_{i=1}^{n}\left[u_i(t) \sum_{e=1}^{n_e} \int_{G_e} \mu(x)\, \partial_x \psi_j(x)\, \partial_x \psi_i(x)\, \mathrm{d}x\right]$$
$$= \sum_{e=1}^{n_e} \int_{G_e} \psi_j(x)\, f(x,t)\, \mathrm{d}x\,. \tag{4.15}$$

The disadvantage of Eq. (4.15) is that each expansion coefficient u_i depends on the integrals over all elements. We can circumvent this problem by choosing local basis functions, i.e. basis functions that are supported by one element only. The discrete equations can then be solved for each element individually. For this we define $N + 1$ basis functions ψ_i^e ($i = 1, \ldots, N + 1$) on each of the n_e elements G_e. The displacement field within the element G_e is then approximated by

$$\bar{u}(x,t)|_{x \in G_e} = \sum_{i=1}^{N+1} u_i^e(t) \psi_i^e(x) \tag{4.16}$$

so that Eq. (4.10) now holds for each element:

$$\sum_{i=1}^{N+1} \ddot{u}_i^e(t) \int_{G_e} \rho(x)\, \psi_j^e(x)\, \psi_i^e(x)\, \mathrm{d}x + \sum_{i=1}^{N+1} u_i^e(t) \int_{G_e} \mu(x)\, \partial_x \psi_j^e(x)\, \partial_x \psi_i^e(x)\, \mathrm{d}x$$
$$= \int_{G_e} \psi_j^e(x)\, f(x,t)\, \mathrm{d}x\,. \tag{4.17}$$

Using matrix notation we write Eq. (4.17) in a more compact form:

$$\mathbf{M}^e \cdot \ddot{\mathbf{u}}^e(t) + \mathbf{K}^e \cdot \mathbf{u}^e(t) = \mathbf{f}^e(t)\,, \quad e = 1, \ldots, n_e\,, \tag{4.18}$$

where $\mathbf{u}^e$, $\mathbf{M}^e$ and $\mathbf{K}^e$ are the local coefficient vector, the local mass matrix and the local stiffness matrix, respectively. The total number of basis functions is now $n_e\,(N+1)$. Since the basis functions are locally supported by one element, the continuity of the discrete displacement $\bar{u}$ at the boundaries between the elements has to be imposed explicitly. To ensure that the wavelengths are sampled nearly uniformly, the size of the elements will usually be chosen proportional to the wave speed $\sqrt{\mu/\rho}$. The integrals in (4.17) can all be treated in the same way if we map each element G_e onto the *standard* or *reference interval* $[-1, 1]$ via an element-specific transformation F_e:

$$F_e : [-1,1] \to G_e\,, \quad x = F_e(\xi)\,, \quad \xi = \xi(x) = F_e^{-1}(x)\,, \quad e = 1, \ldots, n_e\,. \tag{4.19}$$

This transformation is illustrated in Fig. 4.1. Introducing the transformation into Eq. (4.17) gives

$$\sum_{i=1}^{N+1} \ddot{u}_i^e(t) \int_{-1}^{1} \rho[x(\xi)]\, \psi_j^e[x(\xi)]\, \psi_i^e[x(\xi)]\, \frac{\mathrm{d}x}{\mathrm{d}\xi}\, \mathrm{d}\xi$$
$$+ \sum_{i=1}^{N+1} u_i^e(t) \int_{-1}^{1} \mu[x(\xi)]\, \frac{\mathrm{d}\,\psi_j^e[x(\xi)]}{\mathrm{d}\xi}\, \frac{\mathrm{d}\,\psi_i^e[x(\xi)]}{\mathrm{d}\xi} \left(\frac{\mathrm{d}\xi}{\mathrm{d}x}\right)^2 \frac{\mathrm{d}x}{\mathrm{d}\xi}\, \mathrm{d}\xi$$
$$= \int_{-1}^{1} \psi_j^e[x(\xi)]\, f[(x(\xi)), t]\, \frac{\mathrm{d}x}{\mathrm{d}\xi}\, \mathrm{d}\xi\,. \tag{4.20}$$

At this point of the development we specify the basis functions ψ_i. We choose the $N+1$ Lagrange polynomials of degree N that have the corresponding Gauss–Lobatto–Legendre points (GLL points) as collocation points:

$$\psi_i^e[x(\xi)] = \ell_i^{(N)}(\xi)\,, \quad \xi \in [-1, 1]\,. \tag{4.21}$$

This choice is motivated by a number of important results from numerical analysis. We summarise them here and give brief derivations in Appendix A:

(1) Using the GLL points for polynomial interpolation ensures that the absolute value of the Lagrange polynomials $\ell_i^{(N)}$ is smaller than or equal to 1, for any polynomial order (see Sect. A.2.3). This means that Runge's phenomenon can be suppressed.
(2) The GLL points are Fekete points, i.e. they maximise the Vandermonde determinant (see Appendices A.2.1 and A.2.4). Thus, numerical errors right at the collocation points will have the smallest possible effect on the interpolated values between the collocation points.
(3) The Lebesgue constant associated with the GLL points grows slowly – in practice logarithmically – with increasing polynomial order (see Appendix A.2.5). This implies that the interpolation error decreases much more rapidly with increasing polynomial order than in the case of equidistant collocation points – at least when the interpolated function is well behaved.
(4) The GLL points are the collocation points of the GLL quadrature (Sect. A.3.2). One can therefore apply the GLL quadrature formulas to obtain accurate approximations of the integrals in Eq. (4.20) and a *diagonal mass matrix*.

In the interest of a lighter notation we will henceforth omit the superscript $^{(N)}$ in $\ell_i^{(N)}$. Substituting $\ell_i(\xi)$ for $\psi_i[x(\xi)]$ in Eq. (4.20) gives

$$\begin{aligned}
&\sum_{i=1}^{N+1} \ddot{u}_i^e(t) \int_{-1}^{1} \rho'(\xi)\ell_j(\xi)\ell_i(\xi)\,\frac{\mathrm{d}x}{\mathrm{d}\xi}\,\mathrm{d}\xi \\
&+\sum_{i=1}^{N+1} u_i^e(t) \int_{-1}^{1} \mu'(\xi)\,\dot{\ell}_j(\xi)\dot{\ell}_i(\xi)\,\left(\frac{\mathrm{d}\xi}{\mathrm{d}x}\right)^2 \frac{\mathrm{d}x}{\mathrm{d}\xi}\,\mathrm{d}\xi \\
&= \int_{-1}^{1} \ell_j(\xi)\, f'(\xi,t)\,\frac{\mathrm{d}x}{\mathrm{d}\xi}\,\mathrm{d}\xi\,,
\end{aligned} \tag{4.22}$$

where $\dot{\ell}$ denotes the derivative of ℓ with respect to ξ. The transformed density ρ', the elastic modulus μ' and external force f' are defined by

$$\rho'(\xi) := \rho[x(\xi)]\,, \quad \mu'(\xi) := \mu[x(\xi)]\,, \quad f'(\xi) := f[x(\xi)]\,. \tag{4.23}$$

With the GLL quadrature formula (A.74) we can approximate the integral in Eq. (4.22):

$$
\begin{aligned}
&\sum_{i,k=1}^{N+1} \ddot{u}_i^e(t)\, w_k \rho'(\xi)\, \ell_j(\xi)\ell_i(\xi) \frac{\mathrm{d}x}{\mathrm{d}\xi}\bigg|_{\xi=\xi_k} \\
&+ \sum_{i,k=1}^{N+1} w_k u_i^e(t) \mu'(\xi)\, \dot{\ell}_j(\xi)\dot{\ell}_i(\xi) \left(\frac{\mathrm{d}\xi}{\mathrm{d}x}\right)^2 \frac{\mathrm{d}x}{\mathrm{d}\xi}\bigg|_{\xi=\xi_k} \\
&\approx \sum_{k=1}^{N+1} w_k \ell_j(\xi)\, f'(\xi,t) \frac{\mathrm{d}x}{\mathrm{d}\xi}\bigg|_{\xi=\xi_k} .
\end{aligned}
\tag{4.24}
$$

The symbols ξ_k and w_k are the GLL points and their corresponding integration weights, respectively (see Appendix A). The numerical integration in (4.24) is not exact because the integrands are not polynomials of degree $2N-1$ or lower. In the following developments we will nevertheless replace $\approx$ by $=$, keeping in mind that this is an approximation. Recalling the cardinal interpolation property of the Lagrange polynomials, $\ell_i(\xi_k) = \delta_{ik}$, we can simplify (4.24):

$$
\sum_{i=1}^{N+1} M_{ji}^e \ddot{u}_i^e(t) + \sum_{i=1}^{N+1} K_{ji}^e\, u_i^e(t) = f_j^e(t) , \quad e = 1, \dots, n_e , \tag{4.25}
$$

with

$$
M_{ji}^e = w_j \rho'(\xi) \frac{\mathrm{d}x}{\mathrm{d}\xi} \delta_{ij}\bigg|_{\xi=\xi_j} , \tag{4.26}
$$

$$
K_{ji}^e = \sum_{k=1}^{N+1} w_k \mu'(\xi)\, \dot{\ell}_j(\xi)\dot{\ell}_i(\xi) \left(\frac{\mathrm{d}\xi}{\mathrm{d}x}\right)^2 \frac{\mathrm{d}x}{\mathrm{d}\xi}\bigg|_{\xi=\xi_k} , \tag{4.27}
$$

$$
f_j^e(t) = w_j\, f'(\xi,t) \frac{\mathrm{d}x}{\mathrm{d}\xi}\bigg|_{\xi=\xi_j} . \tag{4.28}
$$

Certainly the most advantageous property of the spectral-element discretisation is the diagonality of the local mass matrix $\mathbf{M}^e$. Since the inversion of M_{ji}^e is trivial and computationally inexpensive, we easily obtain an explicit formula for the second-time derivative $\ddot{u}_i(t)$ that can then be discretised.

The numerical integration of Eq. (4.20) has so far been purely local, resulting in linear systems for each element (Eq. 4.25). Related to the individual treatment of the elements is a *local numbering* that allows us to address the GLL points of an element. Points shared between two neighbouring elements are counted twice because they appear in the discretised versions of two integrals. To ensure the continuity of the approximation $\bar{u}$ across the element boundaries, we need to assemble a global system of equations. This is done by introducing a *global numbering* where each GLL point is counted once even when it is shared between elements. Associated which the global numbering is the global displacement vector $\mathbf{u}^{\text{global}}$. Figure 4.1

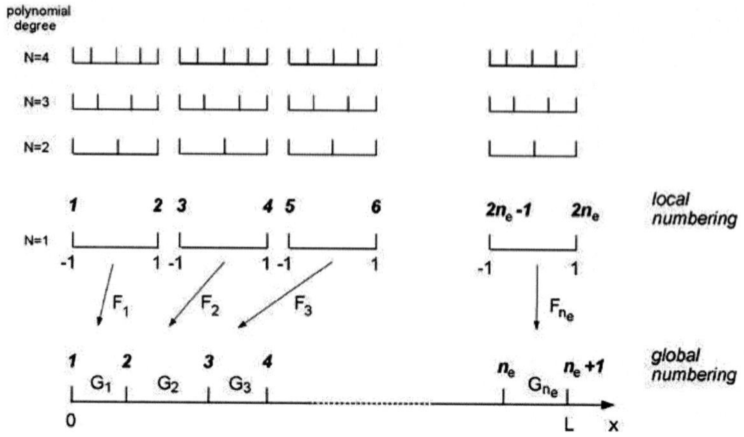

Fig. 4.1 Illustration of the element transformations F_e and the concepts of global and local numbering. The spatial domain $[0, L]$, shown in the lowermost part of the figure, is subdivided into n_e non-overlapping elements. Each element G_e is related to the reference interval $[-1, 1]$ via an element-specific transformation F_e. The local numbering is used to address the node points of the individual elements, some of which are shared between two elements. The global numbering identifies the spatial locations of the node points. This illustration is for the simplest, though unrealistic, case where the elements have no internal node points (polynomial degree 1). Subdivisions of the reference interval for higher polynomial degrees are shown schematically above

illustrates the concept of local and global numbering. The operation of assembling the local mass and stiffness matrices, $\mathbf{M}^e$ and $\mathbf{K}^e$, into their global versions, $\mathbf{M}^{\text{global}}$ and $\mathbf{K}^{\text{global}}$, then simply consists in summing the entries of the local matrices at coincident node points. This leads to a global system of equations,

$$\mathbf{M}^{\text{global}} \cdot \ddot{\mathbf{u}}^{\text{global}}(t) + \mathbf{K}^{\text{global}} \cdot \mathbf{u}^{\text{global}}(t) = \mathbf{f}^{\text{global}}(t) \tag{4.29}$$

that we need to solve for $\mathbf{u}^{\text{global}}$. We note that in practice the stiffness matrix is rarely computed explicitly because only its product with the vector of expansion coefficients is needed.

4.2 Extension to the 3D Case

4.2.1 Mesh Generation

The spectral-element method in more than one dimension starts with the subdivision of the computational domain G into n_e non-overlapping elements G_e such that $G = \bigcup_{i=1}^{n_e} G_e$. This process is referred to as *mesh generation* or *meshing*. The design of the elements should be such that they follow major geologic features including sedimentary basins and faults. To ensure a spatially uniform sampling of

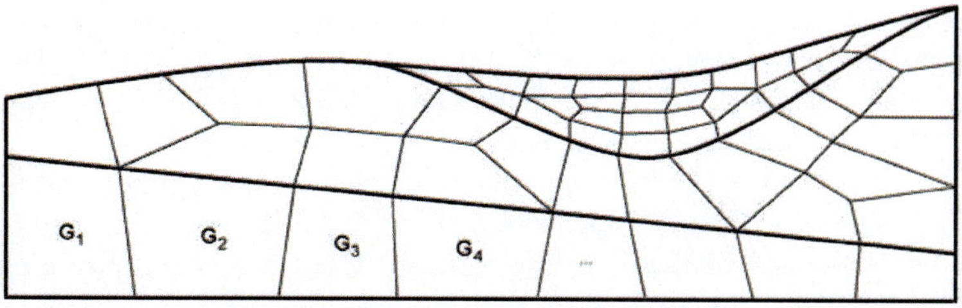

Fig. 4.2 Schematic representation of a mesh in a 2D domain. *Bold lines* indicate structural discontinuities that are honoured by the elements. The seismic velocities in the top layer are small compared to the velocities in the bottom layer where the elements are largest

the seismic waves in different parts of the model, the size of the elements should be proportional to the seismic velocity. Moreover, the faces of the elements should align with structural discontinuities because abrupt changes of the structure and the wave field cannot be represented accurately by the smooth polynomials inside the elements. A mesh is said to *honour discontinuities* when the boundaries of the elements coincide with the discontinuities. A schematic meshing of a 2D domain is shown in Fig. 4.2.

The 3D mesh in Fig. 4.3 was used to analyse near-fault and strong-motion site effects in the Grenoble valley, France (Stupazzini, 2006; Stupazzini et al., 2009). The study region is characterised by strong topographic variations and seismic wave speeds that range between $v_\mathrm{P} \approx 1.5\,\mathrm{km/s}$ and $v_\mathrm{S} \approx 0.3\,\mathrm{km/s}$ at the surface of the alluvial basin (green) and $v_\mathrm{P} \approx 5.9\,\mathrm{km/s}$ and $v_\mathrm{S} \approx 3.4\,\mathrm{km/s}$ in the deep bedrock layers (yellow). This implies that the wavelengths within the model vary by nearly one

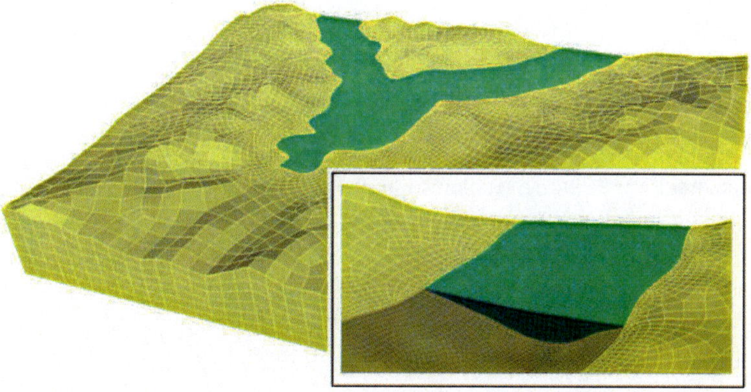

Fig. 4.3 Hexahedral mesh of the Grenoble valley used in a seismic ground motion study (Stupazzini et al., 2009). The wave speeds vary by almost one order of magnitude between the seismically slow alluvial basin (*green*) and the seismically fast bedrock (*yellow*). This variability translates to element sizes that range between 20 m at the surface of the basin and 900 m near the bottom of the model. Note that the seismic discontinuity between basin material and bedrock is honoured by the elements

order of magnitude. The size of the elements in the seismically slow alluvial basin is as small as 20 m and it reaches almost 900 m near the bottom of the model. The total number of elements is 216, 972, and it allows for the propagation of frequencies up to 3 Hz when the polynomial degree is 4.

Today, powerful mesh generation tools, known as meshers, are available (e.g. CUBIT, developed by Sandia National Laboratories or GiD from the International Center for Numerical Methods in Engineering). They greatly simplify the mesh design for complex 3D models.

In order to apply the same numerical quadrature to all the elements G_e, they need to be mapped onto the unit cube $\Lambda = [-1, 1]^3$. Mostly, this transformation will not be given analytically, especially in the case of complex geologic structures. It needs to be approximated instead, and this approximation is based on the concept of *shape functions* and *anchor nodes*. Each element G_e in the physical space is defined by a set of n_a anchor nodes $\mathbf{x}^a$ and their corresponding shape functions N^a. The eight corners of an element are always used as anchor nodes. For elements with straight edges, eight anchor nodes are sufficient to accurately represent the geometry. Additional anchor nodes on the edges or faces may be needed to represent elements with curved edges. A position vector $\mathbf{x}$ in G_e is related to a position vector $\boldsymbol{\xi}$ in the reference cube Λ via the transformation

$$\mathbf{x}(\boldsymbol{\xi}) = \mathbf{F}_e(\boldsymbol{\xi}) = \sum_{a=1}^{n_a} N^a(\boldsymbol{\xi})\,\mathbf{x}^a\,, \tag{4.30}$$

where the shape functions satisfy the condition

$$N^a(\boldsymbol{\xi}^b) = \delta_{ab}\,. \tag{4.31}$$

Equations (4.30) and (4.31) uniquely relate the anchor node $\mathbf{x}^a$ in the physical element G_e to an anchor node $\boldsymbol{\xi}^a$ in the reference cube Λ:

$$\mathbf{x}(\boldsymbol{\xi}^a) = \mathbf{x}^a\,. \tag{4.32}$$

The shape functions N^a are most conveniently defined as products of three Lagrange polynomials, the collocation points of which are the coordinates ξ_i^a of the anchor nodes $\boldsymbol{\xi}^a$ in the reference cube:

$$N^a(\boldsymbol{\xi}) = \ell_{1,a}(\xi_1)\ell_{2,a}(\xi_2)\ell_{3,a}(\xi_3)\,, \tag{4.33}$$

with

$$\ell_{i,a}(\xi_i) = \delta_{ab}\,. \tag{4.34}$$

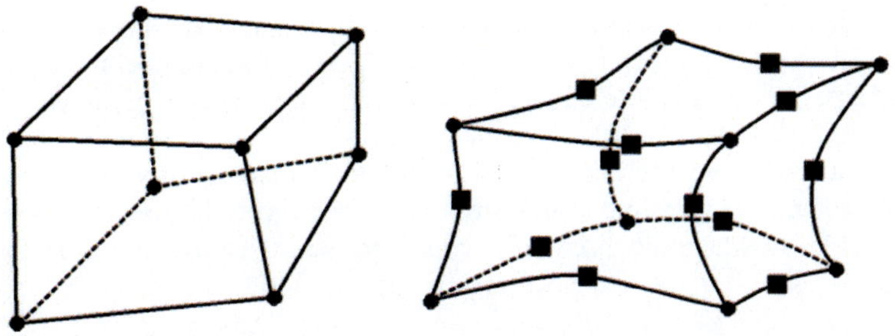

Fig. 4.4 *Left*: The element with *straight edges* is defined by eight anchor nodes and linear shape functions. *Right*: The element with *curved edges* is defined by eight anchor nodes in the corners plus 12 anchor nodes on the edges. The corresponding shape functions are of degree 2. Additional anchor nodes can be defined in the centre and on the faces, leading to an element with 27 anchor nodes

The degree of the Lagrange polynomials depends on the complexity of the element G_e. It is 1 for elements with straight edges and 2 for elements with curved edges. In most applications, the degree of the shape functions is lower than the degree of the polynomials used to interpolate the wave field inside the elements. The transformation F_e is therefore called *subparametric* (Fig. 4.4).

Mesh generation is to some degree an art that cannot be fully automated. The principal difficulty is to design a mesh that honours the complexities of the structural model while producing elements that are as large as possible. Very small elements can of course honour any complexity but they result in prohibitively short time steps because the CFL condition must always be satisfied in the numerical simulations. The choice of tensorised Lagrange polynomials as basis functions in 3D spectral elements furthermore imposes that only hexahedra can be used for the mesh even when tetrahedra may sometimes be better suited from a purely geometric point of view.

Generalisations of the spectral-element method in 2D allow for the combination of quadrangles and triangles within the same mesh (e.g. Komatitsch et al., 2001; Mercerat et al., 2006). In 3D, this approach can be adapted to combine hexahedra and tetrahedra in order to allow for more geometrical flexibility.

Another possible solution of the mesh generation problem consists in the replacement of the original structural model by a long-wavelength equivalent and smooth model (see, for instance, Sect. 13.2.2). The smoothness of the long-wavelength equivalent model alleviates many of the restrictions that complex models with discontinuities impose on the mesh design. Different techniques are available for the construction of long-wavelength equivalent models. Fichtner & Igel (2008) matched dispersion curves of a discontinuous Earth model to the dispersion curves of a smooth Earth model. A more general approach is based on the non-periodic homogenisation of the wave equation (Capdeville et al., 2010a, b; Guillot et al., 2010).

Mesh generation has received much attention in recent years. In practice, however, it can be advantageous to take the proper design of a mesh not too seriously. It is a matter of fact that the precise location and the sharpness of structural discontinuities inside the Earth are uncertain because our data have finite and often insufficient resolution. The implementation of a strict discontinuity in the spectral-element method can therefore lead to strong interface waves that are not observable in the real Earth where the discontinuity may not be as sharp as in our simplified models.

4.2.2 Weak Solution of the Elastic Wave Equation

As a preparatory step towards the spatial discretisation, we first derive the weak form of the elastic wave equation. We base our development on the strong displacement–stress variant of the equations of motion:

$$\rho(\mathbf{x})\,\ddot{\mathbf{u}}(\mathbf{x},t) - \nabla\cdot\boldsymbol{\sigma}(\mathbf{x},t) = \mathbf{f}(\mathbf{x},t) \tag{4.35}$$

$$\boldsymbol{\sigma}(\mathbf{x},t) = \mathbf{C}(\mathbf{x}) : \nabla\mathbf{u}(\mathbf{x},t)\,, \tag{4.36}$$

subject to the boundary and initial conditions

$$\mathbf{n}\cdot\boldsymbol{\sigma}|_{\mathbf{x}\in\partial G} = \mathbf{0}\,,\quad \mathbf{u}|_{t=0} = \dot{\mathbf{u}}|_{t=0} = \mathbf{0}\,. \tag{4.37}$$

For the moment we disregard dissipation, i.e., the time dependence of the elastic tensor **C**. Multiplying Eq. (4.35) by an arbitrary, differentiable, time-independent test function **w** and integrating over G gives

$$\int_G \rho\,\mathbf{w}\cdot\ddot{\mathbf{u}}\,d^3\mathbf{x} - \int_G \mathbf{w}\cdot(\nabla\cdot\boldsymbol{\sigma})\,d^3\mathbf{x} = \int_G \mathbf{w}\cdot\mathbf{f}\,d^3\mathbf{x}\,. \tag{4.38}$$

Invoking the identity

$$\mathbf{w}\cdot(\nabla\cdot\boldsymbol{\sigma}) = \nabla\cdot(\mathbf{w}\cdot\boldsymbol{\sigma}) - \nabla\mathbf{w}:\boldsymbol{\sigma}\,, \tag{4.39}$$

together with Gauss' theorem, yields

$$\int_G \rho\,\mathbf{w}\cdot\ddot{\mathbf{u}}\,d^3\mathbf{x} - \int_{\partial G} \mathbf{w}\cdot\boldsymbol{\sigma}\cdot\mathbf{n}\,d^2\mathbf{x} + \int_G \nabla\mathbf{w}:\boldsymbol{\sigma}\,d^3\mathbf{x} = \int_G \mathbf{w}\cdot\mathbf{f}\,d^3\mathbf{x}\,. \tag{4.40}$$

Upon inserting the free surface boundary condition, Eq. (4.40) condenses to

$$\int_G \rho\,\mathbf{w}\cdot\ddot{\mathbf{u}}\,d^3\mathbf{x} + \int_G \nabla\mathbf{w}:\boldsymbol{\sigma}\,d^3\mathbf{x} = \int_G \mathbf{w}\cdot\mathbf{f}\,d^3\mathbf{x}\,. \tag{4.41}$$

The same procedure is to be repeated for Eq. (4.36), but it does not require any additional transformations. Finding a weak solution to the equations of motion means to find a displacement field $\mathbf{u}$ that satisfies the integral relation (4.41) and

$$\int_G \mathbf{w} \cdot \boldsymbol{\sigma} \, \mathrm{d}^3\mathbf{x} = \int_G \mathbf{w} \cdot \mathbf{C} : \nabla \mathbf{u} \, \mathrm{d}^3\mathbf{x} \tag{4.42}$$

for any test function $\mathbf{w}$ and subject to the initial conditions

$$\int_G \rho \, \mathbf{w} \cdot \mathbf{u}|_{t=0} \, \mathrm{d}^3\mathbf{x} = \int_G \rho \, \mathbf{w} \cdot \dot{\mathbf{u}}|_{t=0} \, \mathrm{d}^3\mathbf{x} = \mathbf{0}\,. \tag{4.43}$$

As in the 1D case we note that the free surface condition is implicit in the weak formulation.

4.2.3 Discretisation of the Equations of Motion

In analogy to the Galerkin method for the 1D case (see Eq. (4.7)) we approximate the p-component u_p of the displacement field $\mathbf{u}$ by a superposition of basis functions

$$\psi_{ijk}(\mathbf{x}) = \psi_{ijk}(x_1, x_2, x_3) \tag{4.44}$$

weighted by expansion coefficients u_p^{ijk}:

$$u_p(\mathbf{x}, t) \approx \bar{u}_p(\mathbf{x}, t) = \sum_{i,j,k=1}^{N+1} u_p^{ijk}(t)\psi_{ijk}(\mathbf{x})\,. \tag{4.45}$$

The corresponding approximation of the stress tensor components σ_{pq} is

$$\sigma_{pq}(\mathbf{x}, t) \approx \bar{\sigma}_{pq}(\mathbf{x}, t) = \sum_{i,j,k=1}^{N+1} \sigma_{pq}^{ijk}(t)\, \psi_{ijk}(\mathbf{x})\,. \tag{4.46}$$

Equations (4.45) and (4.46) already assume that u_p and σ_{pq} are considered inside an element $G_e \subset \mathbb{R}^3$, where they can be represented by $(N+1)^3$ basis functions. We therefore omit the superscript e that we used in Sect. 4.1 to indicate local element-specific quantities.

To find a weak solution in the Galerkin sense, we replace the exact weak formulation from Eqs. (4.41), (4.42) and (4.43) by the requirement that the approximations $\bar{\mathbf{u}}$ and $\bar{\boldsymbol{\sigma}}$ satisfy

$$\int_{G_e} \rho\, \psi_{ijk}\mathbf{e}_p \cdot \ddot{\bar{\mathbf{u}}}\, \mathrm{d}^3\mathbf{x} + \int_{G_e} \nabla(\psi_{ijk}\mathbf{e}_p) : \bar{\boldsymbol{\sigma}}\, \mathrm{d}^3\mathbf{x} = \int_{G_e} \psi_{ijk}\mathbf{e}_p \cdot \mathbf{f}\, \mathrm{d}^3\mathbf{x} \tag{4.47}$$

and

$$\int_{G_e} \psi_{ijk}\mathbf{e}_p \cdot \bar{\sigma}\,\mathrm{d}^3\mathbf{x} = \int_{G_e} \psi_{ijk}\mathbf{e}_p \cdot \mathbf{C} : \nabla\bar{\mathbf{u}}\,\mathrm{d}^3\mathbf{x}\,, \tag{4.48}$$

for all basis functions ψ_{ijk} and for all unit vectors $\mathbf{e}_p$ with $p = 1, 2, 3$. The weak initial conditions are

$$\int_{G_e} \rho\,\psi_{ijk}\mathbf{e}_p \cdot \bar{\mathbf{u}}|_{t=0}\mathrm{d}^3\mathbf{x} = \int_{G_e} \rho\,\psi_{ijk}\mathbf{e}_p\,\dot{\bar{\mathbf{u}}}|_{t=0}\,\mathrm{d}^3\mathbf{x} = \mathbf{0}\,, \tag{4.49}$$

again for all basis functions and unit vectors. To keep the following formulas as readable as possible, we treat the summands in Eqs. (4.47) and (4.48) individually. For the first term on the left-hand side of (4.47) we find

$$\mathbb{F}_{qrs}[\rho\ddot{u}_p] := \int_{G_e} \rho\,\psi_{qrs}\mathbf{e}_p \cdot \ddot{\bar{\mathbf{u}}}\,\mathrm{d}^3\mathbf{x} = \sum_{i,j,k=1}^{N+1} \int_{G_e} \rho(\mathbf{x})\ddot{u}_p^{ijk}(t)\psi_{ijk}(\mathbf{x})\,\psi_{qrs}(\mathbf{x})\,\mathrm{d}^3\mathbf{x}\,. \tag{4.50}$$

The symbol $\mathbb{F}$ is intended to express that $\mathbb{F}_{qrs}[\rho\ddot{u}_p]$ is a discrete local force, averaged over the element G_e. In the next step we relate the elements G_e $(e = 1, \ldots, n_e)$ to the reference cube $\Lambda = [-1, 1]^3$ via invertible and element-specific transformations $\mathbf{F}_e$ that we already made explicit in Eqs. (4.30), (4.31), (4.32) and (4.33):

$$\mathbf{F}_e : [-1, 1]^3 = \Lambda \to G_e\,,\ \mathbf{x} = \mathbf{F}_e(\boldsymbol{\xi})\,,\ \boldsymbol{\xi} = \boldsymbol{\xi}(\mathbf{x}) = \mathbf{F}_e^{-1}(\mathbf{x})\,, e = 1, \ldots, n_e\,. \tag{4.51}$$

The action of the transformations $\mathbf{F}_e$ is illustrated in Fig. 4.5.

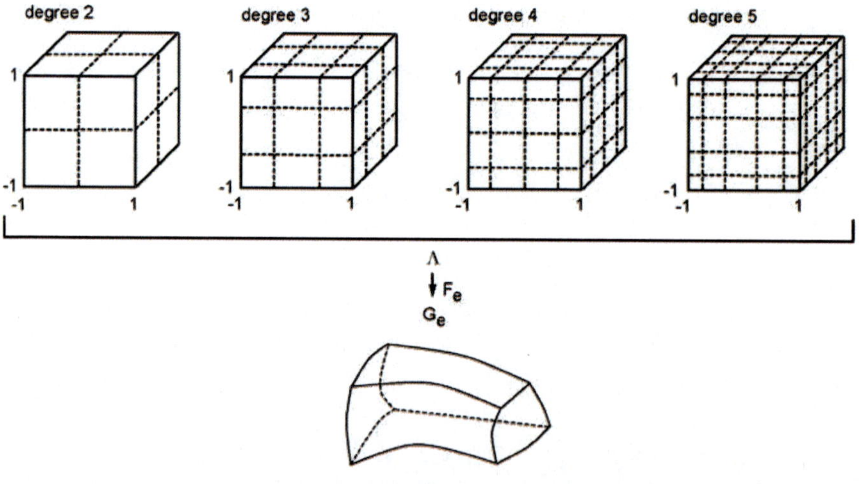

Fig. 4.5 Mapping of the deformed element G_e from the physical space to the reference cube $\Lambda = [-1, 1]^3$. *Dashed lines* indicate the GLL nodal lines for different polynomial degrees

Transforming the integral (4.50) according to the transformation (4.51) yields

$$\mathbb{F}_{qrs}[\rho\ddot{u}_p] = \sum_{i,j,k=1}^{N+1} \int_{\Lambda} \rho[\mathbf{x}(\boldsymbol{\xi})]\, \ddot{u}_p^{ijk}(t)\, \psi_{ijk}[\mathbf{x}(\boldsymbol{\xi})]\, \psi_{qrs}[\mathbf{x}(\boldsymbol{\xi})]\, J(\boldsymbol{\xi})\, \mathrm{d}^3\boldsymbol{\xi}\,, \tag{4.52}$$

where the symbol J in Eq. (4.52) denotes the Jacobian of $\mathbf{F}_e$. It is understood that the elements are chosen such that $J > 0$. For the basis functions $\psi_{ijk}[\mathbf{x}(\boldsymbol{\xi})]$ we choose the product of three Lagrange polynomials collocated at the GLL points:

$$\psi_{ijk}[\mathbf{x}(\boldsymbol{\xi})] = \ell_i(\xi_1)\ell_j(\xi_2)\ell_k(\xi_3)\,. \tag{4.53}$$

The resulting expression for $\mathbb{F}_{qrs}[\rho\ddot{u}_p]$ is then

$$\mathbb{F}_{qrs}[\rho\ddot{u}_p] = \sum_{i,j,k=1}^{N+1} \int_{\Lambda} \rho'(\boldsymbol{\xi})\ddot{u}_p^{ijk}(t)\ell_i(\xi_1)\ell_j(\xi_2)\ell_k(\xi_3)\ell_q(\xi_1)\ell_r(\xi_2)\ell_s(\xi_3)\, J(\boldsymbol{\xi})\, \mathrm{d}^3\boldsymbol{\xi}\,, \tag{4.54}$$

with the transformed density defined by

$$\rho'(\boldsymbol{\xi}) := \rho[\mathbf{x}(\boldsymbol{\xi})]\,. \tag{4.55}$$

Applying the GLL quadrature rule to Eq. (4.54) yields the following simple expression:

$$\mathbb{F}_{qrs}[\rho\ddot{u}_p] = w_q w_r w_s\, \rho'(\boldsymbol{\xi}^{qrs})\ddot{u}_p^{qrs}\, J(\boldsymbol{\xi}^{qrs})\,, \tag{4.56}$$

where we need to evaluate the transformed density and the Jacobian at the GLL points

$$\boldsymbol{\xi}^{qrs} := (\xi_1^q, \xi_2^r, \xi_3^s)\,. \tag{4.57}$$

Equation (4.57) is the 3D equivalent of Eq. (4.26) that specifies the mass matrix for the 1D case. The combination of GLL quadrature and Lagrange polynomials collocated at the GLL points again ensures that the mass matrix in three dimensions is diagonal, meaning that the evaluation of $\mathbb{F}_{qrs}[\rho\ddot{u}_p]$ does not involve any expansion coefficients other than u_p^{qrs}.

We now consider the second term on the left-hand side of Eq. (4.47):

$$\mathbb{F}_{qrs}[(\nabla\cdot\boldsymbol{\sigma})_p] := \int_{G_e} \nabla(\psi_{qrs}\mathbf{e}_p) : \bar{\boldsymbol{\sigma}}\, \mathrm{d}^3\mathbf{x}\,. \tag{4.58}$$

Transforming to the reference cube Λ then gives

$$\mathbb{F}_{qrs}[(\nabla \cdot \boldsymbol{\sigma})_p] = \sum_{n,m=1}^{3} \int_{\Lambda} \frac{\partial \xi_m}{\partial x_n} \frac{\partial}{\partial \xi_m} [\ell_q(\xi_1)\ell_r(\xi_2)\ell_s(\xi_3)]\, \bar{\sigma}'_{np}(\boldsymbol{\xi})\, J(\boldsymbol{\xi})\, \mathrm{d}^3\boldsymbol{\xi}\,, \quad (4.59)$$

where we already substituted the Lagrange polynomials for the general basis functions, as proposed in Eq. (4.53). The transformed stress tensor $\bar{\sigma}'$ is defined as

$$\bar{\boldsymbol{\sigma}}'(\boldsymbol{\xi}) := \bar{\boldsymbol{\sigma}}[\mathbf{x}(\boldsymbol{\xi})]\,. \quad (4.60)$$

Approximating the integral in Eq. (4.59) via the GLL quadrature rule results in a rather lengthy expression:

$$\begin{aligned}\mathbb{F}_{qrs}[(\nabla \cdot \boldsymbol{\sigma})_p] =& \sum_{n=1}^{3}\sum_{i=1}^{N+1} w_i w_r w_s \dot{\ell}_q(\xi_1^i)\, \sigma_{np}^{irs}\, J(\boldsymbol{\xi}^{irs})\, \frac{\partial \xi_1}{\partial x_n}(\boldsymbol{\xi}^{irs}) \\ &+ \sum_{n=1}^{3}\sum_{i=1}^{N+1} w_q w_i w_s \dot{\ell}_r(\xi_2^i)\, \sigma_{np}^{qis}\, J(\boldsymbol{\xi}^{qis})\, \frac{\partial \xi_2}{\partial x_n}(\boldsymbol{\xi}^{qis}) \\ &+ \sum_{n=1}^{3}\sum_{i=1}^{N+1} w_q w_r w_i \dot{\ell}_s(\xi_3^i)\, \sigma_{np}^{qri}\, J(\boldsymbol{\xi}^{qri})\, \frac{\partial \xi_3}{\partial x_n}(\boldsymbol{\xi}^{qri})\,. \end{aligned} \quad (4.61)$$

Equation (4.61) involves a sum over $9(N+1)$ terms which makes it computationally much more expensive than (4.56). Repeating the above procedure for the source term $\mathbf{f}$ in Eq. (4.35) gives

$$\mathbb{F}_{qrs}(f_p) := \int_{G_e} \psi_{qrs}\mathbf{e}_p \cdot \mathbf{f}\, \mathrm{d}^3\mathbf{x} = w_q w_r w_s\, f'_p(\boldsymbol{\xi}^{qrs})\, J(\boldsymbol{\xi}^{qrs})\,, \quad (4.62)$$

with the transformed force density

$$f'_p(\boldsymbol{\xi}) := f_p[\mathbf{x}(\boldsymbol{\xi})]\,. \quad (4.63)$$

Equation (4.62) deserves special attention because its validity is not as obvious as it may initially appear. In fact, we must recall that $\int_{G_e} \psi_{qrs}\mathbf{e}_p \cdot \mathbf{f}\,\mathrm{d}^3\mathbf{x} = w_q w_r w_s\, f'_p(\boldsymbol{\xi}^{qrs})\, J(\boldsymbol{\xi}^{qrs})$ is an approximation. Its accuracy is controlled by the quality of the GLL quadrature that we used to evaluate the projection integral. Since GLL quadrature is exact only for polynomials of degree $2N-1$ and below, the spatial dependence of the source term $\mathbf{f}$ should at least be smooth. In most seismological applications, however, $\mathbf{f}$ is point localised, meaning that its spatial dependence is described by a delta function that is far from being sufficiently smooth. Equation (4.62) is therefore inappropriate for point sources. However, before studying the implementation of point sources in more detail (see Sect. 4.2.4), we continue with the discretisation of the equations of motion.

It remains to consider the approximate weak form of the constitutive relation as specified by Eq. (4.48). For the Galerkin projection of the (mn)-component of the stress tensor $\boldsymbol{\sigma}$, we find

$$\mathbb{F}_{qrs}(\sigma_{mn}) := \int_{G_e} (\psi_{qrs}\mathbf{e}_m \cdot \bar{\boldsymbol{\sigma}})_n \, \mathrm{d}^3\mathbf{x} = w_q w_r w_s \, \sigma_{mn}^{qrs} \, J(\boldsymbol{\xi}^{qrs}) \, . \tag{4.64}$$

Projecting the (mn)-component of the right-hand side in the approximate weak constitutive relation (4.48) gives

$$\begin{aligned}\mathbb{F}_{qrs}[(\mathbf{C} : \nabla\mathbf{u})_{mn}] &= \int_{G_e} (\psi_{qrs}\mathbf{e}_m \cdot \mathbf{C} : \nabla\bar{\mathbf{u}})_n \, \mathrm{d}^3\mathbf{x} \\ &= \sum_{a,b=1}^{3} \sum_{i,j,k=1}^{N+1} \psi_{qrs}(\mathbf{x}) \, C_{mnab} \, \frac{\partial}{\partial x_a} [u_b^{ijk} \, \psi_{ijk}(\mathbf{x})] \, \mathrm{d}^3\mathbf{x} \, .\end{aligned} \tag{4.65}$$

Transforming Eq. (4.65) and inserting the Lagrange polynomials as basis functions yields

$$\begin{aligned}\mathbb{F}_{qrs}[(\mathbf{C} : \nabla\mathbf{u})_{mn}] = \sum_{a,b,p=1}^{3} \sum_{i,j,k=1}^{N+1} \int_{\Lambda} u_b^{ijk} \ell_q(\xi_1)\ell_r(\xi_2)\ell_s(\xi_3) \\ \cdot \, C'_{mnab}(\boldsymbol{\xi}) \, \frac{\partial \xi_p}{\partial x_a} \frac{\partial}{\partial \xi_p} [\ell_i(\xi_1)\ell_j(\xi_2)\ell_k(\xi_3)] \, J(\boldsymbol{\xi}) \, \mathrm{d}^3\boldsymbol{\xi} \, ,\end{aligned} \tag{4.66}$$

with

$$\mathbf{C}'(\boldsymbol{\xi}) := \mathbf{C}[\mathbf{x}(\boldsymbol{\xi})] \, . \tag{4.67}$$

Then following the usual steps, we obtain the Galerkin projection

$$\begin{aligned}\mathbb{F}_{qrs}[(\mathbf{C} : \nabla\mathbf{u})_{mn}] = w_q w_r w_s \, J(\boldsymbol{\xi}^{qrs}) \sum_{a,b=1}^{3} C'_{mnab}(\boldsymbol{\xi}^{qrs}) \\ \cdot \sum_{i=1}^{N+1} \left[u_n^{irs} \dot{\ell}_i(\xi_1^q) \frac{\partial \xi_1}{\partial x_m} + u_n^{qis} \dot{\ell}_i(\xi_2^r) \frac{\partial \xi_2}{\partial x_m} + u_n^{qri} \dot{\ell}_i(\xi_3^s) \frac{\partial \xi_3}{\partial x_m} \right] .\end{aligned} \tag{4.68}$$

At this point we can assemble the discrete equations of motions:

$$[w_q w_r w_s \, \rho'(\boldsymbol{\xi}^{qrs}) \, J(\boldsymbol{\xi}^{qrs})] \, \ddot{u}_p^{qrs} = \mathbb{F}_{qrs}(f_p) - \mathbb{F}_{qrs}[(\nabla \cdot \boldsymbol{\sigma})_p] \, , \tag{4.69}$$

$$[w_q w_r w_s \, J(\boldsymbol{\xi}^{qrs})] \, \sigma_{mn}^{qrs} = \mathbb{F}_{qrs}[(\mathbf{C} : \nabla\mathbf{u})_{mn}] \, . \tag{4.70}$$

The above equations are of general validity. They are useful when deformed elements are needed in order to mesh a structural discontinuity or topography. Equation (4.70) gives the expansion coefficients σ_{mn}^{qrs} of $\boldsymbol{\sigma}$ in terms of the expansion coefficients u_n^{qrs} of $\mathbf{u}$. The former can then be used in Eq. (4.69) to compute the acceleration $\ddot{u}_p^{qrs}$. As a result of using GLL node points together with GLL quadrature, we obtain σ_{mn}^{qrs} and $\ddot{u}_p^{qrs}$ without inverting a large matrix. This is, as already mentioned in Sect. 4.1, one of the principal advantages of the spectral-element method.

Since the explicit computation of the mass and stiffness matrices is unnecessary in practice, we work directly with Eq. (4.69) which is a local, i.e. element-wise, discrete force balance. The corresponding global force balance can be constructed as in the 1D case.

4.2.4 Point Source Implementation

In applications that focus on the far-field properties of the seismic wave field, the seismic source can be represented by either a single-force point source,

$$\mathbf{f}(\mathbf{x}, t) = \mathbf{s}(t)\,\delta(\mathbf{x} - \mathbf{x}^{\mathrm{s}})\,, \tag{4.71}$$

or a moment tensor point source,

$$\mathbf{f}(\mathbf{x}, t) = -\nabla \cdot [\mathbf{M}(t)\,\delta(\mathbf{x} - \mathbf{x}^{\mathrm{s}})]\,, \tag{4.72}$$

located at $\mathbf{x} = \mathbf{x}^{\mathrm{s}}$. The symbols $\mathbf{s}$ and $\mathbf{M}$ denote a vectorial source time function and the moment tensor, respectively. As we already noted in the previous paragraph, the implementation of point sources deserves special attention. This is because the approximation of the projection integral $\mathbb{F}_{qrs}(f_p) := \int_{G_e} \psi_{qrs}\mathbf{e}_p \cdot \mathbf{f}\,\mathrm{d}^3\mathbf{x}$ (Eq. 4.62) by GLL quadrature is inappropriate when the integrand is far from being a polynomial of degree $2N-1$. There are two possible, and in fact fully equivalent, solutions: (1) exact integration and (2) the polynomial approximation of the delta function.

4.2.4.1 Exact Integration

The most straightforward, though physically less insightful, approach is to avoid GLL quadrature altogether. This is possible because integrals over delta functions can be trivially evaluated without any approximations. For the single-force point source from Eq. (4.71), we find

$$\mathbb{F}_{qrs}(f_p) = \int_{G_e} \psi_{qrs}\mathbf{e}_p \cdot \mathbf{f}\,\mathrm{d}^3\mathbf{x} = \int_{G_e} s_p(t)\,\psi_{qrs}(\mathbf{x})\,\delta(\mathbf{x} - \mathbf{x}^{\mathrm{s}})\,\mathrm{d}^3\mathbf{x} = s_p(t)\,\psi_{qrs}(\mathbf{x}^{\mathrm{s}})\,, \tag{4.73}$$

and for the moment tensor source (4.72)

$$\mathbb{F}_{qrs}(f_p) = -\int_{G_e} \psi_{qrs}(\mathbf{x})\, \mathbf{e}_p \cdot \nabla \cdot \left[\mathbf{M}(t)\, \delta(\mathbf{x} - \mathbf{x}^{\mathrm{s}})\right] \mathrm{d}^3\mathbf{x} = \mathbf{e}_p \cdot \mathbf{M}(t) \cdot \nabla \psi_{qrs}(\mathbf{x}^{\mathrm{s}})\,. \tag{4.74}$$

In the reference cube Λ, the basis functions ψ_{qrs} are Lagrange polynomials with the GLL points as collocation points. It follows that more than one coefficient $\mathbb{F}_{qrs}(f_p)$ will be non-zero unless the source location $\mathbf{x}^{\mathrm{s}}$ coincides with a grid point. The numerical point source is therefore potentially non-local. This is reminiscent of the non-local point source approximations used in finite-difference modelling (Sect. 3.4).

4.2.4.2 Polynomial Approximation of the Delta Function

An interesting alternative to the exact integration is the approximation of the delta function by Lagrange polynomials that effectively correspond to a low-pass-filtered version of the point source (Faccioli et al., 1997). For this we assume that $\delta(\mathbf{x}-\mathbf{x}^{\mathrm{s}})$ in the source-bearing element G_e can be approximated in terms of the basis functions $\psi_{ijk}(\mathbf{x})$:

$$\delta(\mathbf{x} - \mathbf{x}^{\mathrm{s}}) \approx \bar{\delta}(\mathbf{x} - \mathbf{x}^{\mathrm{s}}) = \sum_{i,j,k=1}^{N+1} \delta^{ijk}\, \psi_{ijk}(\mathbf{x})\,. \tag{4.75}$$

We now determine the polynomial coefficients δ^{ijk} such that $\bar{\delta}$ reproduces the defining property of δ:

$$\psi(\mathbf{x}^{\mathrm{s}}) = \int_{G_e} \bar{\delta}(\mathbf{x} - \mathbf{x}^{\mathrm{s}})\, \psi(\mathbf{x})\, \mathrm{d}^3\mathbf{x}\,, \tag{4.76}$$

for any test function ψ that can itself be represented by $\psi_{ijk}(\mathbf{x})$. Transforming Eq. (4.76) to the reference cube Λ and substituting Eq. (4.75) gives

$$\psi(\mathbf{x}^{\mathrm{s}}) = \sum_{i,j,k=1}^{N+1} \int_{\Lambda} \delta^{ijk}\, \psi_{ijk}[\mathbf{x}(\boldsymbol{\xi})]\, \psi[\mathbf{x}(\boldsymbol{\xi})]\, J(\boldsymbol{\xi})\, \mathrm{d}^3\boldsymbol{\xi}\,. \tag{4.77}$$

We then insert the Lagrange polynomials as basis functions:

$$\psi(\mathbf{x}^{\mathrm{s}}) = \sum_{i,j,k=1}^{N+1} \int_{\Lambda} \delta^{ijk}\, \ell_i(\xi_1)\ell_j(\xi_2)\ell_k(\xi_3)\, \psi[\mathbf{x}(\boldsymbol{\xi})]\, J(\boldsymbol{\xi})\, \mathrm{d}^3\boldsymbol{\xi}\,. \tag{4.78}$$

As a particular test function we choose

$$\psi[\mathbf{x}(\boldsymbol{\xi})] = \ell_q(\xi_1)\ell_r(\xi_2)\ell_s(\xi_3)\,. \tag{4.79}$$

Inserting (4.79) into (4.78) and approximating the integral with GLL quadrature gives

$$\ell_q(\xi_1^s)\ell_r(\xi_2^s)\ell_s(\xi_3^s) = w_q w_r w_s \, \delta^{qrs} \, J(\boldsymbol{\xi}^{qrs}) \,, \tag{4.80}$$

where

$$(\xi_1^s, \xi_2^s, \xi_3^s) = \boldsymbol{\xi}^s = \mathbf{F}_e(\mathbf{x}^s) \tag{4.81}$$

denotes the source position in the reference coordinate system. The polynomial coefficients δ^{ijk} are thus given by

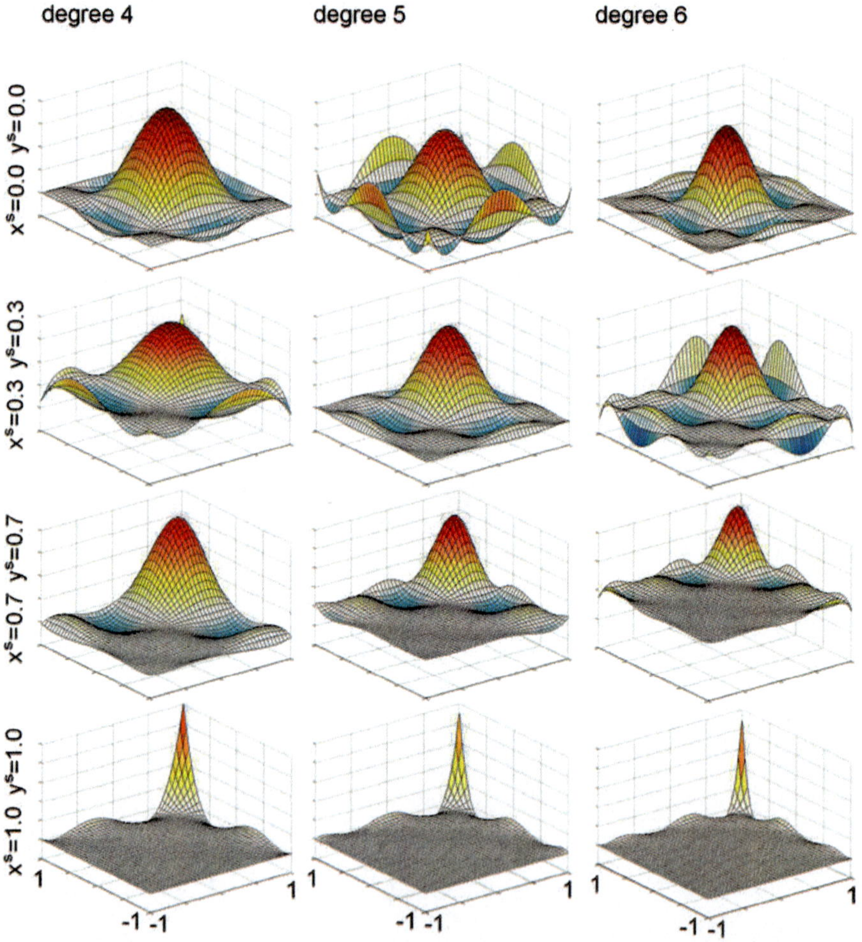

Fig. 4.6 Polynomial approximations of the δ function for different polynomial degrees and different point source localisations inside the reference cube

$$\delta^{ijk} = \frac{\ell_i(\xi_1^{\rm s})\ell_j(\xi_2^{\rm s})\ell_k(\xi_3^{\rm s})}{w_i w_j w_k \, J(\boldsymbol{\xi}^{ijk})} \,. \tag{4.82}$$

A graphical representation of the polynomial approximant $\bar{\delta}$ in two dimensions with $J = 1$ is shown in Fig. 4.6. Evidently, $\bar{\delta}$ is a low-pass-filtered, and therefore non-local, version of δ. The dominant wavelength of $\bar{\delta}$ depends on the polynomial degree and on the location inside the reference cube. It is, in any case, much shorter than the minimum wavelength of the seismic waves, which should be around twice the width of an element. We note that $\bar{\delta}$ integrates to 1 over one element:

$$1 = \int_{G_e} \bar{\delta}(\mathbf{x} - \mathbf{x}^{\rm s}) \, d^3\mathbf{x} \,. \tag{4.83}$$

For moment tensor point sources, it is most convenient to add the polynomial approximation of $\mathbf{M}(t)\,\delta(\mathbf{x} - \mathbf{x}^{\rm s})$ directly to the stress tensor $\boldsymbol{\sigma}$, prior to the computation of the Galerkin projection of $\nabla \cdot \boldsymbol{\sigma}$.

A direct consequence of using a low-pass-filtered version of a point source is that the wave field is inaccurate in the immediate vicinity of the physical source location. Based on numerical experiments, however, Faccioli et al. (1997) and Nissen-Meyer et al. (2007a) conclude that the numerical solutions become hardly distinguishable from the analytical one at distances of more than one element from the true point source. Thus, in most applications, the inaccuracy caused by the low-pass filtering will be very small compared to the error that would result from a source mislocation.

4.3 Variants of the Spectral-Element Method

Chebyshev spectral elements: The spectral-element method, as developed originally in fluid mechanics (Patera, 1984), was based on the element-wise expansion of the solution in terms of Chebyshev polynomials. This choice was motivated by their good approximation properties. Applications of the Chebyshev variant of the spectral-element method to acoustic and elastic wave propagation have been presented by Priolo et al. (1994), Seriani et al. (1995) and Seriani (1998). In contrast to the spectral-element method based on GLL points, the Chebyshev variant does not produce a diagonal mass matrix. The time integration of the equations of motion therefore requires the numerical inversion of a sparse matrix.

Fluid–solid models: The fluid regions of the Earth, including the oceans and the outer core, play an important role in numerous applications. Numerically propagating waves through a fluid–solid model is problematic. Simply setting the shear modulus to zero inside the fluid regions results in spurious oscillations because the displacement formulation of the wave equation has too many degrees of freedom. Therefore, several authors (Komatitsch et al., 2000; Komatitsch & Tromp, 2002; Chaljub et al., 2003, 2007; Chaljub & Valette, 2004) proposed to employ a velocity potential formulation of the equations of motion inside the fluid regions that is coupled to the displacement formulation in the solid part of the Earth model. The potential formulation naturally eliminates the excessive degrees of freedom.

An alternative to the spectral-element solution of the wave equation in the fluid region was presented by Capdeville et al. (2003a) who coupled normal mode solutions inside the spherically symmetric core to spectral-element solutions in the potentially heterogeneous mantle. The coupling is achieved using a Dirichlet-to-Neumann operator. Since the normal mode solution is computationally inexpensive, the overall resource requirements are substantially reduced.

2D spectral elements for radially symmetric Earth models: The applicability of numerical wave propagation to practically relevant problems can be seriously limited by the available computational resources. Doubling the dominant frequency increases the computational costs at least by a factor of 16 when the spatial dimension is 3. However, the restriction to radially symmetric Earth models allows us to work effectively with a 2D computational domain while solving the 3D wave equation (Nissen-Meyer et al., 2007a, 2008; see Fig. 4.7). This is accomplished by utilising analytical expressions for the azimuthal dependencies for all source types in this spherically symmetric case. This greatly reduces the necessary resources, and global wave propagation can easily be handled on desktop computers. It moreover presents an efficient basis for the computation of sensitivity kernels at arbitrary resolution (Nissen-Meyer et al., 2007b).

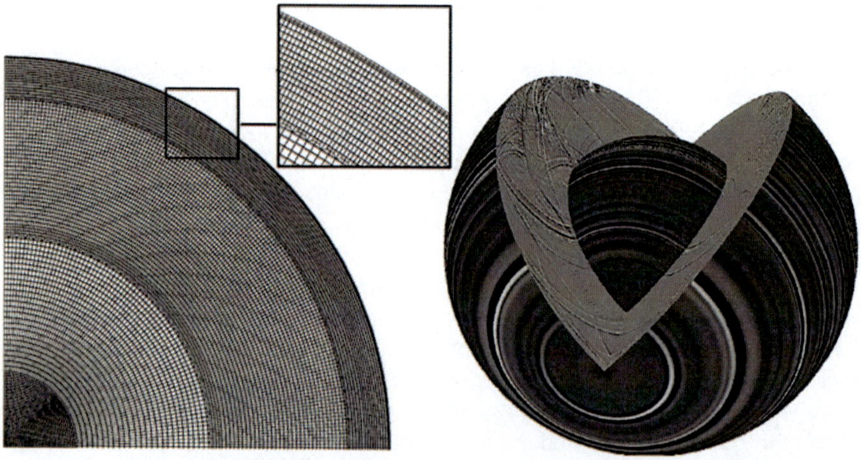

Fig. 4.7 *Left*: Elemental mesh for the 1D Earth model PREM (Dziewonski & Anderson, 1981) that produces accurate waveforms down to periods of 15 s. *Right*: Snapshot of the 3D seismic wave field after 20 min of propagation, for a full-moment tensor earthquake source at dominant period of 10 s. This simulation has been undertaken on a desktop and took 1 h CPU time

4.4 Accuracy and Efficiency

The accuracy of spectral-element solutions is controlled by both the size of the elements and the degree of the Lagrange polynomials. Decreasing the size of the elements and increasing the polynomial degree will mostly lead to higher accuracy. Exceptions are those cases where the numerical dispersion resulting from the low-order time discretisation dominates over the numerical error of the spatial discretisation.

In general, the GLL quadrature is not exact even in the case of a homogeneous medium. This is because the integrands are products of two polynomials of degree N, and GLL quadrature is exact only for polynomials of degree $(2N-1)$ and lower. This deficiency can become important when the material properties vary strongly inside an element.

To ensure that variable material properties are properly accounted for by the mesh, one may increase the polynomial degree. However, the maximum degree is very limited by the CFL stability condition and the available computational resources. This is because the distance between the first two GLL points decreases as $O(N^{-2})$, i.e. quadratically with increasing degree. Thus, choosing $N \gtrapprox 8$ usually results in unreasonably small step lengths. On the other hand, by choosing $N < 4$, one sacrifices much of the accuracy of the spectral-element method. A compromise based on experience is to use polynomial degrees between 4 and 7.

Also based on experience is the rule of thumb that one should use at least five grid points per minimum wavelength in order to obtain accurate solutions when the propagation distance is on the order of 10–50 wavelengths. It is important to keep in mind that numerical dispersion is cumulative, meaning that the numerical error grows steadily as the length of the simulation increases. Thus, a larger number of grid points per wavelength may become necessary when the propagation distance is significantly larger than 50 wavelengths.

Analytical estimates of numerical dispersion in spectral-element modelling have become available only recently. Based on a plane wave analysis in 2D, De Basabe & Sen (2007) found that the relative phase velocity error is below 0.2% when four to five grid points per wavelength are used and when the polynomial degree is 4 or higher. These results have been confirmed by Seriani & Oliveira (2008), who showed that numerical dispersion is negligible when five grid points per wavelength are used together with the polynomial degree 8. Since analytical dispersion analysis usually assumes homogeneous and unbounded media, it is important to keep in mind that additional dispersion errors may arise from boundary conditions or structural heterogeneities.

The term *accurate solution* is subjective to some degree. The required level of accuracy always depends on the specific application. It is recommended to always assess the quality of the synthetic seismograms for a particular application. This can be done using either convergence tests or comparisons to analytical solutions, if available. The commonly used rules of thumb are derived from a small number of well-behaved examples and they need not always be valid.

Chapter 5
Visco-elastic Dissipation

Seismic waves propagating through the Earth are attenuated due to the conversion of elastic energy into heat. While being important, we have ignored this effect in our previous analyses, mostly in the interest of simplicity. In the following paragraphs we show that attenuation can easily be incorporated into any time-domain solver of the elastic wave equation with the help of memory variables.

Based on the work of Robertsson et al. (1994) and Blanch et al. (1995), we start with a specification of the visco-elastic rheology in terms of a superposition of standard linear solids that are controlled by a set of relaxation parameters. The resulting constitutive relation involves a time convolution that can be transformed into a numerically more convenient system of first-order differential equations. We then derive a relation between the relaxation parameters and the quality factor Q that can be inferred from seismological measurements. The chapter concludes with the construction of Q models that are constant within a given frequency range.

5.1 Memory Variables

Under the assumption that all seismologically relevant modes of attenuation can be modelled by linear mechanisms, we restrict ourselves to linear visco-elastic theory. In the interest of a light notation we let the scalars σ, C and ε be representatives of some particular components of the tensors $\boldsymbol{\sigma}$, $\mathbf{C}$ and $\boldsymbol{\varepsilon}$, respectively, and we omit all spatial dependencies. The scalar version of the linear visco-elastic stress–strain relation (2.14) is then defined by

$$\sigma(t) = (C * \dot{\varepsilon})(t) = \int_{-\infty}^{t} C(t - t')\dot{\varepsilon}(t')\,\mathrm{d}t' . \tag{5.1}$$

Many processes including thermal diffusion, grain boundary relaxation, solid-phase transformations and partial melting contribute to the absorption of elastic energy. The combined effect of these mechanisms can be modelled with a stress relaxation function $C(t)$ that is a superposition of N standard linear solids (e.g. Liu et al., 1976; Kennett, 2001; Aki & Richards, 2002):

A. Fichtner, *Full Seismic Waveform Modelling and Inversion*, Advances in Geophysical and Environmental Mechanics and Mathematics,
DOI 10.1007/978-3-642-15807-0_5,

$$C(t) = C_\mathrm{r}\left[1 - \frac{1}{N}\sum_{n=1}^{N}\left(1 - \frac{\tau_{\varepsilon,n}}{\tau_{\sigma,n}}\right)\mathrm{e}^{-t/\tau_{\sigma,n}}\right]H(t)\,, \tag{5.2}$$

where $\tau_{\varepsilon,n}$ and $\tau_{\sigma,n}$ are the strain and stress relaxation times of the nth standard linear solid, respectively. The symbol $H(t)$ denotes the Heaviside function. When the strain has the form of a unit step equal to $H(t)$, the relaxed modulus $C_\mathrm{r} = \lim_{t\to\infty} C(t)$ is equal to the equilibrium stress. The corresponding initial stress $C(0) = C_\mathrm{u}$ is the unrelaxed modulus. Equation (5.2) is very general so that different sets of relaxation times can give nearly identical relaxation functions $C(t)$. To reduce this subjectively undesirable non-uniqueness we limit the number of free parameters by defining a dimensionless variable τ:

$$\tau := \frac{\tau_{\varepsilon,n}}{\tau_{\sigma,n}} - 1\,. \tag{5.3}$$

This gives

$$C(t) = C_\mathrm{r}\left[1 + \frac{\tau}{N}\sum_{n=1}^{N}\mathrm{e}^{-t/\tau_n}\right]H(t)\,, \tag{5.4}$$

where we renamed $\tau_{\sigma,n}$ into τ_n for notational brevity. The new parameter τ measures the relative difference between the relaxed modulus C_r and the unrelaxed modulus C_u:

$$\tau = (C_\mathrm{u} - C_\mathrm{r})/C_\mathrm{r}\,. \tag{5.5}$$

Differentiating (5.4) with respect to t and introducing the result into Eq. (5.1) yields

$$\dot\sigma(t) = C_\mathrm{r}(1+\tau)\,\dot\varepsilon(t) + C_\mathrm{r}\tau\sum_{n=1}^{N} M_n(t)\,, \tag{5.6}$$

where the *memory variables* M_n are defined by

$$M_n(t) := -\frac{1}{N\tau_n}\int_{-\infty}^{\infty}\mathrm{e}^{-(t-t')/\tau_n}\,H(t-t')\,\dot\varepsilon(t')\,\mathrm{d}t'\,. \tag{5.7}$$

Equation (5.7) is computationally inconvenient because the solution of the integral requires knowledge of the strain rate $\dot\varepsilon$ at all times between t_0 and t. This implies that the strain field would need to be stored in iterative time-domain solutions of the wave equation. Fortunately, the differentiation of (5.7) with respect to time allows us to write the memory variables as solutions of the following first-order differential equations:

$$\dot{M}_n = -\frac{1}{N\tau_n}\,\dot{\varepsilon} - \frac{1}{\tau_n}\,M_n\,, \qquad n = 1, \ldots, N. \tag{5.8}$$

Visco-elastic dissipation can thus be modelled by simultaneously solving the momentum Eq. (2.11), the modified stress–strain relation (5.6) and a set of N ordinary differential equations for the memory variables M_n (5.8). The memory variables are formally independent of the elastic parameter C_r. This formulation, proposed by Moczo & Kristek (2005), gives more accurate results in the case of strong attenuation heterogeneities than the earlier treatment by Blanch et al. (1995).

5.2 *Q* Models

Equations (5.6) and (5.8) are easily translate to the tensorial case where every C_{ijkl} becomes time dependent. In practice, however, only dissipation related to the shear modulus μ and the bulk modulus $\kappa = \lambda + \frac{2}{3}\mu$ can be estimated reliably using seismic observations. These estimates are mostly based on measurements of the amplitude reduction of seismic waves that result from the transformation of elastic into thermal energy. For this reason, seismology has traditionally emphasised the quality factor Q instead of particular stress relaxation functions. The definition of Q is based on the complex modulus

$$\tilde{C}(\nu) = \mathbf{i}\,\nu \int_{-\infty}^{\infty} C(t)\,\mathrm{e}^{-\mathrm{i}\nu t}\,\mathrm{d}t\,, \tag{5.9}$$

with the complex frequency $\nu = \omega + \mathbf{i}\gamma$ and $\gamma < 0$. Then

$$Q(\omega) := \frac{\mathfrak{Re}\,\tilde{C}(\omega)}{\mathfrak{Im}\,\tilde{C}(\omega)}\,. \tag{5.10}$$

The seismologist's interest in Q rests on its interpretation in terms of the energy that is dissipated per oscillation cycle, E_{diss}, relative to the mean elastic energy in the absence of dissipation, E_{mean}:

$$\frac{E_{\text{diss}}}{E_{\text{mean}}} = \frac{4\pi}{Q}\,. \tag{5.11}$$

For the stress relaxation function defined in Eq. (5.4) the complex modulus is given by

$$\tilde{C}(\omega) = C_{\mathrm{r}} + \frac{\mathbf{i}\nu\tau C_{\mathrm{r}}}{N}\sum_{n=1}^{N}\frac{\tau_n}{1+\mathbf{i}\nu\tau_n}\,, \tag{5.12}$$

and the resulting expression for Q is

$$Q(\omega) = \frac{\sum_{n=1}^{N}\left(1+\frac{\omega^2\tau_n^2\tau}{1+\omega^2\tau_n^2}\right)}{\sum_{n=1}^{N}\left(\frac{\omega\tau_n\tau}{1+\omega^2\tau_n^2}\right)}. \tag{5.13}$$

The frequency dependence of Q has been studied extensively. While seismological observations (e.g. Flanagan & Wiens, 1998; Cheng & Kennett, 2002; Shito et al., 2004) and laboratory experiments (e.g. Jackson, 1971, 2000) indicate that Q can depend significantly on ω, the constant-Q approximation remains popular because it explains many surface and body wave data sets adequately. See Anderson & Archambeau (1964) for an early example.

It is, in any case, non-trivial to find τ as well as a suitable set of relaxation times τ_n that approximate a pre-defined $Q_0(\omega)$ within a relevant frequency band $[\omega_1, \omega_2]$. The determination of τ and τ_n generally involves the solution of a non-linear optimisation problem (e.g. Blanch et al., 1995; Bohlen, 2002; Hestholm et al., 2006).

In the special case of a large and nearly constant Q we can approximate τ analytically. For this we note that $Q \gg 1$ results from $\tau^{-1} \gg 1$. Inserting $\tau^{-1} \gg 1$ into Eq. (5.13) yields

$$Q^{-1}(\omega) \approx \frac{\tau}{N}\sum_{n=1}^{N}\left(\frac{\omega\tau_n}{1+\omega^2\tau_n^2}\right). \tag{5.14}$$

Since Q is assumed to be approximately constant within $[\omega_1, \omega_2]$, we furthermore find

$$K(\omega) := \sum_{n=1}^{N}\left(\frac{\omega\tau_n}{1+\omega^2\tau_n^2}\right) \approx \text{const}, \quad \text{for } \omega \in [\omega_1, \omega_2], \tag{5.15}$$

and

$$Q^{-1} \approx \frac{\tau K}{N}. \tag{5.16}$$

Equation (5.16) approximates τ provided that the relaxation times τ_n are known. This allows us to rewrite (5.4) in a physically more interpretable form:

$$C(t) = C_r\left[1+\frac{1}{KQ}\sum_{n=1}^{N} e^{-t/\tau_n}\right] H(t). \tag{5.17}$$

The problem of constructing a nearly constant $Q(\omega)$ is now reduced to finding the relaxation times τ_n such that $K(\omega)$ is nearly constant. Since the optimal constant value of $K(\omega)$ is not known a priori, we propose to find τ_n such that the norm of the frequency derivative of $K(\omega)$,

Table 5.1 Dissipation parameters τ and τ_n for an ideally constant Q_0 of 100 in the frequency band from 0.02 to 0.2 Hz. The right column gives the mean deviation of the constructed $Q(\omega)$ from 100

n	τ	τ_1 [s]	τ_2 [s]	τ_3 [s]	τ_4 [s]	τ_5 [s]	ΔQ
2	0.0334	9.11	1.02				6.86
3	0.0451	10.28	1.26	0.10			6.65
4	0.0417	29.01	7.29	1.42	0.40		1.14
5	0.0493	32.06	8.83	1.97	0.46	0.11	0.78

$$J_2(\tau_n) := \int_{\omega_1}^{\omega_2} \left[\frac{d}{d\omega} K(\omega) \right]^2 \, d\omega \,, \tag{5.18}$$

is minimised. This method is robust and reliable when global search algorithms, such as simulated annealing (Kirkpatrick et al., 1983), are used to solve the optimisation problem. Once a suitable set of τ_n is found, τ can be approximated either by Eq. (5.16) or by a second non-linear optimisation stage based on Eq. (5.13).

Effectively, τ determines the absolute values of Q and τ_n determines the frequency range over which it is nearly constant. Table 5.1 lists τ and the relaxation times τ_n for different numbers of relaxation mechanisms for an ideally constant Q_0 of 100 in the frequency band from 0.02 to 0.2 Hz. The corresponding approximations $Q(\omega)$ are shown in Fig. 5.1.

Figure 5.1 reveals that a larger number of relaxation mechanisms gives better constant-Q approximations, especially for frequencies higher than the actual upper frequency limit. For frequencies lower than the lower frequency limit, the superposition of standard linear solids performs poorly in general. As the frequency approaches zero, Q tends to very large values, meaning that the low-frequency

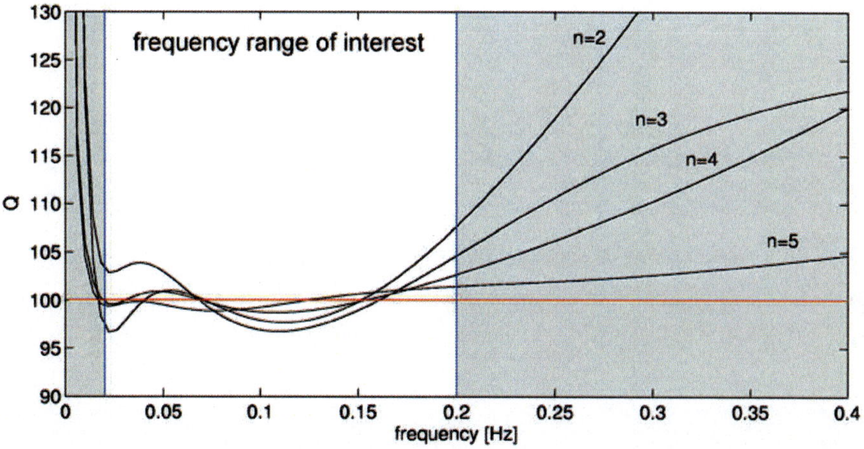

Fig. 5.1 $Q(\omega)$ for different numbers of relaxation mechanisms in the frequency band from 0.0 to 0.4 Hz. The frequency band from 0.02 to 0.2 Hz, for which $Q(\omega)$ is constructed to be approximately constant, is marked by blue lines. A red line indicates the ideal constant Q value of 100

components of the wave field are hardly affected by visco-elastic dissipation. It is a rule of thumb to use two or three relaxation mechanisms per order of magnitude in the frequency range in order to achieve a good constant-Q approximation.

An interesting and useful aspect of Eq. (5.13) is that the relaxation times τ_n always appear in the form of a product with the frequency ω. Therefore, shifting the frequency band from $[\omega_1, \omega_2]$ to $[\gamma\omega_1, \gamma\omega_2]$, with some $\gamma > 0$, merely requires to change τ_n to τ_n/γ in order to achieve a constant-Q approximation within $[\gamma\omega_1, \gamma\omega_2]$.

Chapter 6
Absorbing Boundaries

Restricting the computational domain to only a part of the true physical domain in the interest of computational efficiency introduces reflecting boundaries that do not exist in the true Earth. If not treated adequately, reflections from these artificial boundaries pollute the solution and dominate the numerical error. The most common methods used to suppress the undesired reflections fall into two categories: *absorbing boundary conditions* and *absorbing boundary layers*.

Absorbing boundary conditions, covered in Sect. 6.1, prescribe the wave field and its derivatives along the artificial boundary. *Low-order conditions*, involving first and second derivatives of the wave field, were originally derived from *paraxial approximations* of the wave equation (e.g. Engquist & Majda, 1977; Clayton & Engquist, 1977). Their efficiency depends strongly on the angle of incidence. The absorption is perfect only for waves propagating normal to the boundary. *High-order absorbing boundary conditions* (e.g. Keys, 1985; Higdon, 1991) partly overcome this deficiency.

Absorbing boundary layers are based on the introduction of a thin region along the artificial boundary where the original wave equation is modified such that the amplitudes of incident waves decay rapidly. The *Gaussian taper method* (Cerjan et al., 1985), described in Sect. 6.2, is the simplest and most robust absorbing boundary layer technique. It consists in the iterative multiplication of the incident wave field by numbers smaller than 1. The *perfectly matched layer* (PML) method, introduced by Bérenger (1994), is both more efficient and numerically more involved than the Gaussian taper. It is based on variants of the wave equation that produce waves with exponentially decreasing amplitude. We outline the PML approach in Sect. 6.3.

6.1 Absorbing Boundary Conditions

Absorbing boundary conditions prescribe the space and time derivatives along the artificial boundary. Historically, absorbing boundary conditions were derived from paraxial approximations of the wave equation (e.g. Claerbout & Johnson, 1971). Paraxial approximations, described in Sect. 6.1.1, only allow for propagation around

A. Fichtner, *Full Seismic Waveform Modelling and Inversion*, Advances in Geophysical and Environmental Mechanics and Mathematics,
DOI 10.1007/978-3-642-15807-0_6, © Springer-Verlag Berlin Heidelberg 2011

a pre-defined coordinate axis (e.g. Claerbout & Johnson, 1971). Imposing paraxial approximations along the artificial boundary leads to the absorption of incident waves (e.g. Engquist & Majda, 1977; Clayton & Engquist, 1977; Sect. 6.1.2). While the absorption is perfect at normal incidence, waves propagating at small angles to the boundary are almost perfectly reflected. This deficiency motivates the design of high-order conditions that absorb waves at nearly arbitrary angles of incidence (e.g. Keys, 1985; Higdon, 1991; Sects. 6.1.3 and 6.1.4). Critical issues related to absorbing boundary conditions are numerical stability, practical implementation and efficiency in geologically relevant Earth models (Sect. 6.1.5).

6.1.1 Paraxial Approximations of the Acoustic Wave Equation

The occurrence of reflections at boundaries and interfaces rests on the existence of waves with a propagation direction that is opposite to the propagation direction of the incident wave. This observation motivates the derivation of *paraxial approximations* to the wave equation that only allow for nearly unidirectional propagation. Paraxial approximations, frequently referred to as *one-way wave equations*, have received considerable attention in the context of seismic migration (Claerbout & Johnson, 1971; Gazdag, 1978; Stolt, 1978; Larner & Beasley, 1987) and absorbing boundaries (e.g. Clayton & Engquist, 1977, 1980; Dai et al., 1996). To illustrate the concept, we consider the acoustic wave equation in a homogeneous and source-free full space:

$$\ddot{u} - v^2(\partial_x^2 + \partial_y^2 + \partial_z^2)\, u = 0\,. \tag{6.1}$$

Solutions to Eq. (6.1) are weighted superpositions of plane waves

$$u = \mathrm{e}^{\mathbf{i}(\mathbf{k}\cdot\mathbf{x}-\omega t)}\,, \tag{6.2}$$

where the angular frequency ω, the wave number vector $\mathbf{k} = (k_x, k_y, k_z)$ and the acoustic wave speed v are related via the dispersion relation

$$\frac{\omega^2}{v^2} - k_x^2 - k_y^2 - k_z^2 = 0\,, \tag{6.3}$$

or equivalently

$$k_z = \pm\frac{\omega}{v}\sqrt{1 - \frac{v^2}{\omega^2}\,(k_x^2 + k_y^2)}\,. \tag{6.4}$$

Positive vertical wave numbers, $k_z > 0$, correspond to waves propagating in positive z-direction, and negative vertical wave numbers, $k_z < 0$, correspond to waves propagating in negative z-direction. Our goal is to modify the dispersion relation

(6.4) such that it permits only for waves propagating towards increasing z and then to derive the corresponding wave equation.

The obvious modification of the dispersion relation consists in the restriction to the positive branch of the square root

$$k_z = \frac{\omega}{v}\sqrt{1 - \frac{v^2}{\omega^2}(k_x^2 + k_y^2)} \tag{6.5}$$

that corresponds to the wave equation

$$\left[\partial_z - \mathbf{i}\frac{\omega}{v}\sqrt{1 - \frac{v^2}{\omega^2}(k_x^2 + k_y^2)}\right] u(k_x, k_y, z, \omega) = 0\,. \tag{6.6}$$

Equation (6.6) determines the wave field u in a mixed space–frequency–wave number domain. Its translation to the space–time domain via the inverse Fourier transform results in an integro-differential equation (Engquist & Majda, 1977) that is impractical from a computational point of view. To facilitate the transformation of (6.6) into a numerically more convenient partial differential equation in time and space, we expand the square root from (6.5) into a truncated Taylor series, which results in the following approximations:

$$\frac{vk_z}{\omega} = 1 + \mathcal{O}\left(\left|\frac{v^2}{\omega^2}(k_x^2 + k_y^2)\right|\right), \tag{6.7}$$

$$\frac{vk_z}{\omega} = 1 - \frac{1}{2}\left[\frac{v^2}{\omega^2}(k_x^2 + k_y^2)\right] + \mathcal{O}\left(\left|\frac{v^2}{\omega^2}(k_x^2 + k_y^2)\right|^2\right). \tag{6.8}$$

Following Clayton & Engquist (1977), the approximations (6.7) and (6.8) are commonly referred to as the A1 and A2 *paraxial approximations*. Figure 6.1 illustrates the approximations compared to the dispersion relation of the complete acoustic wave equation. Since (6.7) and (6.8) can be written as polynomials in ω, k_x, k_y and k_z, we easily find the corresponding one-way wave equations:

$$\dot{u} + v\,\partial_z u = 0\,, \tag{6.9}$$

$$\ddot{u} + v\,\partial_z \dot{u} - \frac{1}{2}v^2(\partial_x^2 + \partial_y^2)\,u = 0\,. \tag{6.10}$$

We note that there are many other approximations of the square root from (6.5) that lead to different one-way wave equations (e.g. Claerbout, 1976; Clayton & Engquist, 1977; Zhang, 1985). Equations (6.9) and (6.10) are, however, most commonly used.

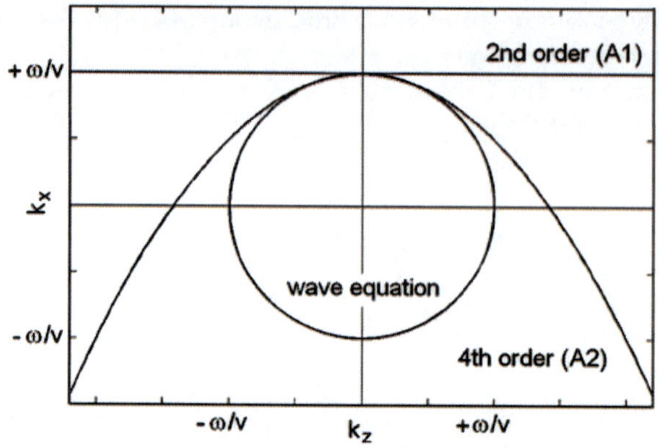

Fig. 6.1 Illustration of the paraxial approximations from (6.7) and (6.8) in the (k_x, k_z)-plane. The circle corresponds to the dispersion relation (6.4) of the acoustic wave equation. In the second-order approximation (A1), the vertical wave number k_z is equal to ω/v for all $k_x \in \mathbb{R}$, thus allowing only for propagation in the positive z-direction. The fourth-order approximation (A2) reproduces the dispersion relation of the wave equation (6.4) for small horizontal wave numbers k_x

6.1.2 *Paraxial Approximations as Boundary Conditions for Acoustic Waves*

Paraxial approximations of the wave equation restrict the propagation directions of waves to a cone surrounding a pre-defined axis. This suggests to use paraxial approximations as absorbing boundary conditions. Assuming, for instance, that the artificial boundary is located along $z = 0$, we may impose the condition

$$\left[\partial_z - \mathbf{i}\frac{\omega}{v}\sqrt{1 - \frac{v^2}{\omega^2}(k_x^2 + k_y^2)}\right] u(k_x, k_y, z, \omega)|_{z=0} = 0 \tag{6.11}$$

that is motivated by Eq. (6.6). To assess the absorption efficiency of condition (6.11), we compute the reflection coefficient R that determines the amplitude of the waves reflected from the artificial boundary. For the solution inside the physical domain we use the superposition of a plane wave travelling in positive z-direction, u^+, and a plane wave travelling in negative z-direction, u^-:

$$u = u^+ + R\,u^-\,, \tag{6.12}$$

with

$$u^+ = \mathrm{e}^{\mathbf{i}(k_x x + k_y y + k_z z - \omega t)}\,, \tag{6.13}$$

and

$$u^- = \mathrm{e}^{\mathbf{i}(k_x x + k_y y - k_z z - \omega t)}\,. \tag{6.14}$$

The wave number vector is given by

$$\mathbf{k} = \frac{\omega}{v} \begin{pmatrix} \cos\phi \, \sin\theta \\ \sin\phi \, \sin\theta \\ \cos\theta \end{pmatrix}, \tag{6.15}$$

where $\phi \in [0, 2\pi)$ is the angle between $\mathbf{k}$ and the positive x-axis, and $\theta \in [0, \pi/2]$ is the angle of incidence between $\mathbf{k}$ and the positive z-axis. Imposing (6.11) upon (6.12) yields

$$R = 0\,, \tag{6.16}$$

meaning that the boundary is perfectly absorbing. Condition (6.11) is given in the (k_x, k_y, z, ω) domain. Its inverse Fourier transform to the space–time domain results in an integro-differential equation that is highly impractical from a computational perspective (e.g. Engquist & Majda, 1977). We therefore replace the perfectly absorbing boundary condition (6.16) by paraxial approximations, that is either by the *A1 condition*

$$\dot{u} + v\, \partial_z u|_{z=0} = 0 \tag{6.17}$$

or the *A2 condition*

$$\ddot{u} + v\, \partial_z \dot{u} - \frac{1}{2} v^2 \,(\partial_x^2 + \partial_y^2)\, u|_{z=0} = 0\,, \tag{6.18}$$

as proposed by Engquist & Majda (1977) and Clayton & Engquist (1977). While being numerically convenient, the boundary conditions (6.17) and (6.18) result in less optimal absorption. The corresponding reflection coefficients are

$$R = -\frac{1-\cos\theta}{1+\cos\theta} \tag{6.19}$$

for the A1 condition (6.17) and

$$R = -\frac{(1-\cos\theta)^2}{(1+\cos\theta)^2} \tag{6.20}$$

for the A2 condition (6.18). Figure 6.2 shows the reflection coefficients of the A1 and A2 conditions as functions of the angle of incidence. Both boundary conditions are perfectly absorbing for plane waves propagating parallel to the z-axis. The reflection coefficient for the A1 condition initially grows more rapidly. However, for angles of incidence larger than $\pi/4$, both conditions result in unacceptably large reflections that may produce errors that are larger than the difference between observed and synthetic seismograms. It is this deficiency that motivates the design of higher order absorbing boundary conditions (e.g. Clayton & Engquist, 1977; Keys, 1985; Higdon, 1991), which are the subject of Sect. 6.1.3.

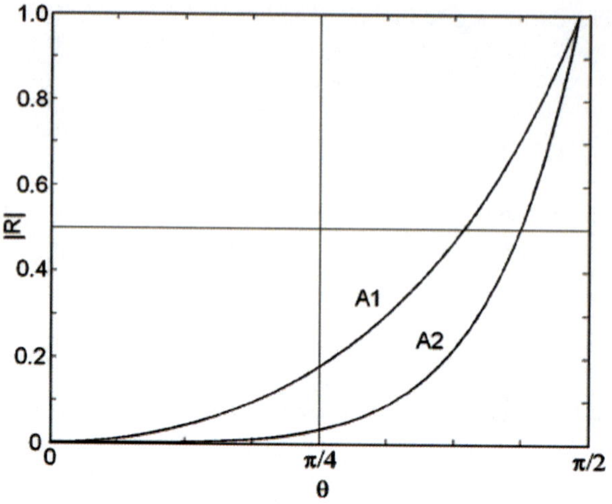

Fig. 6.2 Absolute values of the reflection coefficients of the *A1* (Eq. 6.19) and the *A2* (Eq. 6.20) absorbing boundary conditions as functions of the angle of incidence θ. In both cases the absorption is perfect for normal-incident plane waves, i.e. for $\theta = 0$. The reflection coefficient of the *A1* condition initially grows more rapidly for increasing angles of incidence. However, both conditions generate perfect reflections ($|R| \approx 1$) for angles of incidence close to $\pi/2$

6.1.3 High-Order Absorbing Boundary Conditions for Acoustic Waves

The absorbing boundary conditions considered so far become inefficient for increasing angles of incidence. To derive improved conditions we first return to the A1 condition (6.17) that can be written in the following operator form:

$$A_1 u|_{z=0} = (\cos\theta_1\, \partial_t + v\, \partial_z)\, u|_{z=0} = 0\,, \tag{6.21}$$

with $\theta_1 = 0$. Evaluating $A_1 u$ for the plane wave solutions from (6.12) gives

$$A_1 u = A_1 u^+ + R A_1\, u^- = -\mathrm{i}\omega\,(\cos\theta_1 - \cos\theta)\, u^+ - \mathrm{i}\omega\, R\,(\cos\theta_1 + \cos\theta)\, u^-\,. \tag{6.22}$$

Imposing the A1 boundary condition results in a simple equation for the reflection coefficient:

$$(\cos\theta_1 - \cos\theta) + R\,(\cos\theta_1 + \cos\theta) = 0\,. \tag{6.23}$$

Equation (6.23) implies $R = 0$ for $\theta = \theta_1$, not only when $\theta_1 = 0$ but also for any choice of θ_1 between 0 and $\pi/2$. The boundary condition (6.21) can thus be used to completely annihilate waves with an angle of incidence equal to θ_1.

We now repeat the above exercise and apply a second operator, A_2, to $A_1 u$:

$$\begin{aligned} A_2 A_1 u = &- \omega^2 (\cos\theta_1 - \cos\theta)(\cos\theta_2 - \cos\theta)\, u^+ \\ &- \omega^2 R\, (\cos\theta_1 + \cos\theta)(\cos\theta_2 + \cos\theta)\, u^- \,. \end{aligned} \tag{6.24}$$

Then imposing

$$A_2 A_1 u|_{z=0} = 0 \tag{6.25}$$

again gives an expression for the reflection coefficient R:

$$(\cos\theta_1 - \cos\theta)(\cos\theta_2 - \cos\theta) + R\, (\cos\theta_1 + \cos\theta)(\cos\theta_2 + \cos\theta) = 0\,. \tag{6.26}$$

Since it follows that $R = 0$ for $\theta = \theta_1$ or $\theta = \theta_2$, the absorbing boundary condition (6.25) completely annihilates waves with angles of incidence equal to either θ_1 or θ_2. For the special case $\theta_1 = \theta_2 = 0$, we retrieve the reflection coefficient corresponding to the A2 condition (Eq. 6.20).

The continuation of the above scheme is obvious. To prevent reflections from waves incident at angles $\theta_1, \ldots, \theta_N$, we impose the Nth-order absorbing boundary condition (e.g. Higdon, 1991)

$$\prod_{i=1}^{N} A_i u|_{z=0} = 0\,. \tag{6.27}$$

The corresponding reflection coefficient is then

$$R = \prod_{i=1}^{N} \frac{(\cos\theta_1 - \cos\theta)}{(\cos\theta_1 + \cos\theta)}\,. \tag{6.28}$$

Two examples of reflection coefficients for $N = 3$ and 4 are shown in Fig. 6.3. Higher order absorbing boundary conditions enlarge the range of incidence angles for which almost perfect absorption occurs. They cannot, however, prevent strong reflections from waves propagating nearly parallel to the boundary.

We finally note that the absorption operators A_i can easily be generalised to arbitrarily oriented boundaries:

$$A_i = (\cos\theta_i\, \partial_t + v\, \mathbf{n} \cdot \nabla)\,, \tag{6.29}$$

where $\mathbf{n}$ denotes the unit normal to the boundary ∂G of the computational domain G. Each operator A_i provokes perfect absorption for waves that are incident upon G with an angle of incidence θ_i measured from the normal.

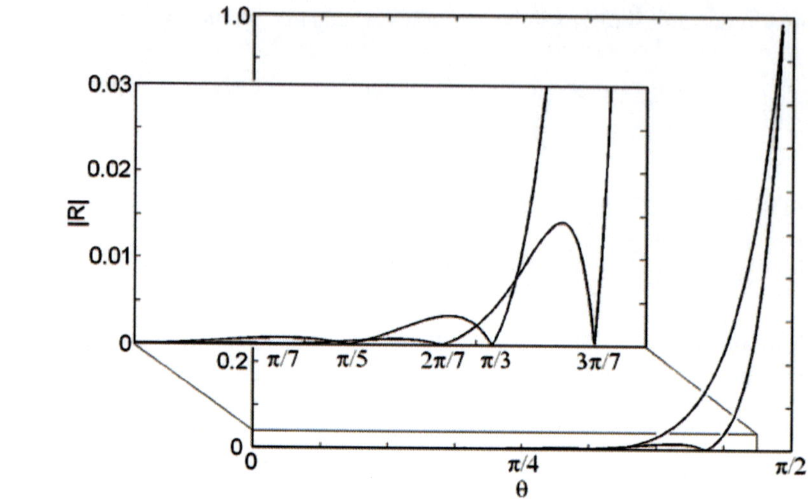

Fig. 6.3 Absolute values of reflection coefficients corresponding to high-order boundary conditions with $(\theta_1, \theta_2, \theta_3) = (0, \pi/5, \pi/3)$ and $(\theta_1, \theta_2, \theta_3, \theta_4) = (0, \pi/7, 2\pi/7, 3\pi/7)$. The reflection coefficients are close to zero for a large range of incident angles θ. However, as θ approaches $\pi/2$, the artificial boundary becomes perfectly reflecting

6.1.4 Generalisation to the Elastic Case

To translate the high-order absorbing boundary condition (6.27) to the elastic case, we need to account for the presence of at least two types of body waves: P waves and S waves propagating at their respective speeds v_P and v_S. For instance, imposing the boundary condition

$$(\partial_t - v_\text{P}\,\partial_z)(\partial_t - v_\text{S}\,\partial_z)\,\mathbf{u}|_{z=0} = \mathbf{0} \tag{6.30}$$

leads to the perfect absorption of normal-incidence P and S waves. The multiplication of (6.30) by v_S/v_P,

$$\left(\frac{v_\text{S}}{v_\text{P}}\,\partial_t - v_\text{S}\,\partial_z\right)(\partial_t - v_\text{S}\,\partial_z)\,\mathbf{u}|_{z=0} = \mathbf{0}\,, \tag{6.31}$$

reveals that the boundary condition (6.30) also annihilates S waves with an angle of incidence equal to $\theta = \operatorname{acos}(v_\text{S}/v_\text{P})$. A more general and high-order version of (6.30), reminiscent of (6.27) is

$$\prod_{i=1}^{N}(c_i\,\partial_t - v_\text{P}\,\partial_z)\mathbf{u}|_{z=0} = \mathbf{0}\,, \tag{6.32}$$

with positive parameters c_i (Higdon, 1991). Choosing $c_i \approx 1$ results in the absorption of normal-incidence P waves, whereas $c_i \approx v_P/v_S$ leads to the absorption of normal-incidence S waves. The crucial point is that each factor in the absorbing boundary condition (6.32) helps to suppress reflections from both P and S waves at different angles of incidence (Higdon, 1990).

6.1.5 Discussion

Absorbing boundary conditions are today widely used in numerical seismic wave propagation. Despite their apparent simplicity, several issues deserve to be discussed in more detail. These include numerical stability, absorption efficiency and the practical implementation.

Stability: Several authors reported numerical instabilities resulting from the application of absorbing boundary conditions. In a series of numerical experiments, Emerman & Stephen (1983) and Mahrer (1986) found that the second-order condition for elastic waves proposed by Clayton & Engquist (1977) becomes unstable for v_S/v_P smaller than ≈ 0.46. Simone & Hestholm (1998) concluded that the occurrence of instabilities depends on the spatial discretisation used to solve the wave equation, and Mahrer (1990) related instabilities to the size of the time increment in time-domain finite-difference modelling. A complete stability analysis of the different absorbing boundary conditions is clearly beyond the scope of this book. It is, however, important to realise that stability is an issue that can become important depending on the numerical scheme and the Earth model.

Absorption efficiency: It is common practice to assess the efficiency of absorbing boundary conditions in series of numerical tests that involve simple Earth models for which analytical solutions to the wave equation are available. Similar tests for heterogeneous and geologically more relevant models are rare. The numerical solutions should therefore be inspected carefully with the experience of a computational seismologist who is able to make the distinction between solution and artefact.

A disadvantageous characteristic of all absorbing boundary conditions is the almost perfect reflection of waves that propagate nearly parallel to the artificial boundary or that have a variable direction of particle motion. This is particularly problematic in shallow models where horizontally propagating surface waves in general and Rayleigh waves in particular may interact with the bottom boundary. Such pollution of the wave field can be avoided by increasing the depth of the model, which implies higher computational costs. The requirement to efficiently absorb surface waves was one the principal motivations for the development of *perfectly matched layer* (PML) techniques, that we will describe in Sect. 6.3.

Implementation of high-order conditions: High-order absorbing boundary conditions, as the ones introduced in Eqs. (6.28) and (6.32), involve time and space derivatives of the wave field that are both unavailable and undesirable in numerical schemes. Time derivatives of order 3 and higher are related to high-frequency wave fields that may not be representable with a given spatio-temporal discretisation.

Space derivatives of high order are numerically expensive and their implementation leads to highly non-local discrete difference operators. Nearly all applications therefore work with either first- or second-order absorbing boundary conditions.

An interesting alternative to the classical high-order conditions for acoustic waves was proposed by Collino (1993). It is based on $N > 0$ auxiliary variables governed by second-order differential equations defined on the boundary. The Nth-order absorbing boundary condition then involves only the first derivatives of the wave field and the auxiliary variables. This scheme overcomes the need to compute high-spatial derivatives, and it is therefore referred to as *local* high-order absorbing boundary condition. For a review of local boundary conditions, the reader is referred to Givoli (2003).

6.2 Gaussian Taper Method

The Gaussian taper method was proposed by Cerjan et al. (1985) in the context of time-domain finite-difference modelling of acoustic and elastic wave propagation. It is the simplest and most robust absorbing boundary layer technique. The general idea of the method is to multiply the wave field inside a narrow strip along the artificial boundaries by a Gaussian taper

$$G(\mathbf{x}) = \mathrm{e}^{-(x_0-x)^2/\gamma^2}\,, \tag{6.33}$$

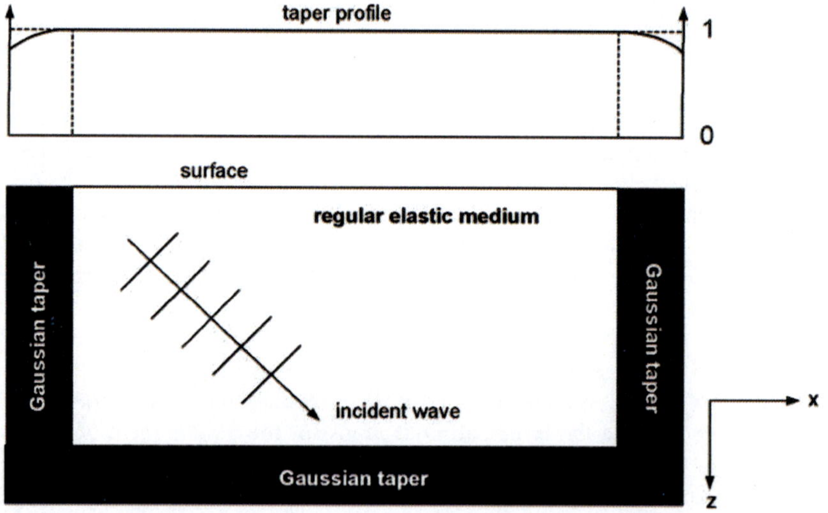

Fig. 6.4 Illustration of the Gaussian taper concept. The regular computation domain is surrounded by an absorbing layer where the wave field is multiplied in each time step by a Gaussian taper. The amplitude of the taper, visualised in the horizontal taper profile, decreases from 1 at the interface between the regular medium and the absorbing layer towards the boundary of the domain

where we assumed that the absorbing layer is orthogonal to the x-direction and that it extends from $x = x_0$ towards increasing values of x. The value of γ is determined empirically such that artificial reflections are minimised. Repeating the tapering procedure in each time step attenuates waves propagating in positive x-direction. This concept is illustrated in Fig. 6.4.

The Gaussian taper method is easily applicable to both acoustic and elastic wave propagation. It is unconditionally stable and attenuates waves at all angles of incidence. Moreover, it requires virtually no changes of pre-existing codes. These advantageous properties explain the extraordinary popularity of this simple technique in the numerical wave propagation community (e.g. Igel et al., 2002; Baig et al., 2003; Furumura & Kennett, 2005).

6.3 Perfectly Matched Layers (PML)

While the Gaussian taper method proves to be efficient for finite-difference methods, it requires unacceptably large boundary layers when high-order methods such as the spectral-element method are used. A more efficient boundary layer technique was introduced by Bérenger (1994), who proposed to modify the electrodynamic wave equation inside a *perfectly matched layer* (PML) such that the solutions decay exponentially without producing reflections from the boundary between the regular medium and the PML. Since Bérenger's original invention, the PML method has been extended to a large variety of wave propagation phenomena. The development of PML techniques for elastic wave propagation was largely motivated by the requirement to efficiently absorb surface waves that propagate nearly parallel to the bottom boundary of the computational domain. Today, reports on variants of the PML method and their respective properties form a vast body of literature. For contributions within a seismological context, the reader is referred to Collino & Tsogka (2001), Komatitsch & Tromp (2003), Festa & Vilotte (2005), Festa et al. (2005), Meza-Fajardo et al. (2008) or Kristek et al. (2009).

To illustrate the general concept, we derive the PML variant known as *anisotropic perfectly matched layers* (APMLs). It was also studied, for example, by Teixeira & Chew (1997) and Zheng & Huang (2002).

6.3.1 General Development

All variants of the PML technique are based on the definition of an absorbing boundary layer – the PML region – where the wave equation is modified such that incident waves decay rapidly. The modification of the wave equation inside the PML region is designed to *perfectly match* the original wave equation, meaning that no artificial reflections occur at the transition of the regular medium to the PML. To derive the PML equations, we consider the velocity–stress formulation of the elastic wave equation in the frequency domain:

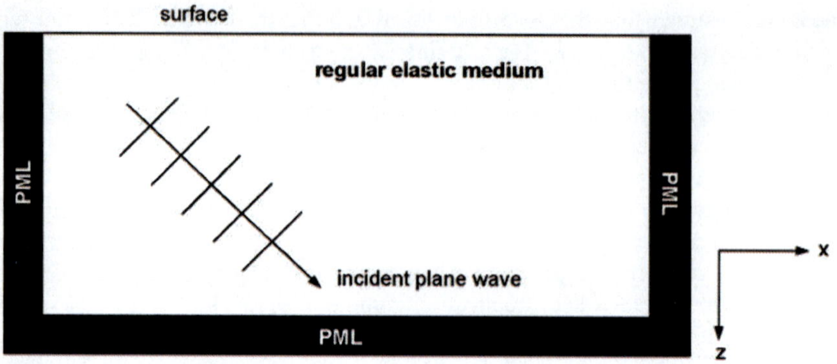

Fig. 6.5 Illustration of the PML concept. The regular computation domain is surrounded by an absorbing layer within which the elastic wave equation is modified such that incident plane waves decay rapidly as they approach the boundary

$$\mathrm{i}\omega\rho\mathbf{v}(\mathbf{x},\omega) = \nabla\cdot\boldsymbol{\sigma}(\mathbf{x},\omega)\,, \tag{6.34}$$

$$\mathrm{i}\omega\boldsymbol{\sigma}(\mathbf{x},\omega) = \mathbf{C}(\mathbf{x},\omega) : \nabla\mathbf{v}(\mathbf{x},\omega)\,, \tag{6.35}$$

$$\mathrm{i}\omega\mathbf{u}(\mathbf{x},\omega) = \mathbf{v}(\mathbf{x},\omega)\,. \tag{6.36}$$

For simplicity we disregard visco-elastic dissipation and we omit the external forces because seismic sources will generally not be placed inside the absorbing layer. To keep the treatment readable, we do not make a notational distinction between time-domain and frequency-domain variables (Fig. 6.5).

In a homogeneous and isotropic medium and when external forces are absent, the elastic wave equation has plane wave solutions of the form

$$\mathbf{u}(\mathbf{x},\omega) = \mathbf{A}\,\mathrm{e}^{-\mathrm{i}\mathbf{k}(\omega)\cdot\mathbf{x}}\,, \tag{6.37}$$

with either

$$\mathbf{A} \,||\, \mathbf{k}(\omega) \quad \text{and} \quad |\mathbf{k}(\omega)|^2 = \frac{\rho\omega^2}{\lambda+2\mu} = \frac{\omega^2}{v_\mathrm{P}^2} \tag{6.38}$$

for the P wave or

$$\mathbf{A} \perp \mathbf{k}(\omega) \quad \text{and} \quad |\mathbf{k}(\omega)|^2 = \frac{\rho\omega^2}{\mu} = \frac{\omega^2}{v_\mathrm{S}^2} \tag{6.39}$$

for the S wave. Our goal is to modify the original equations of motion (6.34), (6.35) and (6.36) such that they allow for plane wave solutions that decay exponentially with distance, say for example, in increasing z-direction. For this, the Cartesian coordinate z is transformed to a new variable $\tilde{z} := \tilde{z}(\mathbf{x})$ such that the resulting solutions are

$$\mathbf{u}(\mathbf{x}, \omega) = \mathbf{A}\, \mathrm{e}^{-\mathbf{i}\mathbf{k}(\omega)\cdot\mathbf{x}}\, \mathrm{e}^{-f(\omega,z)\, z} \tag{6.40}$$

for $z > 0$ and some function f that depends on the particular choice of $\tilde{z}(\mathbf{x})$. Choosing the transformation such that $f(\omega, z)|_{z<0} = 0$ will leave the solution in $z < 0$ unchanged. We use the coordinate transformation

$$\tilde{z} := \int_0^z \gamma_z(z')\, dz' \,. \tag{6.41}$$

At this point of the development we do not specify the integrand γ_z, commonly referred to as *stretching factor*. We merely require $\tilde{z} = z$ or equivalently $\gamma_z(z) = 1$ outside the PML, that is for $z < 0$. To obtain decaying plane waves also in the remaining coordinate directions, the transformations $x \to \tilde{x}$ and $y \to \tilde{y}$ are defined in analogy to (6.41). Replacing the derivatives ∂_x, ∂_y and ∂_z in (6.34) by the derivatives $\partial_{\tilde{x}}$, $\partial_{\tilde{y}}$ and $\partial_{\tilde{z}}$ gives the following set of equations:

$$\mathbf{i}\omega\rho v_x = \gamma_x^{-1}\partial_x\sigma_{xx} + \gamma_y^{-1}\partial_y\sigma_{xy} + \gamma_z^{-1}\partial_z\sigma_{xz}\,, \tag{6.42}$$

$$\mathbf{i}\omega\rho v_y = \gamma_x^{-1}\partial_x\sigma_{yx} + \gamma_y^{-1}\partial_y\sigma_{yy} + \gamma_z^{-1}\partial_z\sigma_{yz}\,, \tag{6.43}$$

$$\mathbf{i}\omega\rho v_z = \gamma_x^{-1}\partial_x\sigma_{zx} + \gamma_y^{-1}\partial_y\sigma_{zy} + \gamma_z^{-1}\partial_z\sigma_{zz}\,. \tag{6.44}$$

Multiplication by $\gamma_x\gamma_y\gamma_z$ yields

$$\mathbf{i}\omega\gamma_x\gamma_y\gamma_z v_x = \gamma_y\gamma_z\partial_x\sigma_{xx} + \gamma_x\gamma_z\partial_y\sigma_{xy} + \gamma_x\gamma_y\partial_z\sigma_{xz}\,, \tag{6.45}$$

$$\mathbf{i}\omega\gamma_x\gamma_y\gamma_z v_y = \gamma_y\gamma_z\partial_x\sigma_{yx} + \gamma_x\gamma_z\partial_y\sigma_{yy} + \gamma_x\gamma_y\partial_z\sigma_{yz}\,, \tag{6.46}$$

$$\mathbf{i}\omega\gamma_x\gamma_y\gamma_z v_z = \gamma_y\gamma_z\partial_x\sigma_{zx} + \gamma_x\gamma_z\partial_y\sigma_{zy} + \gamma_x\gamma_y\partial_z\sigma_{zz}\,. \tag{6.47}$$

The products $\gamma_i\gamma_j$ on the right-hand sides of Eqs. (6.45), (6.46) and (6.47) can be placed under the differentiation because γ_x depends only on x, γ_y only on y and γ_z only on z. We may then write the new version of (6.34) as

$$\mathbf{i}\omega\gamma_x\gamma_y\gamma_z\mathbf{v} = \nabla\cdot\boldsymbol{\sigma}^{\text{pml}}\,, \tag{6.48}$$

with the definition of the PML stress tensor $\boldsymbol{\sigma}^{\text{pml}}$:

$$\boldsymbol{\sigma}^{\text{pml}} := \gamma_x\gamma_y\gamma_z\,\boldsymbol{\sigma}\cdot\boldsymbol{\Lambda}\,, \quad \boldsymbol{\Lambda} := \begin{pmatrix} \gamma_x^{-1} & 0 & 0 \\ 0 & \gamma_y^{-1} & 0 \\ 0 & 0 & \gamma_z^{-1} \end{pmatrix}. \tag{6.49}$$

Since (6.34) and (6.48) are very similar, the PML can be implemented with relative ease by slightly modifying pre-existing codes. It is noteworthy at this point that $\boldsymbol{\sigma}^{\text{pml}}$ is in general not symmetric.

In the next step we consider the constitutive relation. Based on (6.35) and (6.36) we find

$$\sigma_{ij}^{\text{pml}} = \gamma_x\gamma_y\gamma_z\gamma_j^{-1}\sigma_{ij} = \gamma_x\gamma_y\gamma_z\gamma_j^{-1}\sum_{k,l=1}^{3} C_{ijkl}\frac{\partial u_k}{\partial x_l}\,. \tag{6.50}$$

Again replacing ∂_x, ∂_y and ∂_z by $\partial_{\tilde{x}}$, $\partial_{\tilde{y}}$ and $\partial_{\tilde{z}}$, yields

$$\sigma_{ij}^{\text{pml}} = \gamma_x\gamma_y\gamma_z\gamma_j^{-1}\sum_{k,l=1}^{3} C_{ijkl}\gamma_l^{-1}\frac{\partial u_k}{\partial x_l}\,. \tag{6.51}$$

For convenience we define the PML displacement gradient $(\nabla\mathbf{u})^{\text{pml}}$ as

$$(\nabla\mathbf{u})^{\text{pml}} := \gamma_x\gamma_y\gamma_z(\nabla\mathbf{u})\cdot\mathbf{\Lambda}\,. \tag{6.52}$$

We can now assemble the frequency-domain version of the modified wave equation:

$$\mathbf{i}\omega\rho\gamma_x\gamma_y\gamma_z\mathbf{v} = \nabla\cdot\boldsymbol{\sigma}^{\text{pml}}\,, \tag{6.53}$$

$$\mathbf{i}\omega\boldsymbol{\sigma}^{\text{pml}}\cdot\mathbf{\Lambda}^{-1} = \mathbf{C} : (\nabla\mathbf{v})^{\text{pml}}\,. \tag{6.54}$$

To demonstrate that exponentially decaying plane waves are indeed solutions of Eqs. (6.52), (6.53) and (6.54) we consider a homogeneous and isotropic medium. For simplicity, we restrict the analysis to the case where the waves are attenuated when travelling in positive z-direction. This requires $\gamma_x = \gamma_y = 1$ and $\gamma_z = \text{const} \neq 1$. Assuming a plane wave solution of the form

$$\mathbf{u}(\mathbf{x},\omega) = \mathbf{A}\,\mathrm{e}^{-\mathrm{i}(k_x x + k_z z)}\,, \tag{6.55}$$

the problem reduces to finding the dispersion relation $\mathbf{k} = \mathbf{k}(\omega)$. In fact, introducing the ansatz (6.55) into the modified equations of motion (6.53) and (6.54) leads to a matrix equation for the polarisation vector $\mathbf{A}$:

$$\begin{pmatrix} v_{\text{P}}^2k_x^2 + v_{\text{S}}^2k_z^2\gamma_z^{-2} - \omega^2 & 0 & (v_{\text{P}}^2 - v_{\text{S}}^2)(k_xk_z\gamma_z^{-1}) \\ 0 & v_{\text{S}}^2(k_x^2 + k_z^2\gamma_z^{-2}) - \omega^2 & 0 \\ (v_{\text{P}}^2 - v_{\text{S}}^2)(k_xk_z\gamma_z^{-1}) & 0 & v_{\text{P}}^2k_z^2\gamma_z^{-2} + v_{\text{S}}^2k_x^2 - \omega^2 \end{pmatrix}\begin{pmatrix} A_x \\ A_y \\ A_z \end{pmatrix} = \mathbf{0}\,. \tag{6.56}$$

Equation (6.56) is referred to as the *Christoffel equation*. Non-trivial solutions for the polarisation vector $\mathbf{A} = (A_x, A_y, A_z)$ exist only when the determinant of the matrix on the left-hand side of Eq. (6.56) is equal to zero. This requires either

$$k_x^2 + k_z^2\gamma_z^{-2} = \omega^2 v_{\text{P}}^{-2} \tag{6.57}$$

for quasi-P waves or

$$k_x^2 + k_z^2 \gamma_z^{-2} = \omega^2 v_s^{-2} \tag{6.58}$$

for quasi-S waves in the PML region. Equations (6.57) and (6.58) are the dispersion relations inside the PML. The corresponding plane wave solutions are

$$\mathbf{u}_\mathrm{p}(\mathbf{x}, \omega) = \mathbf{A}_p \, \mathrm{e}^{-\mathrm{i}\omega(x \sin\phi + \gamma_z z \cos\phi)/v_\mathrm{P}}, \quad \mathbf{A}_\mathrm{p} \, || \, (k_x, 0, k_z \gamma_z^{-1})^\mathrm{T}, \tag{6.59}$$

and

$$\mathbf{u}_s(\mathbf{x}, \omega) = \mathbf{A}_s \, e^{-\mathrm{i}\omega(x \sin\phi + \gamma_z z \cos\phi)/v_\mathrm{S}}, \quad \mathbf{A}_s \perp (k_x, 0, k_z \gamma_z^{-1})^T. \tag{6.60}$$

The variable ϕ denotes the angle of incidence. Equations (6.59) and (6.60) are valid for any choice of the stretching factor γ_z. In particular, any γ_z with $\mathfrak{Im}\, \gamma_z < 0$ will cause the incident plane wave to decay exponentially as it propagates in positive z-direction. In the following we will consider two classical choices for the stretching factors γ_i that lead to the *standard PML* and the *filtering* or *convolutional PML*.

6.3.2 Standard PML

For the standard PML formulation we choose the stretching factor as

$$\gamma_z(z) := 1 + \frac{a_z(z)}{\mathrm{i}\omega}, \tag{6.61}$$

where the *damping profile* a_z is positive and real. The stretching factors γ_y and γ_x are defined analogously in terms of damping profiles a_y and a_z. To avoid reflections from the interface between the regular domain and the PML, the damping profiles are chosen to smoothly increase from zero right at the interface towards the boundary of the computational domain. Inserting (6.61) into (6.59) and (6.60) yields the plane wave solutions

$$\mathbf{u}(\mathbf{x}, \omega) = \mathbf{A} \, \mathrm{e}^{-\mathrm{i}\omega(x \sin\phi + z \cos\phi)/v} \, \mathrm{e}^{-(a_z z \cos\phi)/v}, \tag{6.62}$$

with v being either v_P or v_S, and $\mathbf{A}$ being either $\mathbf{A}_p$ or $\mathbf{A}_s$. The exponential decay is frequency independent in the case of incident body waves.
With the exception of the corners of the model, the absorbing layers at the x, y and z boundaries do not overlap. We therefore find

$$a_i a_j = a_i^2 \delta_{ij} \tag{6.63}$$

and

$$\gamma_x \gamma_y \gamma_z = \frac{\mathrm{i}\omega + a_x + a_y + a_z}{\mathrm{i}\omega}. \tag{6.64}$$

The resulting equations of motion in the frequency domain are then

$$\rho\,(\mathrm{i}\omega + \mathrm{tr}\,(\mathbf{a}))\,\mathbf{v} = \nabla \cdot \boldsymbol{\sigma}^{\mathrm{pml}}\,, \tag{6.65}$$

$$\boldsymbol{\sigma}^{\mathrm{pml}} \cdot (\mathrm{i}\omega\,\mathbf{I} + \mathbf{a}) = \mathrm{i}\omega\,\mathbf{C} : (\nabla\mathbf{u})^{\mathrm{pml}}\,, \tag{6.66}$$

$$\mathrm{i}\omega\,(\nabla\mathbf{u})^{\mathrm{pml}} = \nabla\mathbf{u} \cdot (\mathrm{i}\omega\,\mathbf{I} + \mathbf{I}\,\mathrm{tr}\,(\mathbf{a}) - \mathbf{a})\,, \tag{6.67}$$

where $\mathbf{I}$ is the 3×3 unit matrix and $\mathbf{a}$ is defined as

$$\mathbf{a} := \begin{pmatrix} a_x & 0 & 0 \\ 0 & a_y & 0 \\ 0 & 0 & a_z \end{pmatrix}. \tag{6.68}$$

Transforming into the time domain finally yields

$$\rho\ddot{\mathbf{u}} + \rho\,\mathrm{tr}\,(\mathbf{a})\,\dot{\mathbf{u}} = \nabla \cdot \boldsymbol{\sigma}^{\mathrm{pml}}\,, \tag{6.69}$$

$$\dot{\boldsymbol{\sigma}}^{\mathrm{pml}} + \boldsymbol{\sigma}^{\mathrm{pml}} \cdot \mathbf{a} = \mathbf{C} : (\nabla\dot{\mathbf{u}})^{\mathrm{pml}}\,, \tag{6.70}$$

$$(\nabla\dot{\mathbf{u}})^{\mathrm{pml}} = \nabla\dot{\mathbf{u}} + \nabla\mathbf{u} \cdot (\mathbf{I}\,\mathrm{tr}\,(\mathbf{a}) - \mathbf{a})\,. \tag{6.71}$$

The choice of γ_i as inversely proportional to ω leads to relatively simple differential equations that can be marched forward in time explicitly.

Figure 6.6 visualises the performance of the anisotropic PMLs in a spectral-element simulation of a shallow earthquake from southern Greece. The width of the absorbing layer is only 10 grid points, which corresponds to two elements. Within the PML the damping profiles a_i increase quadratically from 0 at the interface with the regular medium towards the domain boundaries. The absorption inside the PMLs leads to a continuous decrease of the elastic energy, which, however, never drops to exactly 0. This is due to the finite width of the absorbing layers and the spatial discretisation that causes minor reflections at the interface between the PML and the elastic medium.

6.3.3 Convolutional PML

The construction of the standard PML equations is mostly based on the analysis of body waves propagating in a homogeneous full space. Realistic models, however, include a free surface that allows for the existence of surface waves. To limit the necessary computational resources, the models are usually as shallow as possible. Long-period surface waves may therefore interact with a PML layer that is intended to prevent reflections from the bottom boundary of the domain. In contrast to body waves, surface waves propagate nearly parallel to the PML which leads to their interaction with the bottom of the model (Fig. 6.7). The result of this interaction are long-term instabilities that pollute the solution throughout the complete volume (Festa et al., 2005). A possible solution (Festa & Vilotte, 2005; Festa

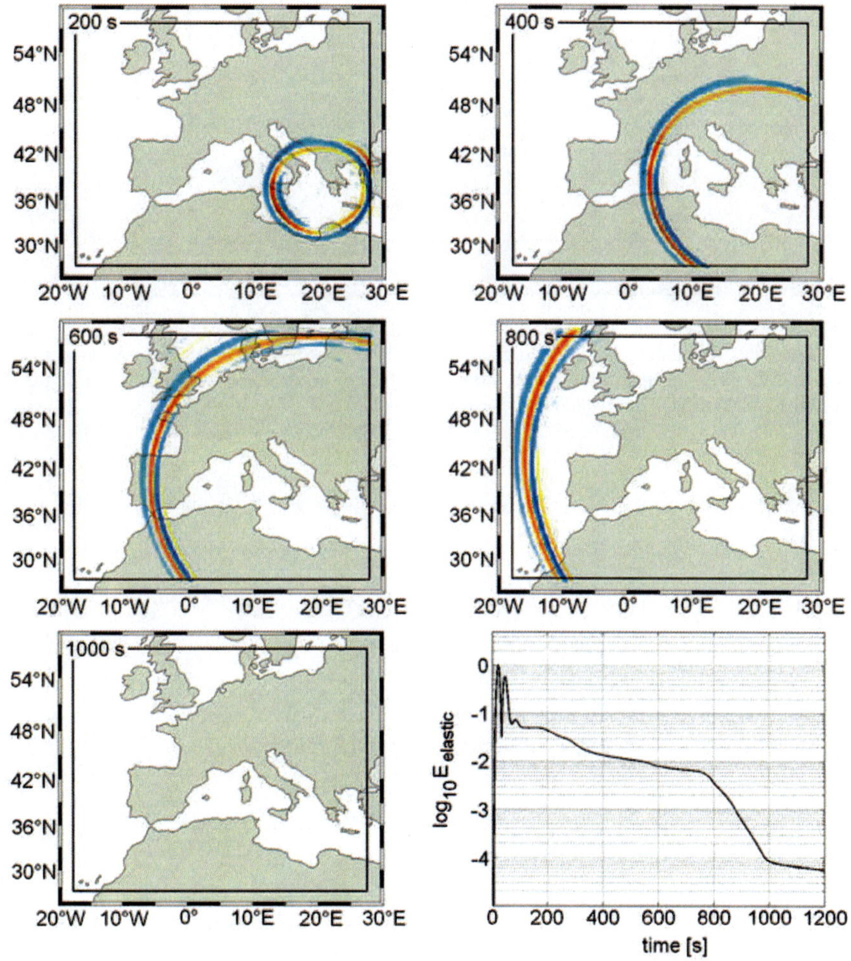

Fig. 6.6 Snapshots of the vertical-component velocity field at 100 km depth taken from a spectral-element simulation (Fichtner et al., 2009). The colour scale saturates at the maximum velocity, the only exception being the snapshot at 1,000 s which is scaled to the maximum amplitude at $t = 800$ s. The wave field is dominated by the high-amplitude Rayleigh waves that emanate from a shallow source in southern Greece. The width of the anisotropic perfectly matched layers, indicated by solid black lines, is two elements, which corresponds to 10 grid points. The normalised elastic energy contained in the model volume is shown in the lower right panel. Following the excitation of the wave field, the energy briefly stabilises between 100 and 200 s. Around 200 s the wavefront reaches the PML at the eastern boundary of the model, resulting in a drop of the elastic energy. After 800 s the wavefront reaches the western boundary, which leads to a further reduction of the elastic energy by two orders of magnitude. The remaining energy is due to the finite width of the absorbing layers and the spatial discretisation that generates minor reflections at the interface between the PML and the regular medium

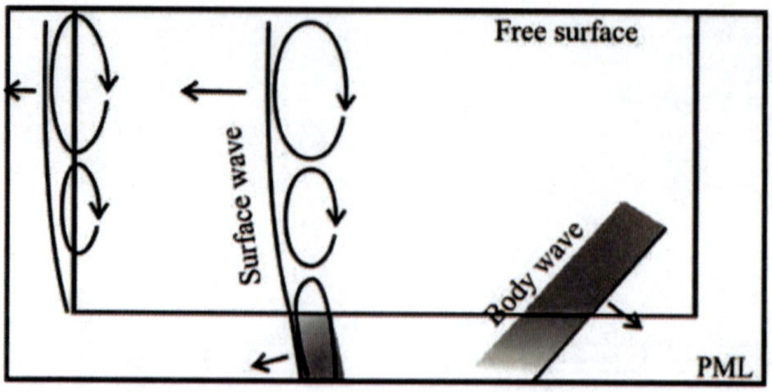

Fig. 6.7 Interaction of surface waves with the bottom PML. Surface waves propagate nearly parallel to the absorbing layer without being attenuated. The penetration of their long-period components into the PML results in long-term instabilities (Figure reproduced from Festa et al. (2005) with permission of the American Geophysical Union)

et al., 2005) is to replace the stretching factor of the standard PML, as defined in Eq. (6.61), by

$$\gamma_z(z) := 1 + \frac{a_z(z)}{\mathbf{i}\omega + \omega_c}\,. \tag{6.72}$$

The pole at $\omega = 0$ from Eq. (6.61) is thus moved to the imaginary axis. Plane wave solutions corresponding to the stretching factor (6.72) are

$$\mathbf{u}(\mathbf{x},\omega) = \mathbf{A}\,\mathrm{e}^{-\mathbf{i}\omega(x\,\sin\phi + z\,\cos\phi)/v}\,\mathrm{e}^{-\frac{\mathbf{i}\omega\omega_c a_z z\cos\phi}{v\,(\omega^2+\omega_c^2)}}\,\mathrm{e}^{-\frac{\omega^2 a_z z\cos\phi}{v\,(\omega^2+\omega_c^2)}}\,. \tag{6.73}$$

Again, v denotes either v_P or v_S, and $\mathbf{A}$ either $\mathbf{A}_\mathrm{p}$ or $\mathbf{A}_\mathrm{S}$. The coordinate transformation adds a frequency-dependent term proportional to

$$-\frac{\omega}{\mathbf{i}\omega + \omega_c} = -\frac{\mathbf{i}\omega\omega_c + \omega^2}{\omega^2 + \omega_c^2} \tag{6.74}$$

to the plane wave solution inside the physical domain. The real part of (6.74), plotted in the left panel of Fig. 6.8, controls the amplitude decay as the wave moves into the PML region, and the imaginary part, shown to the right, introduces a phase shift. For low frequencies the PML behaves as the regular elastic medium. This means, in particular, that the unstable interaction of long-period surface waves with the bottom PML can be avoided (Festa et al., 2005). For increasing frequency, i.e. as $\omega \to \infty$, we retrieve the standard frequency-independent PML.

Introducing the stretching factor from Eq. (6.72) into Eqs. (6.52), (6.53) and (6.54) yields the frequency-domain equations of motion inside the PML:

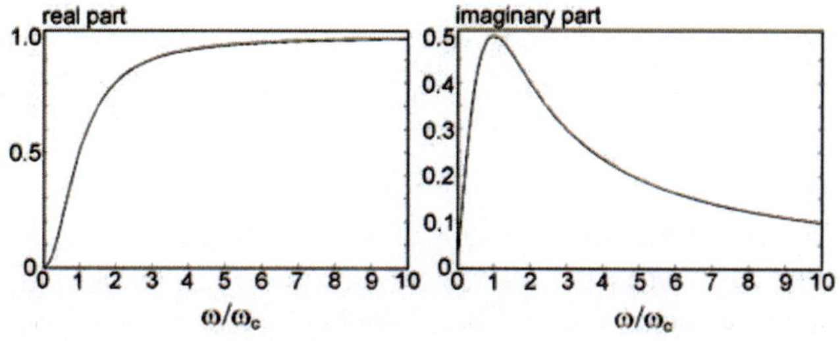

Fig. 6.8 The real and imaginary parts of the decay function from Eq. (6.74). The amplitude decay, controlled by the real part, is frequency dependent. Low-frequency waves are attenuated less than high-frequency waves. For $\omega \to \infty$ we retrieve the standard PML medium

$$\rho\left(\mathbf{i}\omega + \frac{\mathbf{i}\omega\,\mathrm{tr}\,(\mathbf{a})}{\mathbf{i}\omega + \omega_c}\right)\mathbf{v} = \nabla\cdot\boldsymbol{\sigma}^{\mathrm{pml}}\,, \tag{6.75}$$

$$\boldsymbol{\sigma}^{\mathrm{pml}}\cdot[(\mathbf{i}\omega+\omega_c)\,\mathbf{I}+\mathbf{a}] = (\mathbf{i}\omega+\omega_c)\,\mathbf{C}:(\nabla\mathbf{u})^{\mathrm{pml}}\,, \tag{6.76}$$

$$(\mathbf{i}\omega+\omega_c)\,(\nabla\mathbf{u})^{\mathrm{pml}} = \nabla\mathbf{u}\cdot[(\mathbf{i}\omega+\omega_c)\,\mathbf{I}+\mathbf{I}\,\mathrm{tr}\,(\mathbf{a})-\mathbf{a}]\,. \tag{6.77}$$

For $\omega_c = 0$ we retrieve, as expected, the frequency-domain equations of motion in the standard PML, given in Eqs. (6.65), (6.66) and (6.67). The return to the time is straightforward for the PML constitutive relation (6.76) and the definition of the PML displacement gradient (6.77):

$$\dot{\boldsymbol{\sigma}}^{\mathrm{pml}} + (\omega_c\mathbf{I}+\mathbf{a})\,\boldsymbol{\sigma}^{\mathrm{pml}} = \mathbf{C}:(\nabla\dot{\mathbf{u}})^{\mathrm{pml}}\,, \tag{6.78}$$

$$(\nabla\dot{\mathbf{u}})^{\mathrm{pml}} + \omega_c(\nabla\mathbf{u})^{\mathrm{pml}} = \nabla\dot{\mathbf{u}} + \nabla\mathbf{u}\cdot(\omega_c\,\mathbf{I}+\mathbf{I}\,\mathrm{tr}\,(\mathbf{a})-\mathbf{a})\,. \tag{6.79}$$

For the PML momentum balance (6.75) we can follow two different but theoretically equivalent strategies. First we may simply multiply the equation by $(\mathbf{i}\omega+\omega_c)$. This leads to a time-domain momentum balance that involves the numerically inconvenient third time derivative of the displacement field $\mathbf{u}$:

$$\rho\dddot{\mathbf{u}} + \rho\omega_c\ddot{\mathbf{u}} + \mathrm{tr}\,(\mathbf{a})\,\ddot{\mathbf{u}} = \nabla\cdot\boldsymbol{\sigma}^{\mathrm{pml}} + \omega_c\nabla\cdot\boldsymbol{\sigma}^{\mathrm{pml}}\,. \tag{6.80}$$

Alternatively, we can exploit that $(\mathbf{i}\omega+\omega_c)^{-1}$ is the Fourier transform of $H(t)\mathrm{e}^{-\omega_c t}$, where H denotes the Heaviside function. This allows us to write the PML momentum balance in the form of an integro-differential equation that involves the convolution of the acceleration $\ddot{\mathbf{u}}$ and $H(t)\mathrm{e}^{-\omega_c t}$:

$$\rho\ddot{\mathbf{u}} + \mathrm{tr}\,(\mathbf{a})\,\mathbf{M}(t) = \nabla\cdot\boldsymbol{\sigma}^{\mathrm{pml}}\,, \tag{6.81}$$

where the *memory variable* $\mathbf{M}$ is defined by

$$\mathbf{M}(t) := \int_{-\infty}^{\infty} H(t-\tau)\mathrm{e}^{-\omega_c(t-\tau)}\,\ddot{\mathbf{u}}(\tau)\,\mathrm{d}\tau\,. \tag{6.82}$$

Equation (6.82) is similar to the definition of the memory variables that we used in the context of visco-elastic dissipation (Eq. 5.7). Thus, following the recipe from Chap. 5 we can write **M** as the solution of an ordinary differential equation that is numerically more convenient than the convolution in (6.82). For this we simply differentiate the definition of **M** with respect to time:

$$\dot{\mathbf{M}}(t) = \ddot{\mathbf{u}}(t) - \omega_c \mathbf{M}(t)\,. \tag{6.83}$$

The implementation of the convolutional PML therefore requires us to solve the equations of motion (6.78), (6.79) and (6.81), as well as the evolution equation for the memory variable **M** (6.83).

6.3.4 Other Variants of the PML Method

6.3.4.1 Split-Field PML

Our formulation of the PML technique – both in its standard and convolutional forms – is based on the definition of the PML stress tensor (Eq. 6.49) and the PML displacement gradient (Eq. 6.52). One can avoid these definitions by introducing *split fields*. The split-field variant of the PML method is the one classically used in electro- and elasto-dynamics (e.g. Bérenger, 1994; Collino & Tsogka, 2001; Festa & Nielsen, 2003; Festa & Vilotte, 2005). To illustrate the split-field concept, we start with the momentum balance equation and the constitutive relation as they can be written after replacing $\partial/\partial x_i$ by $\gamma_i^{-1}\,\partial/\partial x_i$:

$$\mathbf{i}\omega\rho v_i = \sum_{k=1}^{3} \gamma_k^{-1} \frac{\partial \sigma_{ik}}{\partial x_k}\,, \tag{6.84}$$

$$\mathbf{i}\omega\sigma_{ij} = \sum_{k,l=1}^{3} C_{ijkl}\gamma_l^{-1} \frac{\partial u_k}{\partial x_l}\,. \tag{6.85}$$

Instead of defining a PML stress and a PML displacement gradient, we split the velocity field and the stress tensor according to the derivatives on the right-hand sides of Eqs. (6.84) and (6.85), respectively:

$$v_i^{(k)} := \gamma_k^{-1} \frac{\partial \sigma_{ik}}{\partial x_k}\,, \tag{6.86}$$

$$\mathbf{i}\omega\sigma_{ij}^{(l)} := \sum_{k=1}^{3} C_{ijkl}\gamma_l^{-1}\frac{\partial u_k}{\partial x_l}\,. \tag{6.87}$$

The newly introduced variables $v_i^{(k)}$ and $\sigma_{ij}^{(k)}$ are the *split fields*. It follows directly from Eqs. (6.86) and (6.87) that the complete dynamic fields are the sum of the split fields:

$$v_i = \sum_{k=1}^{3} v_i^{(k)}\,, \tag{6.88}$$

$$\sigma_{ij} = \sum_{l=1}^{3} \sigma_{ij}^{(l)}\,. \tag{6.89}$$

The definitions of the split fields do not involve the sums over the inverse stretching factors γ_i^{-1} that appear in the PML equations of motion (6.84) and (6.85). We can therefore move the γ_i to the left-hand sides of (6.86) and (6.87). Then introducing the stretching factors of the standard PML, as defined in (6.61), leads to the following equations for the split fields:

$$\rho\,(\mathbf{i}\omega + a_k)\,v_i^{(k)} = \frac{\partial\sigma_{ik}}{\partial x_k}\,, \tag{6.90}$$

$$(\mathbf{i}\omega + a_l)\,\sigma_{ij}^{(l)} = \sum_{k=1}^{3} C_{ijkl}\frac{\partial u_k}{\partial x_l}\,. \tag{6.91}$$

These equations can easily be translated to the time domain. To obtain the convolutional variant of the split-field PML, the standard stretching factor (6.61) needs to be replaced by (6.72).

6.3.4.2 PML for Second-Order Systems

The PML technique, as we considered it so far, is based on the velocity–stress formulation of the elastic wave equation, that is, on a system of two partial differential equations that are of first order in time. This prevents the implementation of the PML method in numerical solvers that are based on the displacement formulation which is of second order in time. To overcome this limitation, Komatitsch & Tromp (2003) proposed a split-field PML variant that is applicable to second-order equations. Their formulation merely involves displacement split fields because stress fields are naturally absent. The efficiency of the PML formulation for second-order systems is comparable to the classical PML that was developed for partial differential equations that are of first order in time.

6.3.4.3 Multi-Axial PML

The drawback of the PML method is its long-term instability – noted as early as 1996 by Hu, shortly after Bérenger's original publication (1994). This means that the solutions start to grow indefinitely when the simulations become very long. While this undesirable property of the PMLs has often been ignored in the geophysical literature, it has been intensively studied in the applied mathematics community (e.g. Nehrbass et al., 1996; Abarbanel & Gottlieb, 1997). In the context of elastic wave propagation Bécache et al. (2003) noted that instabilities may be the result of anisotropy, and they derived a set of stability conditions that may, however, not always be met. Meza-Fajardo et al. (2008) introduced a multi-axial PML variant where damping profiles are defined in more than one direction. Their procedure allows the determination of damping profiles that produce stable PML solutions for a given type of anisotropy.

Part II
Iterative Solution of the Full Waveform Inversion Problem

Having solved the forward problem, i.e. the wave equation, it remains to exploit the wealth of waveform information contained in seismograms for the benefit of seismic tomography.

The first step towards this goal is the physically meaningful quantification of the differences between observed waveforms and waveforms that have been computed for an Earth model, $\mathbf{m}$. Mathematically, this waveform difference is expressed in terms of a misfit functional, χ, that generally depends non-linearly on $\mathbf{m}$. Solving the full waveform tomographic problem means to find an optimal Earth model, $\mathbf{m}_{\text{opt}}$, such that $\chi(\mathbf{m}_{\text{opt}})$ is the global minimum of χ. In this sense, full waveform inversion is a special non-linear minimisation problem (Chap. 7). Owing to the large number of model parameters (thousands to millions), the minimisation proceeds iteratively with the help of gradient-based algorithms such as the conjugate gradient or Newton methods.

All gradient-based minimisation algorithms critically rely on the derivative of the misfit functional with respect to the model parameters. The adjoint method (Chaps. 8, 9 and 10) allows us to compute the derivative of χ with optimal efficiency: Starting from the wave equation we can derive the adjoint wave equation, the solution of which is the adjoint wave field. The properties of the adjoint wave field are determined by the adjoint source which is completely specified by the misfit, χ. Correlating the regular and adjoint wave fields yields the derivative of χ with respect to $\mathbf{m}$. A generalisation of the adjoint method allows us to compute the Hessian of χ, which is the carrier of covariance information and which plays a crucial role in the Newton method of non-linear minimisation.

A suitable misfit functional is one of the most important ingredients of a full waveform inversion applied to real data (Chap. 11). The misfit functional should exploit as much information as possible while conforming to the restrictions imposed by the data and the physics of the problem. The L_2 distance between observed and synthetic seismograms is efficient in the localisation of sharp reflectors but often fails to detect long-wavelength Earth structure. Measurements of cross-correlation time shifts and time–frequency misfits have been designed to overcome this deficiency.

Physical intuition is essential for the successful solution of any inverse problem. In the special case of full waveform inversion, our intuition is mostly based on the interpretation of Fréchet or sensitivity kernels, which are the volumetric densities of Fréchet derivatives. For isolated seismic phases such as P, S, sP, etc., Fréchet kernels are generally concentrated around the geometric ray path connecting source and receiver. The width of the kernels is proportional to the frequency bandwidth of the observed waveforms (Chap. 12).

Chapter 7
Introduction to Iterative Non-linear Minimisation

The objective of full waveform inversion in the deterministic sense is to find an optimal Earth model, $\tilde{\mathbf{m}}$, that minimises the misfit functional, $\chi(\mathbf{m})$, used to quantify the discrepancies between the observed seismograms, $\mathbf{u}_0(\mathbf{x}, t)$, and the synthetic seismograms, $\mathbf{u}(\mathbf{m}; \mathbf{x}, t)$ (Sect. 1.2). A model $\mathbf{m}$ may comprise, among other quantities, the spatial distributions of the P wave speed, $v_{\text{P}}(\mathbf{x})$, the S wave speed, $v_{\text{S}}(\mathbf{x})$, and density, $\rho(\mathbf{x})$, that is

$$\mathbf{m}(\mathbf{x}) = [m_1(\mathbf{x}), m_2(\mathbf{x}), m_3(\mathbf{x}), \ldots] = [v_{\text{P}}(\mathbf{x}), v_{\text{S}}(\mathbf{x}), \rho(\mathbf{x}), \ldots] . \tag{7.1}$$

Since χ is generally a complicated non-linear functional of $\mathbf{m}$, we approximate the optimum $\tilde{\mathbf{m}}$ with the help of iterative minimisation algorithms: Starting from a plausible initial model, $\mathbf{m}_0$, iterative minimisation successively updates the current model, $\mathbf{m}_i$, to a new model, $\mathbf{m}_{i+1}$, that explains the data better than its predecessor:

$$\mathbf{m}_{i+1} = \mathbf{m}_i + \gamma_i \, \mathbf{h}_i \quad \text{with} \quad \chi(\mathbf{m}_{i+1}) < \chi(\mathbf{m}_i) . \tag{7.2}$$

The particular choice of the *descent direction*, $\mathbf{h}_i$, and the *step length*, $\gamma_i > 0$, depends on the minimisation scheme used to approximate $\tilde{\mathbf{m}}$. The iterative procedure is repeated until the synthetic data explain the observed data sufficiently well.

Following an introduction to the basic concepts in non-linear minimisation, we discuss the uniqueness of the optimum $\tilde{\mathbf{m}}$, and we review conditions that $\tilde{\mathbf{m}}$ needs to satisfy in order to be considered a minimum of χ (Sect. 7.2). Specific methods for iterative non-linear minimisation are the topic of Sect. 7.3. We conclude with the description of multi-scale inversions and regularisation that ensure the convergence towards physically meaningful models (Sect. 7.4).

The present chapter offers a brief introduction to non-linear minimisation in the specific context of full seismic waveform inversion. For a more comprehensive treatise of both theory and methods the reader is referred to Gill et al. (1981), Fletcher (1987), Polak (1997), Kelley (1999) or Quarteroni et al. (2000).

A. Fichtner, *Full Seismic Waveform Modelling and Inversion*, Advances in Geophysical and Environmental Mechanics and Mathematics,
DOI 10.1007/978-3-642-15807-0_7,

7.1 Basic Concepts: Minima, Convexity and Non-uniqueness

In the interest of readability we define the *model space*, $\mathfrak{M}$, as the space of all admissible Earth models, $\mathbf{m}$. Furthermore, we equip $\mathfrak{M}$ with an L_1 norm

$$||\mathbf{m}||_1 := \sum_k \int_G |m_k(\mathbf{x})| \, d^3\mathbf{x} \tag{7.3}$$

and an L_2 norm

$$||\mathbf{m}||_2 := \sqrt{\sum_k \int_G m_k^2(\mathbf{x}) \, d^3\mathbf{x}} \,. \tag{7.4}$$

Throughout this chapter we assume that the Earth models, $\mathbf{m}$, have been discretised appropriately. This means that any constituent of $\mathbf{m}$, for instance, the P wave speed distribution $v_\mathrm{P}(\mathbf{x})$, is represented as a linear combination of $N < \infty$ orthonormal basis functions, $b_j(\mathbf{x})$:

$$v_\mathrm{P}(\mathbf{x}) = \sum_{j=1}^{N} v_{\mathrm{P},j} \, b_j(\mathbf{x}) \,. \tag{7.5}$$

Popular choices of basis functions are spherical harmonics, blocks or splines. For notational convenience we identify $\mathbf{m}$ as defined in Eq. (7.1) with the vector of expansion coefficients $(v_{\mathrm{P},1}, \ldots, v_{\mathrm{P},N}, v_{\mathrm{S},1}, \ldots)$, that is

$$\mathbf{m} = [v_\mathrm{P}(\mathbf{x}), v_\mathrm{S}(\mathbf{x}), \rho(\mathbf{x}), \ldots] \equiv (v_{\mathrm{P},1}, \ldots, v_{\mathrm{P},N}, v_{\mathrm{S},1}, \ldots) \,. \tag{7.6}$$

This discretisation allows to use matrix notation, as introduced in Sect. 2.1.

7.1.1 Local and Global Minima

Full waveform inversion is characterised by a computationally expensive forward problem solution and a high-dimensional model space that render the application of probabilistic inversion schemes impractical. We are therefore limited to the deterministic solution of the inverse problem, meaning that we wish to find the Earth model $\tilde{\mathbf{m}} \in \mathfrak{M}$ such that the misfit functional $\chi(\mathbf{m})$ is minimal.

We distinguish different types of minima. For this, we consider a neighbourhood $N_r(\tilde{\mathbf{m}})$ of radius r around the optimal model $\tilde{\mathbf{m}}$:

$$N_r(\tilde{\mathbf{m}}) := \{\mathbf{m} \in \mathfrak{M} \mid ||\mathbf{m} - \tilde{\mathbf{m}}||_2 < r\}. \tag{7.7}$$

The model $\tilde{\mathbf{m}} \in \mathfrak{M}$ is a *local minimum* of χ when there exists at least a small neighbourhood $N_r(\tilde{\mathbf{m}})$, where $\chi(\tilde{\mathbf{m}})$ is smaller than or equal to any $\chi(\mathbf{m})$, that is

$$\chi(\tilde{\mathbf{m}}) \leq \chi(\mathbf{m}), \quad \text{for all } \mathbf{m} \in N_r(\tilde{\mathbf{m}}). \tag{7.8}$$

We call $\tilde{\mathbf{m}}$ a *strict local minimum* of χ when $\chi(\tilde{\mathbf{m}})$ is strictly smaller than any $\chi(\mathbf{m})$ at least within a potentially small neighbourhood $N_r(\tilde{\mathbf{m}})$:

$$\chi(\tilde{\mathbf{m}}) < \chi(\mathbf{m}), \quad \text{for all } \mathbf{m} \in N_r(\tilde{\mathbf{m}}). \tag{7.9}$$

A misfit functional may have many local minima, and some of them may produce larger misfits than others. For a *global minimum* $\tilde{\mathbf{m}}$, however, $\chi(\tilde{\mathbf{m}})$ is smaller or equal to $\chi(\mathbf{m})$ for any $\mathbf{m} \in \mathfrak{M}$, not only within a potentially small neighbourhood:

$$\chi(\tilde{\mathbf{m}}) \leq \chi(\mathbf{m}), \quad \text{for all } \mathbf{m} \in \mathfrak{M}. \tag{7.10}$$

The model $\tilde{\mathbf{m}}$ is a *strict global minimum* of χ when

$$\chi(\tilde{\mathbf{m}}) < \chi(\mathbf{m}), \quad \text{for all } \mathbf{m} \in \mathfrak{M}. \tag{7.11}$$

Figure 7.1 illustrates the different types of minima. The distinction between a minimum and a strict minimum is subtle but important because it directly relates to the uniqueness or non-uniqueness of the optimal model. In the case of a strict global minimum, the model $\tilde{\mathbf{m}}$ describes the data better than any other model in $\mathfrak{M}$, and the solution to the inverse problem is said to be *unique*. When the minimum is not strict, then we can find many models that describe the data equally well, meaning that the solution to the inverse problem is *non-unique*.

Using this terminology, we can say that we are interested in finding the strict global minimum of χ, if it exists.

Despite being extremely important, the existence of a strict global minimum has received very little attention in the geophysical literature; and mathematics provides few practical tools that allow us to address this issue in realistic applications. Part of the problem is that χ is not given explicitly, meaning that each evaluation of $\chi(\mathbf{m})$

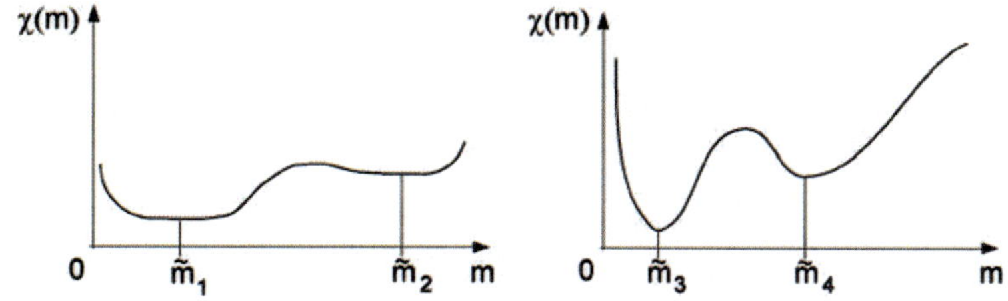

Fig. 7.1 *Left*: Global minimum $\tilde{\mathbf{m}}_1$ and local minimum $\tilde{\mathbf{m}}_2$. *Right*: Strict global minimum $\tilde{\mathbf{m}}_3$ and strict local minimum $\tilde{\mathbf{m}}_4$

for a specific model $\mathbf{m}$ requires the solution of a computationally intense forward problem.

The existence of a global minimum – strict or not – is often conjectural, and the conjecture is based on the assumption that the physics of the problem are modelled sufficiently well.

7.1.2 Convexity: Global Minima and (Non)Uniqueness

The uniqueness and the local or global nature of a minimum depend on the *convexity properties* of the model space $\mathfrak{M}$ and of the misfit functional χ. We call the model space *convex* when the line connecting any two models $\mathbf{m}_1$ and $\mathbf{m}_2$ in $\mathfrak{M}$ is also in $\mathfrak{M}$:

$$(1-\varepsilon)\,\mathbf{m}_1 + \varepsilon\,\mathbf{m}_2 \in \mathfrak{M}\,, \quad \text{for all } \varepsilon \in [0,1] \text{ and for all } \mathbf{m}_1, \mathbf{m}_2 \in \mathfrak{M}\,. \tag{7.12}$$

Provided that $\mathfrak{M}$ is convex, we define *convex misfit functionals* χ as those satisfying the relation

$$\chi[(1-\varepsilon)\,\mathbf{m}_1 + \varepsilon\,\mathbf{m}_2] \le (1-\varepsilon)\,\chi(\mathbf{m}_1) + \varepsilon\,\chi(\mathbf{m}_2)\,, \tag{7.13}$$

for any $\mathbf{m}_1, \mathbf{m}_2 \in \mathfrak{M}$ and $\varepsilon \in [0,1]$. A function defined on the real line is convex when the line connecting any two points on its graph is located above or on the graph. We call χ *strictly convex* when

$$\chi[(1-\varepsilon)\,\mathbf{m}_1 + \varepsilon\,\mathbf{m}_2] < (1-\varepsilon)\,\chi(\mathbf{m}_1) + \varepsilon\,\chi(\mathbf{m}_2)\,, \tag{7.14}$$

for any $\mathbf{m}_1, \mathbf{m}_2 \in \mathfrak{M}$ and $\varepsilon \in (0,1)$. Figure 7.2 shows examples of a convex and a strictly convex function. Now assume that $\tilde{\mathbf{m}}$ is a local minimum of χ and that both χ and $\mathfrak{M}$ are convex. This means that there is a neighbourhood $N_r(\tilde{\mathbf{m}})$ of radius r around $\tilde{\mathbf{m}}$ such that

$$\chi(\tilde{\mathbf{m}}) \le \chi(\mathbf{m})\,, \qquad \text{for all } \mathbf{m} \in N_r(\tilde{\mathbf{m}})\,. \tag{7.15}$$

The convexity of $\mathfrak{M}$ implies that the line $\mathbf{m}(\varepsilon)$ connecting $\tilde{\mathbf{m}}$ to any other element $\mathbf{m}' \in \mathfrak{M}$ is still in $\mathfrak{M}$:

$$\mathbf{m}(\varepsilon) = (1-\varepsilon)\,\tilde{\mathbf{m}} + \varepsilon\,\mathbf{m}' = \tilde{\mathbf{m}} + \varepsilon\,(\mathbf{m}' - \tilde{\mathbf{m}}) \in \mathfrak{M}\,, \tag{7.16}$$

for $\varepsilon \in [0,1]$. By choosing $\varepsilon_r \le r\,||\mathbf{m}' - \tilde{\mathbf{m}}||_2^{-1}$ we ensure that $\mathbf{m}(\varepsilon_r)$ is also in $N_r(\tilde{\mathbf{m}})$, and therefore

$$\chi(\tilde{\mathbf{m}}) \le \chi[\mathbf{m}(\varepsilon_r)] = \chi[(1-\varepsilon_r)\,\tilde{\mathbf{m}} + \varepsilon_r\,\mathbf{m}'] \le (1-\varepsilon_r)\,\chi(\tilde{\mathbf{m}}) + \varepsilon_r\,\chi(\mathbf{m}')\,. \tag{7.17}$$

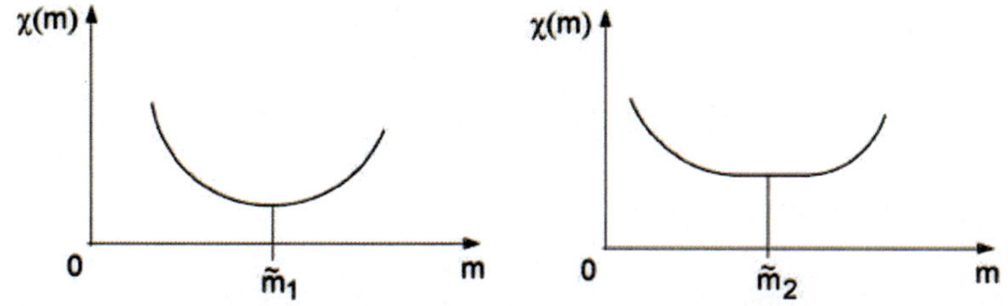

Fig. 7.2 The minimum $\tilde{\mathbf{m}}_1$ of the strictly convex function on the left is global and unique. The minimum $\tilde{\mathbf{m}}_2$ of the convex function on the right is also global but non-unique

Rearranging terms in Eq. (7.18) yields

$$\chi(\tilde{\mathbf{m}}) \leq \chi(\mathbf{m}') . \tag{7.18}$$

Since $\mathbf{m}'$ was chosen arbitrarily from the model space $\mathfrak{M}$, relation (7.18) implies that convexity is sufficient to guarantee that any local minimum is in fact a global minimum. Nevertheless, a global minimum need not be unique. Non-uniqueness occurs, for instance, when χ is constant within a non-empty neighbourhood surrounding $\tilde{\mathbf{m}}$.

The strict convexity of χ, however, ensures that the minimum $\tilde{\mathbf{m}}$ is both global and unique. To see this we assume that $\tilde{\mathbf{m}}'$ is another minimum of χ and that $\tilde{\mathbf{m}}' \neq \tilde{\mathbf{m}}$. As we found before, the convexity of χ implies that $\tilde{\mathbf{m}}$ and $\tilde{\mathbf{m}}'$ are global minima, meaning that $\chi(\tilde{\mathbf{m}}) = \chi(\tilde{\mathbf{m}}')$. Since the model space $\mathfrak{M}$ is also convex, the midpoint between $\tilde{\mathbf{m}}$ and $\tilde{\mathbf{m}}'$ is in $\mathfrak{M}$: $(\tilde{\mathbf{m}}+\tilde{\mathbf{m}}')/2 \in \mathfrak{M}$. Making then use of the assumed strict convexity of χ yields:

$$\chi(\tilde{\mathbf{m}}) \leq \chi\left(\frac{1}{2}\tilde{\mathbf{m}} + \frac{1}{2}\tilde{\mathbf{m}}'\right) < \frac{1}{2}\chi(\tilde{\mathbf{m}}) + \frac{1}{2}\chi(\tilde{\mathbf{m}}') . \tag{7.19}$$

It follows that $\chi(\tilde{\mathbf{m}}) < \chi(\tilde{\mathbf{m}}')$, which contradicts the assumption that $\tilde{\mathbf{m}}'$ is a minimum of χ.

7.1.2.1 Characterisation of Convex Functions

Ideally, we would like the misfit functional, χ, to be strictly convex, because this property ensures that one single Earth model, $\tilde{\mathbf{m}}$, explains the data better than any other model. So far, however, we have no practical means to decide whether χ is strictly convex or not. Working directly with the definition of strict convexity (Eq. 7.14) is clearly impossible because we cannot evaluate the misfit along the line connecting any two elements in the model space. Therefore, we focus our attention on the more handy characterisation of (strictly) convex functions in terms of their first and second derivatives. For this we assume that χ is convex, meaning that

$$\chi[\mathbf{m}_1 + \varepsilon(\mathbf{m}_2 - \mathbf{m}_1)] \leq \chi(\mathbf{m}_1) + \varepsilon[\chi(\mathbf{m}_2) - \chi(\mathbf{m}_1)] , \tag{7.20}$$

for any two models $\mathbf{m}_1, \mathbf{m}_2 \in \mathfrak{M}$. Rearranging (7.20) gives

$$\frac{1}{\varepsilon}\{\chi[\mathbf{m}_1 + \varepsilon(\mathbf{m}_2 - \mathbf{m}_1)] - \chi(\mathbf{m}_1)\} \le \chi(\mathbf{m}_2) - \chi(\mathbf{m}_1), \tag{7.21}$$

and in the limit $\varepsilon \to 0$ we find

$$\chi(\mathbf{m}_2) \ge \chi(\mathbf{m}_1) + (\mathbf{m}_2 - \mathbf{m}_1) \cdot \nabla_m \chi(\mathbf{m}_1). \tag{7.22}$$

The inequality (7.22) implies that the linear approximation

$$\chi(\mathbf{m}_2) \approx \chi(\mathbf{m}_1) + (\mathbf{m}_2 - \mathbf{m}_1) \cdot \nabla_m \chi(\mathbf{m}_1) \tag{7.23}$$

of $\chi(\mathbf{m}_2)$ always underestimates the actual value of $\chi(\mathbf{m}_2)$. In other words, any tangent plane of χ is located below χ itself.

With the help of (7.22) we can relate the Hessian $\mathbf{H}_\chi = \nabla_m \nabla_m \chi$ to the convexity properties of χ: Expanding $\chi(\mathbf{m}_2)$ into a Taylor series about $\mathbf{m}_1$ gives

$$\begin{aligned} \chi(\mathbf{m}_2) = {} & \chi(\mathbf{m}_1) + (\mathbf{m}_2 - \mathbf{m}_1) \cdot \nabla_m \chi(\mathbf{m}_1) \\ & + \frac{1}{2}(\mathbf{m}_2 - \mathbf{m}_1) \cdot \mathbf{H}_\chi(\mathbf{m}_1) \cdot (\mathbf{m}_2 - \mathbf{m}_1) + \mathcal{O}(|(\mathbf{m}_2 - \mathbf{m}_1)|^3). \end{aligned} \tag{7.24}$$

Inserting (7.22) into the previous relation yields

$$0 \le \frac{1}{2}(\mathbf{m}_2 - \mathbf{m}_1) \cdot \mathbf{H}_\chi(\mathbf{m}_1) \cdot (\mathbf{m}_2 - \mathbf{m}_1) + \mathcal{O}(|(\mathbf{m}_2 - \mathbf{m}_1)|^3). \tag{7.25}$$

Since (7.25) holds for any $\mathbf{m}_2$, it follows that

$$0 \le \mathbf{m} \cdot \mathbf{H}_\chi(\mathbf{m}_1) \cdot \mathbf{m}, \quad \text{for all } \mathbf{m}, \mathbf{m}_1 \in \mathfrak{M}. \tag{7.26}$$

This means that the Hessian, $\mathbf{H}_\chi(\mathbf{m}_1)$, of a convex function, χ, is positive semi-definite for any element $\mathbf{m}_1$ from the model space. We note, without proof, that this result can be extended: The Hessian of a strictly convex function is positive definite, and the reverse holds as well.

The requirement of positive definiteness is certainly easier to verify than strict convexity based on its definition (Eq. 7.14). It is, however, still impractical because the calculation of the Hessian for any element in the model space is – at least for realistic problems – numerically too expensive. Nevertheless, the recognition that $\mathbf{H}_\chi$ should ideally be positive definite is of considerable importance from a theoretical point of view, as we shall see in the following paragraphs.

In most realistic applications, our physical intuition tells us that χ can impossibly be strictly convex. In surface wave tomography, for instance, the deep parts of the Earth cannot be resolved, meaning that structural perturbations at sufficiently great depth will not have an influence on χ. Our minimisation problem is in this sense non-unique.

7.1.2.2 Non-uniqueness – An Example in One Dimension

The exploitation of as much information as possible from observed seismograms has always been one of the strongest motivations for full waveform inversion because it is intuitively thought to reduce the non-uniqueness of the tomographic problem at least to some degree. However, the following example illustrates that considerable non-uniqueness may persist even in over-simplified problems, despite using the complete waveform information (e.g. Bamberger et al., 1979): We consider a 1D scalar wave equation as it may be used in seismic reflection experiments:

$$\rho(z)\,\partial_t u(z,t) - \partial_z\left[\mu(z)\,\partial_z u(z,t)\right] = 0\,. \tag{7.27}$$

The variables $z \in [0,\infty)$, ρ and μ signify depth, density and elastic modulus, respectively. We assume that the medium is at rest before $t = 0$ and that a stress $\sigma_0(t)$ prescribed at the surface is responsible for the excitation of waves:

$$\mu(z)\,\partial_z u(z,t)|_{z=0} = \sigma_0(t)\,. \tag{7.28}$$

Our measurement consists in the displacement, $u_0(t)$, recorded at the surface, $z = 0$, and the corresponding synthetic is $u(0,t)$. We now replace the depth variable, z, by a new depth variable, z', that is defined by

$$\frac{\mathrm{d}z'}{\mathrm{d}z} = \phi(z)\,, \quad z'(0) = 0\,, \tag{7.29}$$

where ϕ is any positive function defined on $[0,\infty)$. Rewriting the wave equation (7.27) in terms of z' yields

$$\rho'(z')\,\partial_t u'(z',t) - \partial_{z'}\left[\mu'(z')\,\partial_{z'} u'(z',t)\right] = 0\,, \tag{7.30}$$

with the definitions

$$u'(z',t) = u[z(z'),t]\,, \quad \rho'(z') = \frac{\rho[z(z')]}{\phi[z(z')]}\,, \quad \mu'(z') = \phi[z(z')]\,\mu[z(z')]\,. \tag{7.31}$$

Equation (7.30) demonstrates that $u'(z',t) = u[z(z'),t]$ solves a wave equation with material properties ρ' and μ' that can be different from ρ and μ. By construction we have $z' = 0$ at the surface and therefore

$$u'(z',t)|_{z'=0} = u(z,t)|_{z=0}\,, \tag{7.32}$$

which implies that the change in variables does not affect the solution at the surface. This means that infinitely many models are global minima that explain our observations equally well. Thus, any misfit functional $\chi[u(\rho,\mu)]$ cannot be strictly convex. Let us choose a particular function $\phi(z)$:

$$\phi(z) = \sqrt{\frac{\rho(z)}{\mu(z)}} = \frac{1}{v(z)}, \tag{7.33}$$

where $v(z)$ is the wave propagation speed in the 1D medium. The transformed depth variable

$$z'(z) = \int_0^z \frac{d\xi}{v(\xi)} \tag{7.34}$$

is then the travel time from z to the surface, and the new medium parameters are

$$\rho' = \sqrt{\rho\,\mu} = \eta\,, \quad \mu' = \sqrt{\rho\,\mu} = \eta\,. \tag{7.35}$$

One single medium property, namely the acoustic impedance η as a function of travel time, therefore fully determines the signal recorded at the surface. The individual parameters ρ and μ cannot be inferred independently, and infinitely many combinations of ρ and μ will explain the data equally well.

7.1.2.3 Local Minima and Non-uniqueness in Realistic Problems

The non-uniqueness in the previous example results from the restriction to only one receiver at the surface. Similarly, in realistic 2D and 3D tomographic problems non-uniqueness is related to insufficient data coverage. Seismic sources and receivers are often irregularly distributed, especially in passive experiments on regional to global scales. Some regions of the Earth may therefore not be sampled by the observable wave field. Any change of the Earth model in the unsampled or poorly sampled regions will not or hardly affect the misfit.

Trade-offs between parameters are an additional source of non-uniqueness: The characteristics of seismic waves depend on parameters of different physical nature such as density, wave speeds and relaxation times. Commonly, our data cannot distinguish between variations of one or the other, as it was the case already in our 1D example where the waveforms observed at the surface depend only on the product of ρ and μ.

The presence of multiple local minima is the result of the non-linear relation between Earth models, $\mathbf{m}$, and the corresponding synthetic waveforms, $\mathbf{u}(\mathbf{m})$ (see Deng (1997) for an extreme example). Misfit functionals should ideally be constructed such that they compensate the non-linearity of the forward problem, thus leading to improved convexity. Cross-correlation time shifts (Luo & Schuster, 1991), generalised seismological data functionals (Gee & Jordan, 1992) and time-frequency misfits (Fichtner et al., 2008) are examples of misfit functionals that are designed to compensate the non-linear dependence of seismic waveforms on long-wavelength Earth structure.

The presence of non-uniqueness and multiple local minima requires special measures to be taken in order to ensure the convergence of iterative schemes towards a physically meaningful global minimum. The most commonly adopted strategies are the *multi-scale approach* (Sect. 7.4.1) and *regularisation* (Sect. 7.4.2).

7.2 Optimality Conditions

Optimality conditions that allow us to decide whether an element of the model space is a minimum of χ are essential in the design of iterative minimisation schemes. *First-order conditions* merely involve the first derivative of χ, whereas *second-order conditions* are based on its second derivative.

To construct a necessary first-order optimality condition, we assume that $\tilde{\mathbf{m}}$ is indeed a local minimum of χ. This means that we can find a potentially small neighbourhood $N_r(\tilde{\mathbf{m}})$ around $\tilde{\mathbf{m}}$ such that

$$\chi(\mathbf{m}) \geq \chi(\tilde{\mathbf{m}}), \qquad \text{for all } \mathbf{m} \in N_r(\tilde{\mathbf{m}}). \tag{7.36}$$

Now let $\mathbf{m} \in \mathfrak{M}$ be an arbitrary element of the model space, $\mathfrak{M}$. We then find

$$\mathbf{m} \cdot \nabla_m \chi(\tilde{\mathbf{m}}) = \lim_{\varepsilon \to 0} \frac{1}{\varepsilon}[\chi(\tilde{\mathbf{m}} + \varepsilon\, \mathbf{m}) - \chi(\tilde{\mathbf{m}})] \geq 0, \tag{7.37}$$

because $\varepsilon\, \mathbf{m} \in N_r(\tilde{\mathbf{m}})$ for sufficiently small ε. Relation (7.37) must hold for any $\mathbf{m} \in \mathfrak{M}$ and in particular for $-\mathbf{m} \in \mathfrak{M}$. We therefore have $-\mathbf{m} \cdot \nabla_m \chi(\tilde{\mathbf{m}}) \geq 0$, and it follows that

$$\nabla_m \chi(\tilde{\mathbf{m}}) = \mathbf{0} \tag{7.38}$$

is a necessary first-order optimality condition. The derivative of the misfit functional χ vanishes at the local minimum $\tilde{\mathbf{m}}$. This result suggests that $\nabla_m \chi$ should converge to 0 as the iterative minimisation proceeds.

Equation (7.38) holds not only for minima but also for saddle points and maxima. To unequivocally identify a minimum, we consider the Taylor expansion of χ around $\tilde{\mathbf{m}}$:

$$\chi(\mathbf{m}) - \chi(\tilde{\mathbf{m}}) = \frac{1}{2}(\mathbf{m} - \tilde{\mathbf{m}}) \cdot \mathbf{H}_\chi \cdot (\mathbf{m} - \tilde{\mathbf{m}}) + \mathcal{O}(|\mathbf{m} - \tilde{\mathbf{m}}|^3), \tag{7.39}$$

where we already used the first-order condition (7.38). It follows that $\tilde{\mathbf{m}}$ is a minimum only when the Hessian, $\mathbf{H}_\chi(\tilde{\mathbf{m}})$, is positive definite, that is

$$\mathbf{m} \cdot \mathbf{H}_\chi(\tilde{\mathbf{m}}) \cdot \mathbf{m} > 0, \qquad \text{for all } \mathbf{m} \in \mathfrak{M}. \tag{7.40}$$

While important from a theoretical point of view, the second-order necessary condition (7.40) is rarely used in practice. It is mostly our physical intuition that allows us to make the distinction between minima, saddle points and maxima.

7.3 Iterative Methods for Non-linear Minimisation

Following the introduction of basic concepts and terminology, we now focus our attention on the description of specific methods that can be used to approximate the optimal Earth model $\tilde{\mathbf{m}}$.

7.3.1 *General Descent Methods*

7.3.1.1 Descent Directions

Let $\mathbf{m}_0$ be the initial Earth model, that is, our best guess prior to the inversion. The general goal is to improve $\mathbf{m}_0$ by adding to it an update $\gamma_0\,\mathbf{h}_0$, where $\mathbf{h}_0$ is the descent direction and $\gamma_0 > 0$ is the step length:

$$\mathbf{m}_1 := \mathbf{m}_0 + \gamma_0\,\mathbf{h}_0\,, \qquad \chi(\mathbf{m}_1) < \chi(\mathbf{m}_0)\,. \tag{7.41}$$

For the construction of a suitable descent direction, we require that

$$\chi(\mathbf{m}_1) = \chi(\mathbf{m}_0 + \gamma_0\,\mathbf{h}_0) < \chi(\mathbf{m}_0)\,, \tag{7.42}$$

for all positive step lengths γ_0 that are smaller than a maximum step length γ_{max}. Rearranging (7.42) and taking the limit $\gamma_0 \to 0$ yields

$$\mathbf{h}_0 \cdot \nabla_m \chi(\mathbf{m}_0) = \lim_{\gamma_0 \to 0} \frac{1}{\gamma_0}\left[\chi(\mathbf{m}_0 + \gamma_0\,\mathbf{h}_0) - \chi(\mathbf{m}_0)\right] < 0\,. \tag{7.43}$$

The condition $\mathbf{h}_0 \cdot \nabla_m \chi(\mathbf{m}_0) < 0$ can be considered the definition of a local descent direction, that is a direction indicating where we need to go from $\mathbf{m}_0$ in order to reduce the misfit. It follows immediately, that a descent direction always exists as long as $\nabla_m \chi(\mathbf{m}_0) \neq 0$. To see this, we choose

$$\mathbf{h}_0 = -\nabla_m \chi(\mathbf{m}_0) \tag{7.44}$$

and therefore

$$\mathbf{h}_0 \cdot \nabla_m \chi(\mathbf{m}_0) = -[\nabla_m \chi(\mathbf{m}_0)]^2 < 0\,, \tag{7.45}$$

which means that $\mathbf{h}_0$ from (7.44) satisfies condition (7.43).

Many other descent directions are possible. In fact, when $\mathbf{A}$ is a positive definite matrix, the vector

$$\mathbf{h}_0 = -\mathbf{A} \cdot \nabla_m \chi(\mathbf{m}_0)\,, \tag{7.46}$$

is also a descent direction because

$$\mathbf{h}_0 \cdot \nabla_m \chi(\mathbf{m}_0) = -\nabla_m \chi(\mathbf{m}_0) \cdot \mathbf{A} \cdot \nabla_m \chi(\mathbf{m}_0) < 0\,. \tag{7.47}$$

In an iterative minimisation we successively update the Earth models by going from the current model $\mathbf{m}_i$ along a descent direction $\mathbf{h}_i = -\mathbf{A}_i \cdot \nabla_m \chi(\mathbf{m}_i)$ towards the improved model $\mathbf{m}_{i+1}$. The general descent algorithm can be summarised as follows:

General Descent Method

1. Choose an initial model, $\mathbf{m}_0$, and set $i = 0$.
2. Compute the descent direction

$$\mathbf{h}_i = -\mathbf{A}_i \cdot \nabla_m \chi(\mathbf{m}_i). \tag{7.48}$$

3. Update $\mathbf{m}_i$ according to

$$\mathbf{m}_{i+1} = \mathbf{m}_i + \gamma_i\, \mathbf{h}_i\,,$$

with a suitable step length γ_i that ensures $\chi(\mathbf{m}_{i+1}) < \chi(\mathbf{m}_i)$.

4. Set $i \to i + 1$ and go to Step 2.

This is repeated until the misfit is as small as permitted by the errors in the data.

A critical element of the above algorithm is the gradient $\nabla_m \chi$. It can be computed most efficiently with the help of the adjoint method, which is the topic of Chap. 8. The choice of the positive definite matrices $\mathbf{A}_i$ characterises different descent methods, and it controls the speed of convergence towards the optimum $\tilde{\mathbf{m}}$.

7.3.1.2 The Optimal Step Length

To complete the description of the general descent method, it remains to construct an efficient step length for each iteration. Ideally, we would like to find an optimal γ_i such that $\chi(\mathbf{m}_{i+1}) = \chi(\mathbf{m}_i + \gamma_i\, \mathbf{h}_i)$ is minimal. A necessary condition for the optimality of γ_i is then

$$\frac{\mathrm{d}}{\mathrm{d}\gamma_i}\,\chi(\mathbf{m}_i + \gamma_i\,\mathbf{h}_i) = 0\,. \qquad (7.49)$$

Evaluating the derivative in (7.49) and substituting the linear approximation of $\chi(\mathbf{m}_i + \gamma_i\,\mathbf{h}_i)$ yields

$$0 = \mathbf{h}_i \cdot \nabla_m \chi(\mathbf{m}_i) + \gamma_i\,\mathbf{h}_i \cdot \mathbf{H}_\chi(\mathbf{m}_i) \cdot \mathbf{h}_i\,. \qquad (7.50)$$

Solving for γ_i then gives

$$\gamma_i = -\frac{\mathbf{h}_i \cdot \nabla_m \chi(\mathbf{m}_i)}{\mathbf{h}_i \cdot \mathbf{H}_\chi(\mathbf{m}_i) \cdot \mathbf{h}_i}\,. \qquad (7.51)$$

A commonly used alternative to Eq. (7.51) that allows us to circumvent the computation of the Hessian times vector, $\mathbf{H}_\chi(\mathbf{m}_i) \cdot \mathbf{h}_i$, is a *line search*: For this we choose a small number of trial step lengths, $\gamma_i^{(k)}$ $(k = 1, \ldots, n)$ and evaluate the corresponding misfit functionals $\chi(\mathbf{m}_i + \gamma_i^{(k)}\,\mathbf{h}_i)$. These are then used to approximate $\chi(\mathbf{m}_i + \gamma_i\,\mathbf{h}_i)$ by an interpolating polynomial of degree $n-1$. The minimum of the polynomial is then an approximation to the optimal step length. Numerical experiments reveal that $\chi(\mathbf{m}_i + \gamma_i\,\mathbf{h}_i)$ is often nearly quadratic with respect to γ_i. A second-order polynomial is therefore usually sufficient to obtain a good approximation.

The choice of the trial step lengths is crucial for the success of a line search. The only obvious trial step length is $\gamma_i^{(0)} = 0$ because the corresponding misfit, $\chi(\mathbf{m}_i)$, is already known. Any other $\gamma_i^{(k)}$ can be found most efficiently with the help of intuition that usually provides us with a range of step lengths that lead to physically plausible models.

In principle, the evaluation of χ for each $\gamma_i^{(k)}$ requires the forward simulation of the complete data set. To reduce the numerical costs, Tape et al. (2010) suggest to base the step length optimisation only on a few events that are representative of the whole data set and that sample a large part of the model volume.

7.3.1.3 Termination of the Iterative Minimisation

Given the availability of sufficient computational resources, the minimum of χ can be approximated arbitrarily closely with the help of descent methods. However, $\tilde{\mathbf{m}}$, is likely to explain the observed waveforms so well that the remaining waveform differences are smaller than the uncertainty in the data. The optimal model may therefore contain structure that is not constrained by the observations. To avoid such over-fitting, the iterative minimisation should be terminated before the misfit, $\chi(\mathbf{m})$, drops below a physically plausible threshold ε_χ.

It is important to note that the early termination adds another level of non-uniqueness. This is because the uncertainty in the data does not allow us to make a quantitative distinction between all the models that satisfy $\chi(\mathbf{m}) < \varepsilon_\chi$.

7.3.2 The Method of Steepest Descent

To improve $\mathbf{m}_0$ as much as possible in the first iteration already, we try to find the descent direction, $\mathbf{h}_0$, that leads to the maximum decrease of χ for a small fixed step length, γ_0. For this we determine $\mathbf{h}_0$ with $||\mathbf{h}_0||_2 = 1$ such that

$$\chi(\mathbf{m}_1) - \chi(\mathbf{m}_0) = \chi(\mathbf{m}_0 + \gamma_0\, \mathbf{h}_0) - \chi(\mathbf{m}_0) \approx \gamma_0\, \mathbf{h}_0 \cdot \nabla_m \chi(\mathbf{m}_0) \tag{7.52}$$

is minimal. Assuming that $\mathbf{h}_0$ is a descent direction yields the inequality

$$\gamma_0\, \mathbf{h}_0 \cdot \nabla_m \chi(\mathbf{m}_0) \geq -\gamma_0\, ||\nabla_m \chi(\mathbf{m}_0)||_2 ||\mathbf{h}_0||_2 = -\gamma_0\, ||\nabla_m \chi(\mathbf{m}_0)||_2\,, \tag{7.53}$$

meaning that the quantity that we seek to minimise, $\gamma_0\, \mathbf{h}_0 \cdot \nabla_m \chi(\mathbf{m}_0)$, is always larger than or equal to $-\gamma_0\, ||\nabla_m \chi(\mathbf{m}_0)||_2$. The minimum of $\gamma_0\, \mathbf{h}_0 \cdot \nabla_m \chi(\mathbf{m}_0)$ corresponds to the direction $\mathbf{h}_0$ for which the equal sign holds in (7.53). This is the case for

$$\mathbf{h}_0 = -\frac{\nabla_m \chi(\mathbf{m}_0)}{||\nabla_m \chi(\mathbf{m}_0)||_2}\,, \tag{7.54}$$

which is the *direction of steepest descent*. Our finding that $\mathbf{h}_0$ as defined in (7.54) leads to the most rapid descent for a given small step length γ_0, suggests to repeat this procedure by iteratively moving from a current model $\mathbf{m}_i$ along the local descent direction $-\nabla_m \chi(\mathbf{m}_i)$ towards the updated model $\mathbf{m}_{i+1}$. This is the concept of the steepest descent method:

Method of Steepest Descent

1. Choose an initial model, $\mathbf{m}_0$, and set $i = 0$.
2. Compute the gradient for the current model, $\nabla_m \chi(\mathbf{m}_i)$.
3. Update $\mathbf{m}_i$ according to

$$\mathbf{m}_{i+1} = \mathbf{m}_i - \gamma_i\, \nabla_m \chi(\mathbf{m}_i)\,,$$

   with a suitable step length γ_i that ensures $\chi(\mathbf{m}_{i+1}) < \chi(\mathbf{m}_i)$.
4. Set $i \rightarrow i + 1$, go to Step 2 and repeat until the data are explained sufficiently well.

While being conceptually simple and attractive, the steepest descent method is rarely used in practice because it tends to converge rather slowly towards an acceptable model. This is because a succession of descent directions that are locally optimal may not necessarily be optimal from a global perspective.

7.3.3 Newton's Method and Its Variants

In the steepest descent algorithm, $\mathbf{h}_i$ is determined only from first-derivative information at the current iterate $\mathbf{m}_i$. Newton's method modifies the steepest descent direction with the help of second-derivative information, which leads to much faster convergence. The algorithm exploits the equivalence of finding an extremum of χ and finding a zero of $\nabla_m \chi$. Upon using the necessary condition $\nabla_m \chi(\tilde{\mathbf{m}}) = \mathbf{0}$, the linear approximation of $\nabla_m \chi(\tilde{\mathbf{m}})$ around a nearby model $\mathbf{m}$ reads

$$\mathbf{0} = \nabla_m \chi(\tilde{\mathbf{m}}) \approx \nabla_m \chi(\mathbf{m}) + \mathbf{H}_\chi(\mathbf{m}) \cdot (\tilde{\mathbf{m}} - \mathbf{m}) . \tag{7.55}$$

Formally solving for $\tilde{\mathbf{m}}$ gives

$$\tilde{\mathbf{m}} \approx \mathbf{m} - \mathbf{H}_\chi^{-1}(\mathbf{m}) \cdot \nabla_m \chi(\mathbf{m}) . \tag{7.56}$$

Equation (7.56) suggests the following scheme:

Newton's Method

1. Choose an initial model, $\mathbf{m}_0$, and set $i = 0$.
2. Compute the gradient for the current model, $\nabla_m \chi(\mathbf{m}_i)$.
3. Determine the descent direction, $\mathbf{h}_i$, as the solution of

$$\mathbf{H}_\chi(\mathbf{m}_i) \cdot \mathbf{h}_i = -\nabla_m \chi(\mathbf{m}_i) . \tag{7.57}$$

4. Update $\mathbf{m}_i$ according to

$$\mathbf{m}_{i+1} = \mathbf{m}_i + \mathbf{h}_i . \tag{7.58}$$

5. Set $i \to i + 1$, go to Step 2 and repeat as often as needed.

In each iteration, Newton's method requires the computation of the Hessian times vectors and the solution of the linear system (7.57) which is known as the *Newton equation*. Its solution, the descent direction $\mathbf{h}_i$, can be found, for instance, with the help of iterative matrix solvers (e.g. Quarteroni et al., 2000). Note that Newton's method corresponds to choosing the matrix $\mathbf{A}_i$ from Eq. (7.48) as

$$\mathbf{A}_i = \mathbf{H}_\chi^{-1}(\mathbf{m}_i) . \tag{7.59}$$

Provided that $\mathbf{H}_\chi(\mathbf{m}_i)$ is positive definite, $-\mathbf{H}_\chi^{-1}(\mathbf{m}_i) \cdot \nabla_m \chi(\mathbf{m}_i)$ is a descent direction. To reveal the power of Newton's method, we consider the distance between the new iterate $\mathbf{m}_{i+1}$ and the optimum $\tilde{\mathbf{m}}$, that is $||\mathbf{m}_{i+1} - \tilde{\mathbf{m}}||_2$, relative to the previous distance, $||\mathbf{m}_i - \tilde{\mathbf{m}}||_2$: From Eqs. (7.57) and (7.58) we have

$$\mathbf{m}_{i+1} - \tilde{\mathbf{m}} = \mathbf{m}_i - \tilde{\mathbf{m}} - \mathbf{H}_\chi^{-1}(\mathbf{m}_i) \cdot \nabla_m \chi(\mathbf{m}_i) . \tag{7.60}$$

Inserting the second-order approximation

$$\nabla_m \chi(\mathbf{m}_i) \approx -\mathbf{H}_\chi(\mathbf{m}_i) \cdot (\tilde{\mathbf{m}} - \mathbf{m}_i) - \frac{1}{2}(\tilde{\mathbf{m}} - \mathbf{m}_i) \cdot \nabla_m \mathbf{H}_\chi(\mathbf{m}_i) \cdot (\tilde{\mathbf{m}} - \mathbf{m}_i) \tag{7.61}$$

leads to

$$\mathbf{m}_{i+1} - \tilde{\mathbf{m}} \approx \frac{1}{2} \mathbf{H}_\chi^{-1}(\mathbf{m}_i) \cdot [(\tilde{\mathbf{m}} - \mathbf{m}_i) \cdot \nabla_m \mathbf{H}_\chi(\mathbf{m}_i) \cdot (\tilde{\mathbf{m}} - \mathbf{m}_i)] , \tag{7.62}$$

and therefore

$$||\mathbf{m}_{i+1} - \tilde{\mathbf{m}}||_2 \lessapprox c \, ||\mathbf{m}_i - \tilde{\mathbf{m}}||_2^2 , \tag{7.63}$$

with $c := \frac{1}{2}||\mathbf{H}_\chi^{-1}(\mathbf{m}_i)||_2||\nabla \mathbf{H}_\chi(\mathbf{m}_i)||_2$. Equation (7.63) implies that Newton's method converges quadratically, that is, faster than any other feasible iterative minimisation scheme. The high speed of convergence comes at the prize of having to compute the Hessian times a vector (see Sect. 8.4) and the solution of the Newton equation (7.57) in each iteration.

Newton's method relies critically on the appropriateness of the local approximation from Eq. (7.55). When the current iterate, $\mathbf{m}_i$, is far from the optimum, the Hessian, $\mathbf{H}_\chi(\mathbf{m}_i)$, may have negative eigenvalues and be ill-conditioned or even singular. This can result in very slow convergence, movement in non-descent directions and heavy oscillations that diverge from the solution.

Being unsuitable in its raw form, Newton's method has been modified in many different ways. In the *damped* version of the algorithm, the update $\mathbf{h}_i = -\mathbf{H}_\chi^{-1}(\mathbf{m}_i) \cdot \nabla_m \chi(\mathbf{m}_i)$ is multiplied by a scalar parameter $\alpha \in (0, 1]$:

$$\mathbf{m}_{i+1} = \mathbf{m}_i - \alpha \, \mathbf{H}_\chi^{-1}(\mathbf{m}_i) \cdot \nabla_m \chi(\mathbf{m}_i) . \tag{7.64}$$

Using a conventional line search, α is chosen such that $\mathbf{m}_{i+1}$ reduces χ as much as possible. Since $\mathbf{H}_\chi(\mathbf{m}_i)$ will be nearly singular in most applications, its inversion usually requires regularisation. The *regularised Newton method* therefore adds a scalar multiple of the unit matrix, $\mathbf{I}$, to $\mathbf{H}_\chi(\mathbf{m}_i)$:

$$\mathbf{m}_{i+1} = \mathbf{m}_i - [\mathbf{H}_\chi(\mathbf{m}_i) + \beta \, \mathbf{I}]^{-1} \cdot \nabla_m \chi(\mathbf{m}_i) . \tag{7.65}$$

Just as α, the parameter β may be found with the help of a line search. Note that the regularised Newton method approaches the steepest descent method for large values of β.

A whole class of methods aims to increase the computational efficiency by avoiding the computation of the Hessian. For this we use that χ depends on $\mathbf{m}$ only via the synthetic wave field $\mathbf{u}(\mathbf{m})$, that is $\chi = \chi[\mathbf{u}(\mathbf{m})]$. Differentiating χ twice with respect to $\mathbf{m}$ and making use of the chain rule gives

$$\mathbf{H}_\chi = \nabla_u \nabla_u \chi(\nabla_m \mathbf{u}, \nabla_m \mathbf{u}) + \nabla_u \chi(\nabla_m \nabla_m \mathbf{u}) . \tag{7.66}$$

The first term on the right-hand side of Eq. (7.66) is the *approximate Hessian,*

$$\bar{\mathbf{H}}_\chi := \nabla_u \nabla_u \chi(\nabla_m \mathbf{u}, \nabla_m \mathbf{u}) . \tag{7.67}$$

Since $\bar{\mathbf{H}}_\chi$ involves first derivatives only, it can be used as a computationally less expensive substitute for the *full Hessian* $\mathbf{H}_\chi$. This replacement is easy to justify when $\mathbf{u}$ depends quasi-linearly on $\mathbf{m}$ so that $\nabla_m \nabla_m \mathbf{u}$ is small compared to $\bar{\mathbf{H}}_\chi$. The Newton-like method based on the approximate Hessian,

$$\mathbf{m}_{i+1} = \mathbf{m}_i - \bar{\mathbf{H}}_\chi^{-1}(\mathbf{m}_i) \cdot \nabla_m \chi(\mathbf{m}_i) , \tag{7.68}$$

is known as the *Gauss–Newton method.* Its regularised version

$$\mathbf{m}_{i+1} = \mathbf{m}_i - [\bar{\mathbf{H}}_\chi(\mathbf{m}_i) + \beta\, \mathbf{I}]^{-1} \cdot \nabla_m \chi(\mathbf{m}_i) \tag{7.69}$$

is the *Levenberg method* (Levenberg, 1944). Marquardt (1963) suggested to replace the identity matrix, $\mathbf{I}$, in Eq. (7.69) by the diagonal of $\bar{\mathbf{H}}_\chi(\mathbf{m}_i)$ in order to overcome slow convergence in the directions where the gradient is small. The resulting algorithm

$$\mathbf{m}_{i+1} = \mathbf{m}_i - [\bar{\mathbf{H}}_\chi(\mathbf{m}_i) + \beta \operatorname{diag} \bar{\mathbf{H}}_\chi(\mathbf{m}_i)]^{-1} \cdot \nabla_m \chi(\mathbf{m}_i) \tag{7.70}$$

is referred to as the *Levenberg–Marquardt method.* All modifications of the Newton method sacrifice the quadratic convergence locally near the optimum for an improved global convergence.

The pure Newton method based on the full Hessian is currently not being used in realistic full seismic waveform inversions, because the quadratic convergence does not appear to compensate the numerical costs involved in the computation of the Hessian. An example where Newton's method was used for a 1D synthetic problem can be found in Santosa & Symes (1988). Pratt et al. (1998) demonstrated the benefits of the Gauss–Newton method as compared to the steepest descent method. An application of the Gauss–Newton method for a 3D synthetic problem was presented by Epanomeritakis et al. (2008).

7.3.4 The Conjugate-Gradient Method

To illustrate the conjugate-gradient method, we start with the quadratic misfit functional

$$\chi(\mathbf{m}) = \chi(\tilde{\mathbf{m}}) + \frac{1}{2}(\mathbf{m} - \tilde{\mathbf{m}}) \cdot \mathbf{H}_\chi \cdot (\mathbf{m} - \tilde{\mathbf{m}}) . \tag{7.71}$$

The Hessian, $\mathbf{H}_\chi$, is assumed to be symmetric and positive definite so that $\tilde{\mathbf{m}}$ is the unique minimum of χ. Our interest is in iterative schemes where the descent directions satisfy the following orthogonality conditions:

$$0 = \mathbf{h}_i \cdot \nabla_m \chi(\mathbf{m}_j)\,, \tag{7.72}$$

$$0 = \mathbf{h}_i \cdot \mathbf{H}_\chi \cdot \mathbf{h}_j\,, \tag{7.73}$$

for any $0 \leq i < j \leq n$, where n is the dimension of the model space $\mathfrak{M}$. In the following we will refer to descent directions that fulfil (7.72) and (7.73) as *conjugate directions.*

7.3.4.1 The Benefit of Using Conjugate Descent Directions

To reveal the benefit of conjugate directions, we compute the nth iterate $\mathbf{m}_n$. The repeated application of the general iterative scheme, $\mathbf{m}_{i+1} = \mathbf{m}_i + \gamma_i\,\mathbf{h}_i$, yields

$$\mathbf{m}_n = \mathbf{m}_{j+1} + \sum_{i=j+1}^{n-1} \gamma_i\,\mathbf{h}_i\,, \tag{7.74}$$

for j between 0 and $n-1$. Subtracting $\tilde{\mathbf{m}}$ from both sides of Eq. (7.74) and multiplying by $\mathbf{H}_\chi$ gives

$$\nabla_m \chi(\mathbf{m}_n) = \nabla_m \chi(\mathbf{m}_{j+1}) + \sum_{i=j+1}^{n-1} \gamma_i\,\mathbf{H}_\chi \cdot \mathbf{h}_i\,, \tag{7.75}$$

because $\nabla_m \chi(\mathbf{m}) = \mathbf{H}_\chi \cdot (\mathbf{m} - \tilde{\mathbf{m}})$. With the help of Eq. (7.72) we find

$$\mathbf{h}_j \cdot \nabla_m \chi(\mathbf{m}_n) = \sum_{i=j+1}^{n-1} \gamma_i\,\mathbf{h}_j \cdot \mathbf{H}_\chi \cdot \mathbf{h}_i\,. \tag{7.76}$$

By choosing descent directions that are mutually orthogonal with respect to $\mathbf{H}_\chi$ in the sense of Eq. (7.73), we can force the right-hand side of (7.76) to zero, that is

$$\mathbf{h}_j \cdot \nabla_m \chi(\mathbf{m}_n) = 0\,, \tag{7.77}$$

for any j between 0 and $n-1$. Since the mutually orthogonal descent directions, $\mathbf{h}_0, \mathbf{h}_1, \ldots, \mathbf{h}_{n-1}$, form a basis of the n-dimensional model space $\mathfrak{M}$, it follows from (7.77) that $\nabla_m \chi(\mathbf{m}_n) = \mathbf{0}$. This means that the nth iterate $\mathbf{m}_n$ is equal to the minimum $\tilde{\mathbf{m}}$ of the quadratic misfit functional χ. Clearly, the advantage of using an algorithm that satisfies the orthogonality conditions (7.72) and (7.73) is that the solution is reached after at most n iterations, given that numerical errors are negligible. This is in contrast to the steepest descent method where the minimum may not be reached within a finite number of iterations.

7.3.4.2 The Conjugate-Gradient Method for Quadratic Misfit Functionals

Having shown their advantages, it remains to construct the conjugate directions explicitly. Fortunately, this can be done automatically within an iterative descent scheme that is just a slight modification of the steepest descent algorithm:

Conjugate-Gradient Method

1. Choose an initial model, $\mathbf{m}_0$. Set $i = 0$ and $\mathbf{h}_0 = -\nabla_m \chi(\mathbf{m}_0)$.
2. Compute the optimal step length γ_i according to Eq. (7.51):

$$\gamma_i = -\frac{\mathbf{h}_i \cdot \nabla_m \chi(\mathbf{m}_i)}{\mathbf{h}_i \cdot \mathbf{H}_\chi \cdot \mathbf{h}_i} .$$

3. Update $\mathbf{m}_i$ via

$$\mathbf{m}_{i+1} = \mathbf{m}_i + \gamma_i \, \mathbf{h}_i .$$

4. Compute the gradient for the next iterate, $\nabla_m \chi(\mathbf{m}_{i+1})$.
5. Compute the descent direction for the next iteration, $\mathbf{h}_{i+1}$, according to

$$\beta_i := \frac{||\nabla_m \chi(\mathbf{m}_{i+1})||_2^2}{||\nabla_m \chi(\mathbf{m}_i)||_2^2} , \tag{7.78}$$

$$\mathbf{h}_{i+1} = -\nabla_m \chi(\mathbf{m}_{i+1}) + \beta_i \, \mathbf{h}_i . \tag{7.79}$$

6. Set $i \to i + 1$, go to Step 2 and repeat as often as needed.

We can proof by induction that the above procedure indeed generates conjugate descent directions: In the interest of readability we introduce the notation $\mathbf{g}_i := \nabla_m \chi(\mathbf{m}_i)$, and we note two useful relations that are essential for the proof:

$$\nabla_m \chi(\mathbf{m}_i) = \mathbf{g}_i = \mathbf{H}_\chi \cdot (\mathbf{m}_i - \tilde{\mathbf{m}}) \tag{7.80}$$

and

$$\mathbf{g}_{i+1} - \mathbf{g}_i = \mathbf{H}_\chi \cdot (\mathbf{m}_{i+1} - \tilde{\mathbf{m}}) - \mathbf{H}_\chi \cdot (\mathbf{m}_i - \tilde{\mathbf{m}}) = \gamma_i \mathbf{H}_\chi \cdot \mathbf{h}_i . \tag{7.81}$$

The induction is with respect to the index j as used in (7.72) and (7.73). For $j = 1$ we find

$$\mathbf{h}_0 \cdot \mathbf{g}_1 = \mathbf{h}_0 \cdot [\mathbf{H}_\chi \cdot (\mathbf{m}_0 + \gamma_0 \mathbf{h}_0 - \tilde{\mathbf{m}})] = -\mathbf{h}_0 \cdot \mathbf{h}_0 + \gamma_0 \mathbf{h}_0 \cdot \mathbf{H}_\chi \cdot \mathbf{h}_0 = 0 \tag{7.82}$$

and

$$\begin{aligned}\mathbf{h}_0 \cdot \mathbf{H}_\chi \cdot \mathbf{h}_1 &= \mathbf{h}_0 \cdot \mathbf{H}_\chi \cdot (\beta_0 \mathbf{h}_0 - \mathbf{g}_1) = \gamma_0^{-1}(\mathbf{g}_1 - \mathbf{g}_0) \cdot (-\mathbf{g}_1 - \beta_0 \mathbf{g}_0) \\ &= \gamma_0^{-1}(-\mathbf{g}_1^2 + \beta_0 \mathbf{g}_1 \cdot \mathbf{h}_0 - \mathbf{h}_0 \cdot \mathbf{g}_1 + \beta_0 \mathbf{g}_0^2) = \gamma_0^{-1}(-\mathbf{g}_1^2 + \beta_0 \mathbf{g}_0^2) \\ &= 0\,. \end{aligned} \tag{7.83}$$

Equations (7.82) and (7.83) mean that the orthogonality conditions (7.72) and (7.73) are satisfied in the first iteration. Now assuming that (7.72) and (7.73) hold for any $0 \le i < j$, we have

$$\mathbf{h}_i \cdot \mathbf{g}_{j+1} = \mathbf{h}_i \cdot (\mathbf{g}_j + \gamma_j \mathbf{H}_\chi \cdot \mathbf{h}_j) = 0 \tag{7.84}$$

and

$$\begin{aligned}\mathbf{h}_i \cdot \mathbf{H}_\chi \cdot \mathbf{h}_{j+1} &= \mathbf{h}_i \cdot \mathbf{H}_\chi \cdot (-\mathbf{g}_{j+1} + \beta_j \mathbf{h}_j) = -\mathbf{h}_i \cdot \mathbf{H}_\chi \cdot \mathbf{g}_{j+1} \\ &= \gamma_i^{-1}(\mathbf{g}_i - \mathbf{g}_{i+1}) \cdot \mathbf{g}_{j+1} = \gamma_i^{-1}(\mathbf{h}_{i+1} + \beta_i \mathbf{h}_i - \mathbf{h}_i - \beta_{i-1} \mathbf{h}_{i-1}) \cdot \mathbf{g}_{j+1} \\ &= 0\,. \end{aligned} \tag{7.85}$$

Similarly we can show that $\mathbf{h}_j \cdot \mathbf{g}_{j+1} = 0$ and $\mathbf{h}_j \cdot \mathbf{H}_\chi \cdot \mathbf{h}_{j+1} = 0$. This concludes the proof.

7.3.4.3 The Conjugate-Gradient Method for General Non-linear Functionals

For purely quadratic functionals χ with positive definite Hessian, the conjugate-gradient method is effectively a direct matrix solver that converges to the exact solution after at most n iterations.

When χ is a general non-linear functional, the orthogonality relations do not hold and the algorithm may need more than n iterations to reach an acceptable result. Also, since the Hessian is mostly not available, a suitable step length γ_i must be determined with the help of a line search. The conjugate-gradient algorithm as outlined above for the quadratic case but applied to an arbitrary χ is commonly referred to as *Fletcher–Reeves method* (Fletcher & Reeves, 1964). Popular variants are the *Polak–Ribière method* (Polak & Ribière, 1969) that defines β_i as

$$\beta_i := \frac{\nabla_m \chi(\mathbf{m}_{i+1}) \cdot [\nabla_m \chi(\mathbf{m}_{i+1}) - \nabla_m \chi(\mathbf{m}_i)]}{\nabla_m \chi(\mathbf{m}_i) \cdot \nabla_m \chi(\mathbf{m}_i)} \tag{7.86}$$

and the *Hestenes–Stiefel method* (Hestenes & Stiefel, 1952) where β_i is given by

$$\beta_i := \frac{\nabla_m \chi(\mathbf{m}_{i+1}) \cdot [\nabla_m \chi(\mathbf{m}_{i+1}) - \nabla_m \chi(\mathbf{m}_i)]}{\mathbf{h}_i \cdot [\nabla_m \chi(\mathbf{m}_{i+1}) - \nabla_m \chi(\mathbf{m}_i)]}\,. \tag{7.87}$$

All definitions of β_i are equivalent for quadratic χ. In the more general case, the preferred formula appears to be a matter of personal preference.

7.3.4.4 Convergence and Pre-conditioning Sensu Stricto

Usually, the model space is so large that the minimisation is terminated after a number of iterations that is much smaller than n. In this case, the convergence rate of the algorithm becomes relevant. As a general rule, the closer $\mathbf{H}_\chi$ is to the unit matrix $\mathbf{I}$, the faster the method converges towards $\tilde{\mathbf{m}}$. In fact, when χ is quadratic with $\mathbf{H}_\chi = \mathbf{I}$, one single iteration is sufficient to reach the minimum, $\tilde{\mathbf{m}}$, independent of the initial model, $\mathbf{m}_0$.

Although in practice we have no control over $\mathbf{H}_\chi$, we can modify the iterative scheme such that it effectively solves a minimisation problem where the Hessian is closer to $\mathbf{I}$ than in the original problem. For this we note that the minimisation of the quadratic $\chi(\mathbf{m})$ as defined in Eq. (7.71) is fully equivalent to the minimisation of

$$\chi'(\mathbf{m}') := \chi(\tilde{\mathbf{m}}) + \frac{1}{2}(\mathbf{m}' - \tilde{\mathbf{m}}') \cdot \mathbf{H}'_\chi \cdot (\mathbf{m}' - \tilde{\mathbf{m}}'), \tag{7.88}$$

with

$$\mathbf{m}' := \mathbf{P}^{-1} \cdot \mathbf{m}, \tag{7.89}$$

$$\tilde{\mathbf{m}}' := \mathbf{P}^{-1} \cdot \tilde{\mathbf{m}}, \tag{7.90}$$

$$\mathbf{H}'_\chi := \mathbf{P} \cdot \mathbf{H}_\chi \cdot \mathbf{P} \tag{7.91}$$

and any positive definite *pre-conditioning matrix* $\mathbf{P}$ that is designed to approach $\mathbf{H}'_\chi$ as closely as possible to the unit matrix. In the ideal case we have $\mathbf{P} = \mathbf{H}_\chi^{-1/2}$. More realistic choices are mostly based on heuristics and physical intuition. The conjugate-gradient method applied to (7.88) converges towards the minimum $\tilde{\mathbf{m}}'$ of χ', that is related to the minimum $\tilde{\mathbf{m}}$ of χ via Eq. (7.90).

In the general non-linear case the local Hessian $\mathbf{H}_\chi(\mathbf{m}_i)$ is usually not known explicitly so that its pre-conditioned version $\mathbf{H}'_\chi(\mathbf{m}_i)$ cannot be computed. Yet, based on the quadratic approximations

$$\chi(\mathbf{m}_i) \approx \chi(\tilde{\mathbf{m}}) + \frac{1}{2}(\mathbf{m}_i - \tilde{\mathbf{m}}) \cdot \mathbf{H}_\chi(\tilde{\mathbf{m}}) \cdot (\mathbf{m}_i - \tilde{\mathbf{m}}) \tag{7.92}$$

and

$$\chi'(\mathbf{m}'_i) \approx \chi(\tilde{\mathbf{m}}) + \frac{1}{2}(\mathbf{m}'_i - \tilde{\mathbf{m}}') \cdot \mathbf{H}'_\chi(\tilde{\mathbf{m}}') \cdot (\mathbf{m}'_i - \tilde{\mathbf{m}}'), \tag{7.93}$$

we find

$$\nabla_{m'}\chi(\mathbf{m}'_i) \approx \mathbf{P} \cdot \nabla_m \chi(\mathbf{m}_i). \tag{7.94}$$

Equation (7.94) allows us to apply the conjugate-gradient algorithm to the minimisation of $\chi'(\mathbf{m}'_i)$ using only the gradient of χ, that is $\nabla_m \chi(\mathbf{m}_i)$:

Pre-conditioned Conjugate–Gradient Method (Fletcher–Reeves Variant)

1. Choose an initial model, $\mathbf{m}_0$. Set $i = 0$, $\mathbf{m}'_0 = \mathbf{P}^{-1} \cdot \mathbf{m}_0$ and $\mathbf{h}_0 = -\mathbf{P} \cdot \nabla_m \chi(\mathbf{m}_0)$.
2. Determine the optimal step length γ_i using a line search.
3. Update $\mathbf{m}'_i$ according to

$$\mathbf{m}'_{i+1} = \mathbf{m}'_i + \gamma_i \, \mathbf{h}_i \,.$$

4. Determine $\mathbf{m}_{i+1}$ via $\mathbf{m}_{i+1} = \mathbf{P} \cdot \mathbf{m}'_{i+1}$.
5. Compute the gradient for the next iterate, $\nabla_m \chi(\mathbf{m}_{i+1})$.
6. Compute the descent direction for the next iteration, $\mathbf{h}_{i+1}$, according to

$$\beta_i := \frac{||\mathbf{P} \cdot \nabla_m \chi(\mathbf{m}_{i+1})||_2^2}{||\mathbf{P} \cdot \nabla_m \chi(\mathbf{m}_i)||_2^2} \,, \tag{7.95}$$

$$\mathbf{h}_{i+1} = -\mathbf{P} \cdot \nabla_m \chi(\mathbf{m}_{i+1}) + \beta_i \, \mathbf{h}_i \,. \tag{7.96}$$

7. Set $i \to i + 1$, go to Step 2 and repeat as often as needed.

7.3.4.5 Gradient Pre-conditioning

In the context of full waveform inversion, pre-conditioning is often not understood in its strict mathematical form where a minimisation problem is replaced by an equivalent one for the benefit of faster convergence. Instead, the conjugate-gradient (or the steepest descent) method is used in its original form, the only difference being that the local gradient, $\nabla_m \chi(\mathbf{m}_i)$, is manipulated in a physically plausible way. *Gradient pre-conditioning* in this sense is effectively a form of regularisation that forces the solution towards desirable models that reflect our prior knowledge or expectation.

A very common manipulation is smoothing, meaning that $\nabla_m \chi(\mathbf{m}_i)$ is low-pass filtered before it is used to compute the next descent direction. Smoothing results in the removal of short-wavelength structure that we do not expect to be resolvable with a given set of data. Additional pre-conditioning can be necessary because the gradient tends to be extremely large in the immediate vicinity of the sources and receivers (see also Chap. 12). To prevent these large contributions from becoming dominant features in the tomographic images, the raw gradient is often cut so that its maximum is below a reasonable threshold. A similar approach was taken by Igel et al. (1996) who multiplied the gradient by the inverse geometric spreading.

There are, in principle, no limits to gradient pre-conditioning as long as it leads to accelerated convergence towards physically meaningful models. Specific pre-conditioners are usually found empirically with the help of synthetic inversions.

The beauty of the conjugate gradient method is that it tends to converge much faster than the steepest descent algorithm, and the accelerated convergence comes at nearly no additional computational costs. This explains the popularity of conjugate gradients in full waveform inversion, where they have been used early on (e.g. Mora, 1987, 1988). For recent applications of conjugate-gradient-type methods the reader is referred to Fichtner et al. (2009) and Tape et al. (2010).

7.4 Convergence

Realistic inverse problems are characterised by the presence of multiple local minima and the non-uniqueness of the global minimum. To avoid the convergence towards potentially meaningless local minima, special inversion strategies, such as the multi-scale approach and regularisation, need to be adopted.

The multi-scale approach, described in Sect. 7.4.1, helps to guide the iterative inversion towards the basin of attraction of the global minimum. Regularisation (Sect. 7.4.2) forces the iteration towards models that are consistent with prior knowledge or personal preference.

7.4.1 The Multi-Scale Approach

By construction, descent methods march steadily downhill on the misfit surface, approaching a minimum, $\tilde{\mathbf{m}}$, of χ. This implies that all iterates, $\mathbf{m}_i$, are confined to one basin of attraction that may not necessarily correspond to the global minimum. To avoid the convergence towards a local minimum, the initial model, $\mathbf{m}_0$, must be within the basin of attraction of the global minimum, the location of which is usually not known a priori.

A very successful strategy that greatly reduces the requirements concerning the initial model is illustrated in Fig. 7.3. It exploits the observation that the complexity of the misfit functional is directly proportional to the dominant length scales in the Earth model. Loosely speaking, rough Earth models generate rough misfit functionals with numerous local minima. However, the misfit functionals corresponding to smooth models tend to be smooth with fewer local minima. The iterative inversion should therefore start with long-period data that are used to constrain the long-wavelength structure. A sufficiently good initial model for this first inversion stage can usually be obtained from classical traveltime tomography (e.g. Pratt & Goulty, 1991; Zhou et al., 1995; Korenaga et al., 1997; Dessa & Pascal, 2003; Bleibinhaus et al., 2007; Fichtner et al., 2010). When the low-frequency data are well explained, the iteration is stopped and the final model is used as initial model for the next round of iterations for shorter wavelength structure, based on shorter period data. This procedure – sometimes referred to as *multi-scale approach* – is repeated until the data with the highest frequencies have been inverted for the short-

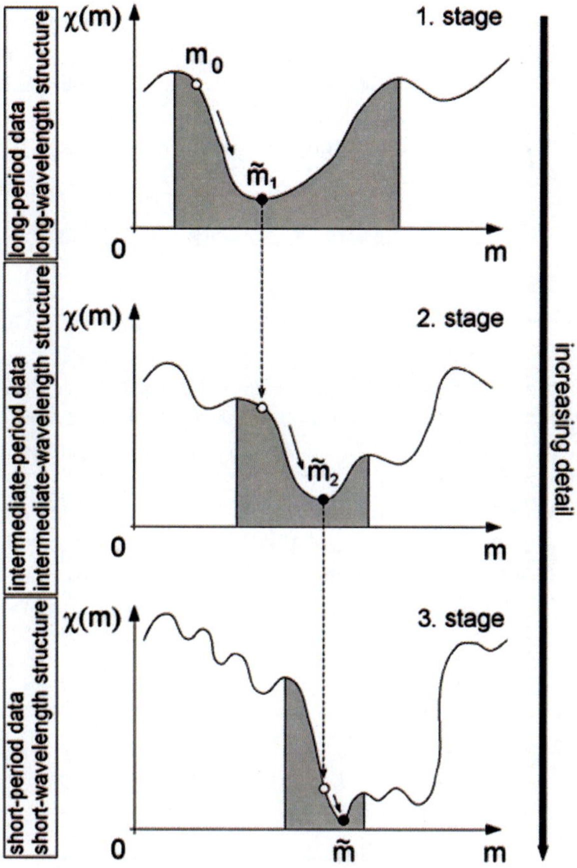

Fig. 7.3 Illustration of the multi-scale approach for the case of a three-stage inversion. *Top*: The first stage of the inversion starts from the initial model $\mathbf{m}_0$ (*empty circle*, ∘). The iterative minimisation is based on long-period data and attempts to find the optimal long-wavelength model (*filled circle*, •). The objective functional, $\chi(\mathbf{m})$, is well behaved with a wide basin of attraction (*shaded area*) around the long-wavelength optimal model, $\tilde{\mathbf{m}}_1$. *Centre*: The best model from the first stage is used as initial model in the second stage that works with intermediate-period data. Since the admissible wavelengths in the model are decreased by relaxing the regularisation, the misfit functional is rougher and the basin of attraction around the optimum $\tilde{\mathbf{m}}_2$ is narrower. *Bottom*: Model $\tilde{\mathbf{m}}_2$ is used as initial model in the third stage where short-period data are used to retrieve the short-wavelength components in the model. The basin of attraction around the global minimum $\tilde{\mathbf{m}}$ is particularly narrow. As the inversion proceeds from one stage to the next, increasingly many details appear in the tomographic images (see Fig. 14.6 for an example)

wavelength structure (e.g. Pica et al., 1990; Bunks et al., 1995; Ravaut et al., 2004; Bleibinhaus et al., 2007; Fichtner et al., 2010).

The importance of the multi-scale approach becomes apparent already in simplistic synthetic inversions such as the one shown in Fig. 7.4. There we consider the propagation of scalar waves in a 100 km by 100 km two-dimensional domain where all boundaries are perfectly reflecting. The target model, $\mathbf{m}_t$, used to generate artificial data, $u(\mathbf{m}_t, \mathbf{x}^r, t)$, is shown on the left. It consists of a 14 km by 14 km positive velocity perturbation embedded within a homogeneous background. Nine sources and the receivers are co-located at the positions $\mathbf{x}^r$ ($r = 1, \ldots, 9$) indicated by the black triangles, and the misfit functional is the cumulative L_2 distance

$$\chi(\mathbf{m}) = \sum_{r=1}^{9} \int_T [u(\mathbf{m}_t; \mathbf{x}^r, t) - u(\mathbf{m}; \mathbf{x}^r, t)]^2 \, dt \,. \tag{7.97}$$

The upper right panel shows the result of an inversion with a single frequency band that corresponds to a dominant wavelength of $\lambda = 14$ km, which is equal to the size of the velocity perturbation. Using a steepest descent algorithm (Sect. 7.3.2), the iteration arrived at a model that explains the artificial data very well, but fails to retrieve the target model. Clearly, the model shown in the upper right panel of Fig. 7.4 is a local minimum of χ.

The multi-scale approach, visualised in the lower right panel of Fig. 7.4, can prevent the convergence towards a potentially meaningless local minimum. The inversion starts in the first stage with long-period waveforms that correspond to a dominant wavelength of $\lambda = 40$ km. The final model from the first stage is used

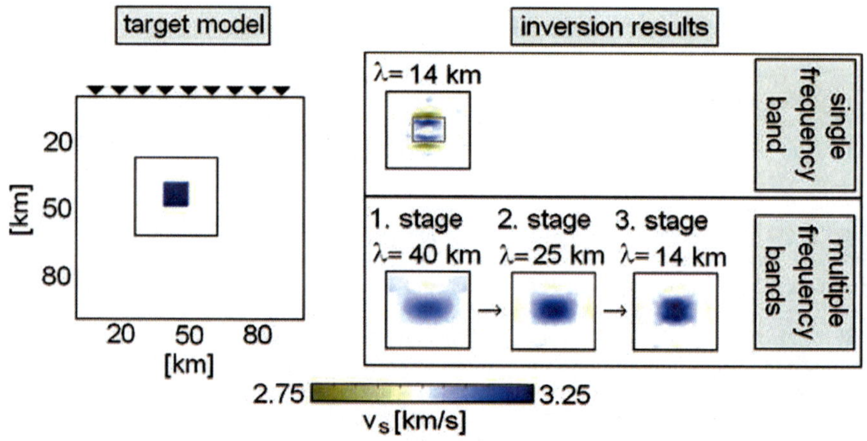

Fig. 7.4 Illustration of the multi-scale approach in a synthetic inversion. The minimisation was based on the method of steepest descent (Sect. 7.3.2), and the initial model was homogeneous. *Left*: The target model used to generate the artificial data consists of a positive velocity perturbation embedded within a homogeneous background. *Top right*: The final model after a single frequency band inversion is a local minimum of χ that explains the artificial data very well without being close to the target model. (Only the central part of the final model is displayed.) *Bottom right*: Iterative minimisation with successively shorter periods. The dominant wavelengths of the waves propagating through the model decrease from $\lambda = 40$ km in the first stage to $\lambda = 14$ km in the third stage. Displayed are the central parts of the respective final models. After the third stage the data are perfectly explained and the optimal model resembles the target model. The detailed setup of this numerical experiment is explained in the text

as initial model for the second stage where the dominant wavelength is reduced to 25 km. After the third stage, the shortest period data are explained slightly better than in the single frequency band inversion, and the optimal model is close to the target $\mathbf{m}_t$.

There are two important things to note: First, the success of the multi-scale approach is empirical, meaning that there is no guarantee that it works under different circumstances. Second, the extent to which a single frequency band inversion fails depends strongly on the misfit functional, and the one chosen in (7.97) is rather inappropriate for full waveform inversion, as we will discuss in Chap. 11.

All iterative minimisation schemes described in the following paragraphs fall into the class of descent methods, meaning that the misfit is steadily reduced by moving within one basin of attraction along local descent directions. Each method comes with advantages and disadvantages that need to be weighted in the light of a particular application.

7.4.2 Regularisation

To ensure the convergence of iterative minimisation schemes towards a physically meaningful optimum, $\tilde{\mathbf{m}}$, the inverse problem needs to be *regularised*. Regularisation means to give preference to models that reflect prior knowledge and/or a philosophical concept such as the principle that models should be simplistic in a certain sense (Occam's razor).

The choice of a regularisation method depends very much on the nature of a specific application and on the particular aspects of the Earth's structure that one wishes to constrain. Regularisation is therefore subjective to some degree.

In the following paragraphs we summarise regularisation approaches that are commonly used in seismic tomography and that we consider to be of special relevance in full waveform inversion:

1. *Tikhonov regularisation*: The idea of Tikhonov regularisation, named after its original inventor (Tikhonov, 1963), is to base the inversion on a regularised misfit functional, χ_{reg}, which is equal to χ plus a quadratic term:

$$\chi_{\text{reg}}(\mathbf{m}) := \chi(\mathbf{m}) + \gamma\, ||\mathbf{R} \cdot (\mathbf{m} - \mathbf{m}_{\text{ref}})||_2^2\,, \tag{7.98}$$

where $\mathbf{R}$ and $\mathbf{m}_{\text{ref}}$ denote the *Tikhonov matrix* and a reference model, respectively. Typically, $\mathbf{m}_{\text{ref}}$ is a laterally homogeneous or otherwise simplistic model of the Earth. The strictly positive parameter γ balances the contributions of the pure misfit, $\chi(\mathbf{m})$ and the regularisation term $||\mathbf{R} \cdot (\mathbf{m} - \mathbf{m}_{\text{ref}})||_2^2$.

Different choices for the Tikhonov matrix are possible: When $\mathbf{R}$ is equal to the identity matrix, that is $\mathbf{R} = \mathbf{I}$, the regularisation is referred to as *norm damping*. Preference is then given to models that are close to $\mathbf{m}_{\text{ref}}$ in the sense that the differences $\mathbf{m} - \mathbf{m}_{\text{ref}}$ are small. This regularisation effect is generally strongest in regions of the Earth where the structure is poorly constrained by the data.

In the extreme case where χ bears no information at all on the structure of the Earth, that is when $\chi(\mathbf{m}) = \text{const}$, the optimal model $\tilde{\mathbf{m}}$ is equal to $\mathbf{m}_{\text{ref}}$.

A common alternative to norm damping is *derivative damping*, where the optimal model is forced to have smooth variations. The Tikhonov matrix is then $\mathbf{R} = \mathbf{D}$, where $\mathbf{D}$ is a finite-difference operator acting on the model space. Typically, $\mathbf{D}$ is a finite-difference approximation to the Laplacian operator. The discrete derivative emphasises sharp features so that the regularisation term is large for rough models.

Tikhonov regularisation has been studied extensively in the context of linearised tomographic problems. For a summary the reader is referred to Nolet (2008).

2. *Total variation regularisation*: Norm and derivative damping are the classic regularisation techniques in seismic tomography despite some well-known deficiencies. While enforcing low-amplitude model variations, norm damping still allows for small-scale oscillations that may not be resolvable. Derivative damping eliminates small-scale oscillations, but it also suppresses the sharp interfaces that one would like to image with the help of full waveform inversion.

   Originally developed in the context of image processing (e.g. Vogel & Oman, 1996), total variation regularisation combines the advantages of norm and derivative damping while avoiding most of their drawbacks. Total variation regularisation is particularly effective in recovering sharp material interfaces.

   To illustrate the concept, we consider a scalar function $y = f(x)$ defined on the interval $[0, 1]$. Imagine a point travelling along the graph of f. The cumulative distance travelled in y-direction is the total variation of f, denoted $\mathcal{V}(f)$. When f is differentiable, $\mathcal{V}(f)$ can be computed explicitly via

$$\mathcal{V}(f) = \int_0^1 |\partial_x f(x)| \, dx = ||\partial_x f||_1 \,, \tag{7.99}$$

   meaning that the total variation is equal to the L_1 norm of $\partial_x f$. To understand what $\mathcal{V}(f)$ effectively measures, we look at the examples shown in Fig. 7.5.

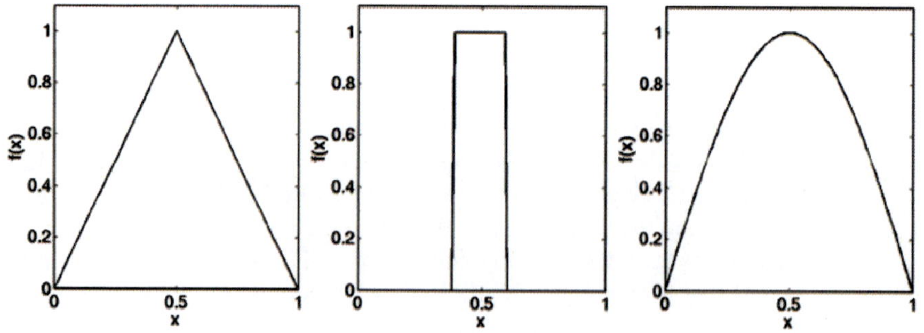

Fig. 7.5 Exemplary scalar functions defined on the interval $[0, 1]$. The total variation, $\mathcal{V}(f)$, of all functions is equal to 2, despite the very different smoothness properties

While the smoothness properties of the three functions are very different, they share the same total variation, $\mathcal{V}(f) = 2$. This is because the point on the graph travels the same distance in y-direction. The total variation depends on the cumulative deviation from zero, but not on the presence of sharp features such as kinks or jumps.

This example suggests to regularise χ by adding the total variation of the model, $\mathbf{m}$:

$$\chi_{\text{reg}}(\mathbf{m}) := \chi(\mathbf{m}) + \gamma\, ||\bar{\nabla}\mathbf{m}||_1\,, \tag{7.100}$$

where $\bar{\nabla}$ is a discrete approximation to the gradient operator ∇. The parameter γ plays the same role as in Tikhonov regularisation. Total variation regularisation gives preference to models that are piecewise smooth and small in the sense that there is little cumulative deviation from zero. Unlike derivative damping, total variation regularisation allows for the presence of sharp interfaces. This property is referred to as *edge preservation*.

Despite being a promising approach, total variation regularisation has not yet been applied to real data full waveform inversion. Synthetic examples in one and two dimensions can be found in Askan et al. (2007), Askan & Bielak (2008) and Burstedde & Ghattas (2009).

3. *Regularisation through parameterisation*: The basis functions used to discretise the space-continuous Earth models (see Eq. 7.5) play a key role in the regularisation of the inverse problem, because they pre-determine the admissible classes of models. Choosing, for instance, a spectral basis such as spherical harmonics, imposes smoothness constraints that do not permit models with sharp features.

   Since full waveform inversion aims to recover strong contrasts in material properties, local basis functions should be used. For optimal regularisation, the support of the basis functions needs to be adapted to our prior estimates of the resolving lengths. This helps to prevent the occurrence of unresolved short-wavelength oscillations in the tomographic images.

   Methods for such adaptive discretisation on a global scale have been presented by several authors (e.g. Debayle & Sambridge, 2004; Nolet & Montelli, 2005). On a regional scale and in the particular context of full waveform inversion, it appears most natural to use the discretisation from the numerical modelling also for the representation of the Earth model. Tape et al. (2010), for example, used the Lagrange polynomial basis of their spectral-element simulations. This approach automatically adapts the parameterisation to the wavelength of the seismic waves, which roughly coincides with the expected resolution length, at least in the sufficiently covered regions.

4. *Regularisation through finite number of iterations*: A very subtle and hardly quantifiable form of regularisation is introduced by the finite number of iterations used to approximate an optimum model, $\tilde{\mathbf{m}}$. As the non-linear inversion proceeds from one model to the next (see Eq. 7.2), the misfit decreases continuously because more and more aspects of the data can be explained. In this process, outliers in the data can become dominant, thus introducing large artefacts

in the tomographic images. The iteration must therefore be stopped sufficiently early, and it is up to the careful tomographer to decide what 'sufficient' means in practice.

While often disregarded, the finite number of iterations is likely to be a major player in the regularisation of non-linear inverse problems.

Chapter 8
The Time-Domain Continuous Adjoint Method

The adjoint method is a mathematical tool that allows us to compute the gradient of an objective functional with respect to the model parameters very efficiently. In this chapter we derive a general formulation of the adjoint method that is independent of a particular physical problem. We introduce important concepts such as time reversal, adjoint sources, adjoint fields and Fréchet or sensitivity kernels. An extension of the adjoint method to second derivatives naturally leads to Hessian kernels. We conclude this chapter with an application to the elastic wave equation including anisotropy and attenuation.

8.1 Introduction

The derivatives of an observable with respect to the parameters of the theory used for its prediction play a fundamental role in all physical sciences. Derivatives are indispensable in sensitivity analysis, numerous non-linear optimisation methods (see Sect. 7.3) and in the development of the physical intuition that is crucial for the efficient and meaningful solution of inverse problems.

The adjoint method as a particularly efficient tool for the computation of the partial derivatives of an objective functional seems to have originated in the field of control theory. In control theory one considers an observable **u** that is the output of a dynamical system. The behaviour of the system depends on model parameters **m** that are linked to **u** via a physical theory symbolised by an operator **L**,

$$\mathbf{L}(\mathbf{u},\mathbf{m}) = \mathbf{f}\,, \tag{8.1}$$

where **f** represents external forces. The goal is to choose the model parameters **m** as a function of the currently observed output, such that the system operates optimally. In mathematical terms optimality means that a problem-specific objective functional $\chi(\mathbf{m}) = \chi[\mathbf{u}(\mathbf{m})]$ is minimal. A change of χ in response to a change from $\mathbf{m}$ to $\mathbf{m} + \varepsilon\,\delta\mathbf{m}$ is approximated by the derivative of χ with respect to $\mathbf{m}$ in the direction $\delta\mathbf{m}$:

$$\nabla_m \chi(\mathbf{m})\,\delta\mathbf{m} = \lim_{\varepsilon\to 0}\frac{1}{\varepsilon}\left[\chi(\mathbf{m}+\varepsilon\,\delta\mathbf{m}) - \chi(\mathbf{m})\right]. \tag{8.2}$$

A. Fichtner, *Full Seismic Waveform Modelling and Inversion*, Advances in Geophysical and Environmental Mechanics and Mathematics,
DOI 10.1007/978-3-642-15807-0_8, © Springer-Verlag Berlin Heidelberg 2011

In many applications, including seismic tomography, the number of model parameters is large and the solution of the forward problem $\mathbf{L}(\mathbf{u}, \mathbf{m}) = \mathbf{f}$ is computationally expensive. This renders a finite-difference approximation of (8.2) for all possible directions $\delta\mathbf{m}$ practically impossible. This is where the adjoint method comes into play.

The adjoint state of a hyperbolic differential equation, such as the wave equation, can be found as early as 1968 in the book by J.-L. Lions on the optimal control of systems governed by partial differential equations. It was one of Lions students, G. Chavent, who may have been the first to use the adjoint method for the determination of distributed parameters. His thesis entitled *Analyse Fonctionelle et Identification de Coefficients Répartis dans les Équations aux Dérivées Partielles* dates from 1971.

Probably one of the first geoscientific applications of control theory in conjunction with the adjoint method was presented by Chavent and his co-workers Dupuy and Lemonnier (1975). In the framework of petroleum engineering they 'determined the permeability distribution by matching the history of pressure in a single-phase field, given flow production data'. To highlight some of the basic concepts still found in modern applications, we briefly expand on their approach: Chavent, Dupuy and Lemonnier considered a reservoir Ω with a permeable boundary $\partial\Omega$ from which a fluid was extracted at NW production wells. The pressure p as a function of time and space was assumed to be governed by a diffusion equation and influenced by a transmissivity coefficient b, a known storage coefficient a and the production rate q_i of the wells. The goal was to find the transmissivity of the reservoir by matching the history of the computed pressure p at the wells to the truly observed pressure history, denoted p^0. As in optimal control theory, this was stated in the form of a minimisation problem involving an objective functional, $\chi(b, p^0, p)$, that quantifies the misfit between data, p^0, and the computed pressure estimate, p, as a function of b. The minimisation of χ with respect to b was achieved using a steepest descent method, and the direction of steepest descent was obtained via the adjoint method. Finite differencing would have been prohibitively expensive even though the discretised transmissivity model merely comprised 171 grid points.

Following a series of theoretical studies on the 1D seismic inverse problem (Bamberger et al., 1977, 1979), Bamberger et al. (1982) presented what is likely to be the first application of the adjoint method for seismic imaging purposes. The *inversion of normal incidence seismograms* was indeed set as an optimal control problem.

Since the mid-1980s, the adjoint method has been used in a variety of physical sciences, including meteorology (e.g. Talagrand & Courtier, 1987), ground water modelling (e.g. Sun, 1994) and geodynamics (e.g. Bunge et al., 2003). Examples of seismological applications of the adjoint method may be found in Tarantola (1988), Tromp et al. (2005), Fichtner et al. (2006a, b), Sieminski et al. (2007a, b), Liu & Tromp (2008), Stich et al. (2009), Tape et al. (2009, 2010) and Fichtner et al. (2009, 2010).

There are two distinct approaches towards the practical solution of an optimisation problem that are closely linked to different variants of the adjoint method. One may first develop the optimisation scheme based on the continuous forward problem and then discretise the equations. This approach, known as *first optimise then discretise (OTD)*, is related to the *continuous adjoint method*, which is the topic of this chapter. The continuous adjoint method is very general, and it involves basic notions of functional analysis. Alternatively, one may *first discretise then optimise (DTO)*, which means that the development of the optimisation scheme is based on a forward problem that has been discretised previously. Different variants of the *discrete adjoint method* correspond to the DTO approach. They involve simple matrix algebra and their specific form depends on the numerical method used to discretise the forward problem. In Chap. 10 we describe the frequency-domain discrete adjoint method. Both OTD and DTO are equivalent provided that the discretisation of the forward problem produces sufficiently accurate solutions.

8.2 General Formulation

We consider a physical observable $\mathbf{u}$ that depends on the position vector $\mathbf{x} \in G \subset \mathbb{R}^3$, time $t \in T = [t_0, t_1]$ and on model parameters $\mathbf{m} \in \mathfrak{M}$:

$$\mathbf{u} = \mathbf{u}(\mathbf{m}; \mathbf{x}, t)\,. \tag{8.3}$$

The model space $\mathfrak{M}$ contains all admissible parameters $\mathbf{m}$, and the semicolon in Eq. (8.3) indicates that $\mathbf{u}$ evolves in space and in time, whereas the model parameters are assumed to be fixed for a given realisation of $\mathbf{u}$. In seismology $\mathbf{u}$ represents an elastic wave field that is linked via the wave equation, symbolically written as $\mathbf{L}(\mathbf{u}, \mathbf{m}) = \mathbf{f}$, to external sources $\mathbf{f}$ and parameters $\mathbf{m}$ such as the mass density ρ and the elastic tensor $\mathbf{C}$ in the Earth (see Sect. 2.2).

It is commonly not $\mathbf{u}$ itself, but a scalar objective functional $\chi(\mathbf{m}) = \chi[\mathbf{u}(\mathbf{m})]$ that we are interested in. Objective functionals serve two closely related purposes: First, they can represent a measurement process that translates a pure physical entity $\mathbf{u}$, such as a seismic wave, to a secondary observable, for instance, the energy at a receiver position $\mathbf{x} = \mathbf{x}^{\mathrm{r}}$:

$$\chi(\mathbf{m}) = \int_T \int_G \dot{\mathbf{u}}^2(\mathbf{m}; \mathbf{x}, t)\, \delta(\mathbf{x} - \mathbf{x}^{\mathrm{r}})\, \mathrm{d}t\, \mathrm{d}^3\mathbf{x}\,. \tag{8.4}$$

Second, when observed data $\mathbf{u}^0(\mathbf{x}, t)$ are available, χ can be used to quantify the discrepancy between the observation and the theoretical prediction $\mathbf{u}(\mathbf{m}; \mathbf{x}, t)$. The objective functional then plays the role of a misfit functional that may, for instance, quantify the L_2 distance between observed and predicted seismograms at $\mathbf{x} = \mathbf{x}^{\mathrm{r}}$:

$$\chi(\mathbf{m}) = \frac{1}{2} \int_T \int_G [\mathbf{u}(\mathbf{m}; \mathbf{x}, t) - \mathbf{u}^0(\mathbf{x}, t)]^2\, \delta(\mathbf{x} - \mathbf{x}^{\mathrm{r}})\, \mathrm{d}t\, \mathrm{d}^3\mathbf{x}\,. \tag{8.5}$$

While Eqs. (8.4) and (8.5) are special cases of objective functionals, the appearance of an integral over space and time is of surprising generality not only in seismological applications. We therefore write χ in the form

$$\chi(\mathbf{m}) = \int_T \int_G \chi_1[\mathbf{u}(\mathbf{m}; \mathbf{x}, t)]\, \mathrm{d}t\, \mathrm{d}^3\mathbf{x} = \langle \chi_1(\mathbf{m}) \rangle\,, \tag{8.6}$$

where we introduced $\langle\, .\, \rangle$ as a short notation for the integral over $T \times G$. The derivative $\nabla_m \chi\, \delta\mathbf{m}$ of $\chi[\mathbf{u}(\mathbf{m})]$ with respect to $\mathbf{m}$ in a direction $\delta\mathbf{m}$ follows from the chain rule:

$$\nabla_m \chi\, \delta\mathbf{m} = \nabla_u \chi\, \delta\mathbf{u} = \langle \nabla_u \chi_1\, \delta\mathbf{u} \rangle \tag{8.7}$$

where

$$\delta\mathbf{u} := \nabla_m \mathbf{u}\, \delta\mathbf{m} \tag{8.8}$$

denotes the derivative of $\mathbf{u}$ with respect to $\mathbf{m}$ in the direction $\delta\mathbf{m}$. The difficulty of Eq. (8.7) lies in the appearance of $\delta\mathbf{u}$ which is often hard to evaluate numerically. For a first-order finite-difference approximation of $\nabla_m \chi$ one needs to determine $\mathbf{u}(\mathbf{m} + \varepsilon\, \delta\mathbf{m})$ for each possible direction $\delta\mathbf{m}$. This, however, becomes infeasible in the case of numerically expensive forward problems and large model spaces. Consequently, we may not be able to compute $\nabla_m \chi$ unless we manage to eliminate $\delta\mathbf{u}$ from Eq. (8.7). For this purpose we differentiate the theoretical relationship $\mathbf{L}(\mathbf{u}, \mathbf{m}) = \mathbf{f}$ with respect to $\mathbf{m}$. Again invoking the chain rule for differentiation gives

$$\nabla_m \mathbf{L}\, \delta\mathbf{m} + \nabla_u \mathbf{L}\, \delta\mathbf{u} = \mathbf{0}\,. \tag{8.9}$$

The right-hand side of Eq. (8.9) vanishes because the external sources $\mathbf{f}$ do not depend on the model parameters $\mathbf{m}$. We now multiply (8.9) by an arbitrary test function $\mathbf{u}^\dagger$ and then apply the integral $\langle\, .\, \rangle$:

$$\langle \mathbf{u}^\dagger \cdot \nabla_m \mathbf{L}\, \delta\mathbf{m} \rangle + \langle \mathbf{u}^\dagger \cdot \nabla_u \mathbf{L}\, \delta\mathbf{u} \rangle = 0\,. \tag{8.10}$$

Adding Eqs. (8.7) and (8.10) gives

$$\nabla_m \chi\, \delta\mathbf{m} = \langle \nabla_u \chi_1\, \delta\mathbf{u} \rangle + \langle \mathbf{u}^\dagger \cdot \nabla_u \mathbf{L}\, \delta\mathbf{u} \rangle + \langle \mathbf{u}^\dagger \cdot \nabla_m \mathbf{L}\, \delta\mathbf{m} \rangle\,. \tag{8.11}$$

We can rewrite Eq. (8.11) using the adjoint operators $\nabla_u \chi_1^\dagger$ and $\nabla_u \mathbf{L}^\dagger$ which are defined by

$$\langle \nabla_u \chi_1\, \delta\mathbf{u} \rangle = \langle \delta\mathbf{u} \cdot \nabla_u \chi_1^\dagger \rangle \tag{8.12}$$

and

$$\langle \mathbf{u}^\dagger \cdot \nabla_u \mathbf{L}\, \delta\mathbf{u} \rangle = \langle \delta\mathbf{u} \cdot \nabla_u \mathbf{L}^\dagger\, \mathbf{u}^\dagger \rangle\,, \tag{8.13}$$

for any $\delta\mathbf{u}$ and $\mathbf{u}^\dagger$. We then obtain

$$\nabla_m \chi\, \delta\mathbf{m} = \langle \delta\mathbf{u} \cdot (\nabla_u \chi_1^\dagger + \nabla_u \mathbf{L}^\dagger \mathbf{u}^\dagger) \rangle + \langle \mathbf{u}^\dagger \cdot \nabla_m \mathbf{L}\, \delta\mathbf{m} \rangle\,. \tag{8.14}$$

We may now eliminate $\delta\mathbf{u}$ from Eq. (8.14) if we can determine a field $\mathbf{u}^\dagger$ to satisfy

$$\nabla_u \mathbf{L}^\dagger \mathbf{u}^\dagger = -\nabla_u \chi_1^\dagger\,. \tag{8.15}$$

Equation (8.15) is referred to as the *adjoint equation* of (8.1), and $\mathbf{u}^\dagger$ and $-\nabla_u \chi_1^\dagger$ are the *adjoint field* and the *adjoint source*, respectively. When the solution $\mathbf{u}^\dagger$ of the adjoint equation is found, then the derivative of the objective functional reduces to

$$\nabla_m \chi\, \delta\mathbf{m} = \langle \mathbf{u}^\dagger \cdot \nabla_m \mathbf{L}\, \delta\mathbf{m}\rangle\,. \tag{8.16}$$

By construction, $\nabla_m \chi\, \delta\mathbf{m}$ can now be computed for any differentiation direction $\delta\mathbf{m}$ without the explicit knowledge of $\delta\mathbf{u}$. This advantage comes at the price of having to find the adjoint operator $\nabla_u \mathbf{L}^\dagger$ and a solution of the adjoint problem (8.15).

Equation (8.15) can be simplified when the operator $\mathbf{L}$ is linear in $\mathbf{u}$. It then follows that $\mathbf{L}^\dagger$ is also linear and therefore

$$\mathbf{L}^\dagger(\mathbf{u}^\dagger) = -\nabla_u \chi_1^\dagger\,. \tag{8.17}$$

8.2.1 Fréchet Kernels

Much of our physical intuition is based on the interpretation of sensitivity or Fréchet kernels which are defined as the volumetric densities of the Fréchet derivative $\nabla_m \chi$:

$$K_m := \frac{\mathrm{d}}{\mathrm{d}V}\,\nabla_m \chi = \int_T \mathbf{u}^\dagger \cdot \nabla_m \mathbf{L}\,\mathrm{d}t\,. \tag{8.18}$$

Using the notion of sensitivity kernels, we can recast Eq. (8.16) as follows:

$$\nabla_m \chi\, \delta\mathbf{m} = \int_G K_m(\mathbf{x})\, \delta\mathbf{m}(\mathbf{x})\, \mathrm{d}^3\mathbf{x}\,. \tag{8.19}$$

The sensitivity kernels $K_m(\mathbf{x})$ reveal how the objective functional $\chi(\mathbf{m})$ is affected by model parameter changes at position $\mathbf{x}$ in the Earth. It is the study of $K_m(\mathbf{x})$ for different types of seismic waves and different objective functionals that allows us to design efficient inversion schemes and to interpret the results in a physically meaningful way. A collection of sensitivity kernels for different combinations of objective functionals and Earth model parameters is presented in Chap. 12.

8.2.2 Translation to the Discretised Model Space

In most applications, the model space, $\mathfrak{M}$, is discretised, meaning that the components, m_i, of the space-continuous model

$$\mathbf{m}(\mathbf{x}) = [m_1(\mathbf{x}), m_2(\mathbf{x}), m_3(\mathbf{x}), \dots] \in \mathfrak{M} \tag{8.20}$$

are expressed as a linear combination of $N < \infty$ basis functions, $b_j(\mathbf{x})$:

$$m_i(\mathbf{x}) = \sum_{j=1}^{N} \mu_{ij}\, b_j(\mathbf{x})\,. \tag{8.21}$$

Commonly, the basis functions are spherical harmonics, blocks, wavelets or splines. With the representation (8.21), the model, $\mathbf{m}$, and the objective functional, χ, are fully determined by the coefficients or model parameters μ_{ij}. We are therefore interested in the partial derivatives $\partial\chi/\partial\mu_{ij}$. Using the definition of the classical derivative, we find

$$\begin{aligned}\frac{\partial \chi}{\partial \mu_{ij}} &= \lim_{\varepsilon \to 0} \frac{1}{\varepsilon}\left[\chi(\dots, \mu_{ij} + \varepsilon, \dots) - \chi(\dots, \mu_{ij}, \dots)\right] \\ &= \lim_{\varepsilon \to 0} \frac{1}{\varepsilon}\left[\chi(\dots, m_i + \varepsilon b_j, \dots) - \chi(\dots, m_i, \dots)\right] \\ &= \nabla_{m_i}\chi\, b_j = \int_G K_{m_i}(\mathbf{x})\, b_j(\mathbf{x})\, \mathrm{d}^3\mathbf{x}\,.\end{aligned} \tag{8.22}$$

It follows from Eq. (8.22) that the gradient in the classical sense, $\partial\chi/\partial\mu_{ij}$, is given by the projection of the sensitivity kernel K_{m_i} onto the basis function, b_j.

8.2.3 Summary of the Adjoint Method

Owing to the outstanding importance of the adjoint method in the context of full waveform inversion, we briefly summarise the key equations:

Forward problem:

$$\mathbf{L}(\mathbf{u}, \mathbf{m}) = \mathbf{f}\,.$$

Objective functional:

$$\chi(\mathbf{m}) = \int_T \int_G \chi_1[\mathbf{u}(\mathbf{m}; \mathbf{x}, t)]\, \mathrm{d}t\, \mathrm{d}^3\mathbf{x} = \langle \chi_1 \rangle\,.$$

Adjoint equation:

$$\nabla_u \mathbf{L}^\dagger \mathbf{u}^\dagger = -\nabla_u \chi_1^\dagger\,.$$

Derivative of the objective functional:

$$\nabla_m \chi \, \delta\mathbf{m} = \langle \mathbf{u}^\dagger \cdot \nabla_m \mathbf{L} \, \delta\mathbf{m} \rangle = \int_G K_m \, \delta\mathbf{m} \, d^3\mathbf{x} \, .$$

Fréchet kernel:

$$K_m = \int_T \mathbf{u}^\dagger \cdot \nabla_m \mathbf{L} \, \delta\mathbf{m} \, dt \, .$$

8.3 Derivatives with Respect to the Source

Using the method developed so far, we can compute the derivative of the objective functional χ with respect to the model parameters $\mathbf{m}$, given that the adjoint problem can be solved. Equation (8.1) states that the observable $\mathbf{u}$ does depend not only on $\mathbf{m}$, $\mathbf{x}$ and t, but also on the right-hand side $\mathbf{f}$. This is an obvious matter of fact in seismology where the properties of the seismic wave field are strongly influenced by the characteristics of the source. This dependence can formally be expressed as

$$\mathbf{u} = \mathbf{u}(\mathbf{f}, \mathbf{m}; \mathbf{x}, t) \, , \tag{8.23}$$

and one may ask how the augmented objective functional $\chi = \chi(\mathbf{m}, \mathbf{f})$ changes as the right-hand side of (8.1) passes from $\mathbf{f}$ to $\mathbf{f} + \varepsilon\delta\mathbf{f}$. Differentiating χ with respect to $\mathbf{f}$ in the direction $\delta\mathbf{f}$ yields

$$\nabla_f \chi \, \delta\mathbf{f} = \nabla_u \chi \, \delta\mathbf{u} = \langle \nabla_u \chi_1 \, \delta\mathbf{u} \rangle \, , \tag{8.24}$$

with $\delta\mathbf{u}$ now defined by

$$\delta\mathbf{u} = \nabla_f \chi \, \delta\mathbf{f} \, . \tag{8.25}$$

Again, we need to eliminate $\delta\mathbf{u}$ from Eq. (8.24). For this, we define

$$\tilde{\mathbf{L}}(\mathbf{u}, \mathbf{m}, \mathbf{f}) = \mathbf{L}(\mathbf{u}, \mathbf{m}) - \mathbf{f} \, . \tag{8.26}$$

The equation

$$\tilde{\mathbf{L}}(\mathbf{u}, \mathbf{m}, \mathbf{f}) = \mathbf{0} \tag{8.27}$$

is now fully equivalent to $\mathbf{L}(\mathbf{u}, \mathbf{m}) = \mathbf{f}$. The derivative of $\tilde{\mathbf{L}}$ with respect to $\mathbf{f}$ in the direction $\delta\mathbf{f}$ is given by

$$\nabla_f \tilde{\mathbf{L}}\, \delta\mathbf{f} + \nabla_u \tilde{\mathbf{L}}\, \delta\mathbf{u} = \mathbf{0}\,. \tag{8.28}$$

Following the recipe of the previous section, we multiply (8.28) by an arbitrary test function $\mathbf{u}^\dagger$ and apply the integral $\langle\,.\,\rangle$:

$$\langle \mathbf{u}^\dagger \cdot \nabla_f \tilde{\mathbf{L}}\, \delta\mathbf{f} \rangle + \langle \mathbf{u}^\dagger \cdot \nabla_u \tilde{\mathbf{L}}\, \delta\mathbf{u} \rangle = 0\,. \tag{8.29}$$

Adding Eq. (8.29) to Eq. (8.24) then yields

$$\begin{aligned} \nabla_f \chi\, \delta\mathbf{f} &= \langle \nabla_u \chi_1\, \delta\mathbf{u} \rangle + \langle \mathbf{u}^\dagger \cdot \nabla_f \tilde{\mathbf{L}}\, \delta\mathbf{f} \rangle + \langle \mathbf{u}^\dagger \cdot \nabla_u \tilde{\mathbf{L}}\, \delta\mathbf{u} \rangle \\ &= \langle \nabla_u \chi_1\, \delta\mathbf{u} \rangle - \langle \mathbf{u}^\dagger \cdot \delta\mathbf{f} \rangle + \langle \mathbf{u}^\dagger \cdot \nabla_u \mathbf{L}\, \delta\mathbf{u} \rangle\,. \end{aligned} \tag{8.30}$$

Assuming that the adjoints $\nabla_u \chi_1^\dagger$ and $\nabla_u \mathbf{L}^\dagger$ exist allows us to rearrange (8.30):

$$\nabla_f \chi\, \delta\mathbf{f} = \langle \delta\mathbf{u} \cdot (\nabla_u \chi_1^\dagger + \nabla_u \mathbf{L}^\dagger \mathbf{u}^\dagger) \rangle - \langle \mathbf{u}^\dagger \cdot \delta\mathbf{f} \rangle\,. \tag{8.31}$$

Again, we can eliminate $\delta\mathbf{u}$ when a field $\mathbf{u}^\dagger$ can be found such that the adjoint equation

$$\nabla_u \mathbf{L}^\dagger \mathbf{u}^\dagger = -\nabla_u \chi_1^\dagger \tag{8.32}$$

is satisfied. The derivative $\nabla_f \chi\, \delta\mathbf{f}$ then simplifies to

$$\nabla_f \chi\, \delta\mathbf{f} = -\langle \mathbf{u}^\dagger \cdot \delta\mathbf{f} \rangle\,. \tag{8.33}$$

The adjoint equation for derivatives with respect to the right-hand side $\mathbf{f}$ (Eq. 8.32) is the same as for derivatives with respect to the model parameters $\mathbf{m}$ (Eq. 8.15). We can thus compute any derivative from one adjoint field: $\mathbf{u}^\dagger$.

In Sect. 9.2 we present a selection of first derivatives with respect to source parameters, including source location, initiation time and moment tensor components. Moreover, we demonstrate the close relation between time-reverse imaging and the adjoint method.

8.4 Second Derivatives

Following the description of the adjoint method for the first derivative, it is natural to seek an extension to second derivatives. For this we consider the objective functional differentiated with respect to $\mathbf{m}$ in a differentiation direction $\delta\mathbf{m}_1$, that is $\nabla_m \chi\, \delta\mathbf{m}_1$. Differentiating $\nabla_m \chi\, \delta\mathbf{m}_1$ once more but in a direction $\delta\mathbf{m}_2$ yields

$$\nabla_m (\nabla_m \chi\, \delta\mathbf{m}_1)\, \delta\mathbf{m}_2 = \nabla_m \nabla_m \chi\, (\delta\mathbf{m}_1, \delta\mathbf{m}_2)\,. \tag{8.34}$$

The second derivative $H_\chi := \nabla_m \nabla_m \chi$, also referred to as the *Hessian* of χ, is linear in both its arguments, which are the differentiation directions $\delta\mathbf{m}_1$ and $\delta\mathbf{m}_2$.

Our interest in the Hessian is largely motivated by its fundamental role in *resolution analysis*, the *Newton method* of non-linear optimisation (see Sect. 7.3.3) and the computation of *optimal step lengths* in the general descent method (see Sect. 7.3.1). Before delving into the extension of the adjoint method to the computation of second derivatives, we briefly summarise some of the applications where knowledge of the Hessian is crucial.

8.4.1 Motivation: The Role of Second Derivatives in Optimisation and Resolution Analysis

8.4.1.1 Relation Between Hessian and Posterior Covariance

The physical meaning of the Hessian becomes most apparent in the context of probabilistic inverse problems (e.g. Tarantola, 2005), where the probability of the correctness of a model, $\mathbf{m}$, is expressed in terms of a probability density, $\sigma(\mathbf{m})$. In the specific case of a linear forward problem and Gaussian distributions describing prior knowledge and measurement errors, $\sigma(\mathbf{m})$ takes the form

$$\sigma(\mathbf{m}) = \text{const.}\, \mathrm{e}^{-\chi_g(\mathbf{m})}\,, \tag{8.35}$$

with the misfit functional

$$\chi_g(\mathbf{m}) = \frac{1}{2}\, \tilde{C}_M^{-1}\,(\mathbf{m} - \tilde{\mathbf{m}}, \mathbf{m} - \tilde{\mathbf{m}})\,. \tag{8.36}$$

The operator $\tilde{C}_\mathrm{M}$ is the *posterior covariance operator*, and $\tilde{\mathbf{m}}$ is the maximum-likelihood model. All information on the resolution of and trade-offs between model parameters is contained in $\tilde{C}_\mathrm{M}$.

To establish a link between the posterior covariance operator and the Hessian, H_χ, of a functional, χ, we approximate χ quadratically around the optimal model $\tilde{\mathbf{m}}$:

$$\chi(\mathbf{m}) \approx \chi(\tilde{\mathbf{m}}) + \frac{1}{2}\, H_\chi(\tilde{\mathbf{m}})\,(\mathbf{m} - \tilde{\mathbf{m}}, \mathbf{m} - \tilde{\mathbf{m}})\,. \tag{8.37}$$

In Eq. (8.37) we used the fact that the first derivative of χ evaluated at the optimal model $\tilde{\mathbf{m}}$ is equal to zero, i.e. $\nabla_m \chi(\tilde{\mathbf{m}}) = \mathbf{0}$ (see Chap. 7). The comparison of Eqs. (8.35), (8.36) and (8.37) suggests the interpretation of the Hessian, H_χ, in terms of the inverse posterior covariance, $\tilde{C}_\mathrm{M}^{-1}$. Accepting a Gaussian model, we can construct a probability density

$$\sigma(\mathbf{m}) = \text{const.}\, \mathrm{e}^{-\frac{1}{2} H_\chi(\tilde{\mathbf{m}})\,(\mathbf{m}-\tilde{\mathbf{m}}, \mathbf{m}-\tilde{\mathbf{m}})}\,, \tag{8.38}$$

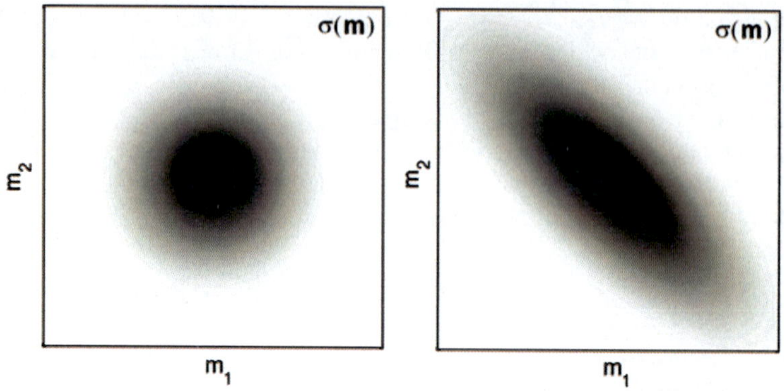

Fig. 8.1 *Left*: Probability density $\sigma(\mathbf{m}) = \sigma(m_1, m_2)$ for a Hessian with zero diagonal elements. The model parameters are independent. *Right*: The same as on the *left* but for a Hessian with large non-zero off-diagonal elements. The model parameters cannot be constrained independently

which can be used to quantify resolution and trade-offs, at least locally in the vicinity of the optimum $\tilde{\mathbf{m}}$. Large diagonal elements of the Hessian indicate that the corresponding model parameters are well resolved, meaning that the misfit χ increases substantially when one individual parameter deviates from its optimum value. In contrast, small diagonal elements suggest low resolution of individual parameters. The off-diagonal elements of the Hessian quantify the extent to which two different model parameters trade off. When off-diagonal elements are large the corresponding model parameters can hardly be constrained independently.

This concept is schematically illustrated in Fig. 8.1 for the case of a two-dimensional model space. The figure on the left shows a probability density, $\sigma(\mathbf{m}) = \sigma(m_1, m_2)$, that corresponds to a Hessian with zero off-diagonal elements. The probability densities of the individual parameters, m_1 and m_2, are independent of each other: They are always Gaussian with identical mean and variance. The probability density on the right-hand side of Fig. 8.1 is for a Hessian with large non-zero off-diagonal elements. The probability for m_1 now depends on m_2 and vice versa, meaning that the two model parameters cannot be constrained independently.

A proper covariance analysis requires knowledge of the posterior covariance operator which is the inverse Hessian. Inverting H_χ will, however, be computationally too expensive for many problems, where the model space is large and the forward problem costly. Still, the probability density defined in Eq. (8.38) provides invaluable information on resolution that goes beyond classical chequerboard tests.

8.4.1.2 Extremal Bounds Analysis

In contrast to probabilistic inversion, *extremal bounds analysis* (e.g. Meju, 2009) addresses resolution in a deterministic sense. Assume that $\tilde{q}$ is the misfit associated with the optimal model, $\tilde{\mathbf{m}}$, that is $\chi(\tilde{\mathbf{m}}) = \tilde{q}$. Our goal is to find an extremal model, $\mathbf{m}_{\text{extr}}$, such that

$$\mathbf{b}\,\mathbf{m}_{\text{extr}} = \text{extr}\,!\,, \tag{8.39}$$

where the projection operator, $\mathbf{b}$, can be used, for instance, to isolate specific model parameters. We constrain the problem by the requirement that the misfit corresponding to $\mathbf{m}_{\text{extr}}$ is equal to a pre-defined value q: $\chi(\mathbf{m}_{\text{extr}}) = q$. The difference $q - \tilde{q}$ is usually chosen to be the error in χ, that is due to inaccurate measurements, noise or an incomplete theory. The Langrange function corresponding to the constrained requirement (8.39) is

$$\mathcal{L} = \mathbf{b}\,\mathbf{m} + \frac{1}{\lambda}\,[\chi(\mathbf{m}) - q]\,, \tag{8.40}$$

with the Lagrange multiplier, λ. Inserting the quadratic approximation (8.37) into (8.40), differentiating and setting the resulting expression to zero yields

$$\mathbf{0} = \mathbf{b} + \frac{1}{\lambda}\,H_\chi(\mathbf{m}_{\text{extr}} - \tilde{\mathbf{m}})\,. \tag{8.41}$$

Rearranging Eq. (8.41) gives

$$\mathbf{m}_{\text{extr}} = \tilde{\mathbf{m}} - \lambda\,H_\chi^{-1}\mathbf{b}\,, \tag{8.42}$$

where the Lagrange multiplier is found to be

$$\lambda = \pm\sqrt{\frac{2\,(q - \tilde{q})}{\mathbf{b}\,H_\chi^{-1}\,\mathbf{b}}}\,. \tag{8.43}$$

Extremal bounds analysis provides quantitative estimates of the maximum possible model perturbations that are still consistent with the data. Its application requires knowledge of the inverse Hessian, H_χ^{-1}, applied to the projection operator, $\mathbf{b}$. This quantity can be computed iteratively using, for instance, conjugate-gradient type algorithms. The need for the inverse Hessian applied to a vector reappears in the Newton method for non-linear minimisation.

8.4.1.3 Optimal Step Length in Gradient-Based Minimisation

Gradient-based minimisation algorithms iteratively minimise the misfit functional χ by updating the current model $\mathbf{m}_i$ with $\gamma_i\,\mathbf{h}_i$, where $\mathbf{h}_i$ is a descent direction computed from $\nabla_m \chi(\mathbf{m}_i)$, and $\gamma_i > 0$ is the step length (see Sect. 7.3.1):

$$\mathbf{m}_{i+1} := \mathbf{m}_i + \gamma_i\,\mathbf{h}_i\,, \quad \chi(\mathbf{m}_{i+1}) < \chi(\mathbf{m}_i)\,. \tag{8.44}$$

The step length γ_i that reduces χ as much as possible within one iteration can be computed with the help of H_χ:

$$\gamma_i = -\frac{\nabla_m \chi(\mathbf{m}_i)\,\mathbf{h}_i}{H_\chi(\mathbf{m}_i)(\mathbf{h}_i, \mathbf{h}_i)} . \tag{8.45}$$

The Hessian therefore allows us to determine efficient step lengths in iterative minimisation schemes.

8.4.1.4 The Newton Method

The Newton method (see Sect. 7.3.3) is a special gradient-based algorithm that iteratively minimises the objective functional χ by updating the current iterate $\mathbf{m}_i$ to the next iterate $\mathbf{m}_{i+1}$ by adding a search direction $\mathbf{h}_i$,

$$\mathbf{m}_{i+1} = \mathbf{m}_i + \mathbf{h}_i \,, \tag{8.46}$$

where $\mathbf{h}_i$ is determined by the *Newton equation*:

$$H_\chi\,\mathbf{h}_i = -\nabla_m \chi(\mathbf{m}_i) . \tag{8.47}$$

Under favourable conditions, the Newton method can be shown to converge quadratically, that is, faster than other optimisation schemes. The size of the model space usually requires that Eq. (8.47) be solved iteratively as well, using, for instance, the *conjugate-gradient method* described in Sect. 7.3.4. The iterative solution of (8.47) for each model iterate $\mathbf{m}_i$ defines an *inner loop* within the optimisation of the objective functional χ, which is itself referred to as the *outer loop*.

Methods of the conjugate-gradient type for the solution of the Newton equation involve the Hessian, H_χ, applied to a model $\delta\mathbf{m}_1$, i.e., $H_\chi\,\delta\mathbf{m}_1$. This means that the explicit Hessian does not need to be computed. This reduces the computational costs of the Newton method to a feasible level. Despite being feasible, the Newton method has so far not enjoyed much popularity in seismic waveform inversion. One application to the 1D wave equation can be found in Santosa & Symes (1988).

8.4.2 Extension of the Adjoint Method to Second Derivatives

To compute $H_\chi \delta\mathbf{m}_1$ we adopt the following strategy: First, we derive an equation that gives $H_\chi(\delta\mathbf{m}_1, \delta\mathbf{m}_2)$ in terms of the adjoint field that we already introduced in Sect. 8.2. Then, in order to isolate $\delta\mathbf{m}_2$, we define a secondary adjoint field that is the solution of the adjoint equation with a right-hand side determined by $\delta\mathbf{m}_1$.

We start with the explicit computation of $H_\chi(\delta\mathbf{m}_1, \delta\mathbf{m}_2)$. For this we first differentiate χ with respect to $\mathbf{m}$ in the direction $\delta\mathbf{m}_1$,

$$\nabla_m \chi\,\delta\mathbf{m}_1 = \langle \nabla_u \chi_1\,\delta_1\mathbf{u}\rangle \,, \tag{8.48}$$

with the derivative of the forward field $\delta_1\mathbf{u} := \nabla_m\mathbf{u}\,\delta\mathbf{m}_1$. Repeating this procedure for a second direction, $\delta\mathbf{m}_2$, yields

$$H_\chi(\delta \mathbf{m}_1, \delta \mathbf{m}_2) = \nabla_m \nabla_m \chi(\delta \mathbf{m}_1, \delta \mathbf{m}_2) = \langle \nabla_u \nabla_u \chi_1(\delta_1 \mathbf{u}, \delta_2 \mathbf{u}) + \nabla_u \chi_1 \, \delta_{12} \mathbf{u} \rangle \,, \tag{8.49}$$

with the first derivative $\delta_2 \mathbf{u} := \nabla_m \mathbf{u} \, \delta \mathbf{m}_2$ and the Hessian of the forward wave field $\delta_{12} \mathbf{u} := \nabla_m \nabla_m \mathbf{u}(\delta \mathbf{m}_1, \delta \mathbf{m}_2)$. The first term on the right-hand side of Eq. (8.49), $\langle \nabla_u \nabla_u \chi_1(\delta_1 \mathbf{u}, \delta_2 \mathbf{u}) \rangle$, is the *approximate Hessian* of the objective functional applied to $\delta \mathbf{m}_1$ and $\delta \mathbf{m}_2$. It merely requires first derivatives that we can already compute efficiently with the help of the standard adjoint method that we described in Sect. 8.2. The approximate Hessian is used as a computationally less expensive substitute of the *full Hessian*, $\nabla_m \nabla_m \chi$, in the Gauss–Newton and the Levenberg–Marquardt methods of non-linear optimisation (e.g. Pratt et al., 1998; Sect. 7.3.3).

Our goal is to eliminate the second derivative $\delta_{12} \mathbf{u}$ from Eq. (8.49) because its computation for all possible directions would require us to solve the adjoint problem N_m times, where N_m is the dimension of the (discretised) model space.
We proceed as in the standard adjoint method for first derivatives and differentiate the forward problem, $\mathbf{L}(\mathbf{u}, \mathbf{m}) = \mathbf{f}$, with respect to $\mathbf{m}$:

$$\mathbf{0} = \nabla_m \mathbf{L} \, \delta \mathbf{m}_1 + \nabla_u \mathbf{L} \, \delta_1 \mathbf{u} \,. \tag{8.50}$$

Differentiating (8.50) once more with respect to $\mathbf{m}$ but in the direction $\delta \mathbf{m}_2$ yields

$$\begin{aligned} \mathbf{0} = {} & \nabla_m \nabla_m \mathbf{L}(\delta \mathbf{m}_1, \delta \mathbf{m}_2) + \nabla_u \nabla_m \mathbf{L}(\delta \mathbf{m}_1, \delta_2 \mathbf{u}) \\ & + \nabla_m \nabla_u \mathbf{L}(\delta_1 \mathbf{u}, \delta \mathbf{m}_2) + \nabla_u \nabla_u \mathbf{L}(\delta_1 \mathbf{u}, \delta_2 \mathbf{u}) + \nabla_u \mathbf{L} \delta_{12} \mathbf{u} \,. \end{aligned} \tag{8.51}$$

In the next step we multiply Eq. (8.51) with an arbitrary test function $\mathbf{u}_1^\dagger$ and integrate over time and space:

$$\begin{aligned} 0 = {} & \langle \mathbf{u}_1^\dagger \cdot \nabla_m \nabla_m \mathbf{L}(\delta \mathbf{m}_1, \delta \mathbf{m}_2) \rangle + \langle \mathbf{u}_1^\dagger \cdot \nabla_u \nabla_m \mathbf{L}(\delta \mathbf{m}_1, \delta_2 \mathbf{u}) \rangle \\ & + \langle \mathbf{u}_1^\dagger \cdot \nabla_m \nabla_u \mathbf{L}(\delta_1 \mathbf{u}, \delta \mathbf{m}_2) \rangle + \langle \mathbf{u}_1^\dagger \cdot \nabla_u \nabla_u \mathbf{L}(\delta_1 \mathbf{u}, \delta_2 \mathbf{u}) \rangle + \langle \mathbf{u}_1^\dagger \cdot \nabla_u \mathbf{L} \delta_{12} \mathbf{u} \rangle \,. \end{aligned} \tag{8.52}$$

Adding (8.52) to (8.49) and rearranging terms gives

$$\begin{aligned} H_\chi(\delta \mathbf{m}_1, \delta \mathbf{m}_2) = {} & \langle (\nabla_u \chi_1 + \mathbf{u}_1^\dagger \cdot \nabla_u \mathbf{L}) \, \delta_{12} \mathbf{u} \rangle + \langle \nabla_u \nabla_u \chi_1(\delta_1 \mathbf{u}, \delta_2 \mathbf{u}) \rangle \\ & + \langle \mathbf{u}_1^\dagger \cdot \nabla_m \nabla_m \mathbf{L}(\delta \mathbf{m}_1, \delta \mathbf{m}_2) \rangle + \langle \mathbf{u}_1^\dagger \cdot \nabla_u \nabla_m \mathbf{L}(\delta \mathbf{m}_1, \delta_2 \mathbf{u}) \rangle \\ & + \langle \mathbf{u}_1^\dagger \cdot \nabla_m \nabla_u \mathbf{L}(\delta_1 \mathbf{u}, \delta \mathbf{m}_2) \rangle + \langle \mathbf{u}_1^\dagger \cdot \nabla_u \nabla_u \mathbf{L}(\delta_1 \mathbf{u}, \delta_2 \mathbf{u}) \rangle \,. \end{aligned} \tag{8.53}$$

We can now eliminate the second derivative of the forward field, $\delta_{12} \mathbf{u}$, from Eq. (8.53) by imposing that the test field $\mathbf{u}_1^\dagger$ be the solution of the adjoint equation

$$\nabla_u \mathbf{L}^\dagger \mathbf{u}_1^\dagger = -\nabla_u \chi_1^\dagger \,. \tag{8.54}$$

The adjoint equation (8.54) is identical to the adjoint equation for first derivatives (8.15), which means that the adjoint field, $\mathbf{u}_1^\dagger$, can in practice be recycled for the computation of the Hessian. We will henceforth refer to $\mathbf{u}_1^\dagger$ as the *primary adjoint field* and to (8.54) as the *primary adjoint equation*.

When the primary adjoint field, $\mathbf{u}_1^\dagger$, satisfies the primary adjoint equation (8.54), then $H_\chi(\delta\mathbf{m}_1, \delta\mathbf{m}_2)$ can indeed be expressed in terms of *first* derivatives with respect to $\mathbf{m}$:

$$\begin{aligned} H_\chi(\delta\mathbf{m}_1, \delta\mathbf{m}_2) = &\langle \nabla_u \nabla_u \chi_1(\delta_1\mathbf{u}, \delta_2\mathbf{u})\rangle + \langle \mathbf{u}_1^\dagger \cdot \nabla_m \nabla_m \mathbf{L}(\delta\mathbf{m}_1, \delta\mathbf{m}_2)\rangle \\ &+ \langle \mathbf{u}_1^\dagger \cdot \nabla_u \nabla_m \mathbf{L}(\delta\mathbf{m}_1, \delta_2\mathbf{u})\rangle + \langle \mathbf{u}_1^\dagger \cdot \nabla_m \nabla_u \mathbf{L}(\delta_1\mathbf{u}, \delta\mathbf{m}_2)\rangle \\ &+ \langle \mathbf{u}_1^\dagger \cdot \nabla_u \nabla_u \mathbf{L}(\delta_1\mathbf{u}, \delta_2\mathbf{u})\rangle . \end{aligned} \tag{8.55}$$

The last term on the right-hand side of Eq. (8.55), involving $\nabla_u\nabla_u\mathbf{L}$, is zero for linear operators, including the wave equation operator. Whether the second term, involving $\nabla_m\nabla_m\mathbf{L}$, is zero or not depends on the specific parameterisation of the model space. The wave equation operator, for instance, is linear in density and the elastic parameters, which lead to a zero second derivative, i.e. $\nabla_m\nabla_m\mathbf{L} = \mathbf{0}$. However, when the model is parameterised in terms of density and seismic wave speeds, quadratic terms appear, and $\nabla_m\nabla_m\mathbf{L}$ is generally non-zero.

To apply conjugate-gradient-type methods to the solution of the Newton equation (8.47) we require the Hessian applied to a model perturbation, that is $H_\chi\,\delta\mathbf{m}_1$ and not $H_\chi(\delta\mathbf{m}_1, \delta\mathbf{m}_2)$. Unfortunately, $\delta\mathbf{m}_2$ appears implicity via $\delta_2\mathbf{u}$ in Eq. (8.55). The next step is therefore the isolation of $\delta\mathbf{m}_2$. For clarity we restrict the following development to operators $\mathbf{L}(\mathbf{u})$ that are linear in $\mathbf{u}$, and we omit the dependence of $\mathbf{L}$ on $\mathbf{m}$ in the notation. This gives

$$\nabla_u\nabla_u\mathbf{L}(\mathbf{u})(\delta_1\mathbf{u}, \delta_2\mathbf{u}) = \mathbf{0}\,, \tag{8.56}$$

$$\nabla_u\nabla_m\mathbf{L}(\mathbf{u})(\delta\mathbf{m}_1, \delta_2\mathbf{u}) = \nabla_m\mathbf{L}(\delta_2\mathbf{u})\,\delta\mathbf{m}_1\,, \tag{8.57}$$

$$\nabla_m\nabla_u\mathbf{L}(\mathbf{u})(\delta_1\mathbf{u}, \delta\mathbf{m}_2) = \nabla_m\mathbf{L}(\delta_1\mathbf{u})\,\delta\mathbf{m}_2\,. \tag{8.58}$$

After slight rearrangements we can now write Eq. (8.55) in the following form:

$$\begin{aligned} H_\chi(\delta\mathbf{m}_1, \delta\mathbf{m}_2) = &\langle \nabla_u\nabla_u\chi_1(\delta_1\mathbf{u}, \delta_2\mathbf{u})\rangle + \langle \mathbf{u}_1^\dagger \cdot \nabla_m\mathbf{L}(\delta_2\mathbf{u})\,\delta\mathbf{m}_1\rangle \\ &+ \langle \mathbf{u}_1^\dagger \cdot \nabla_m\nabla_m\mathbf{L}(\mathbf{u})(\delta\mathbf{m}_1, \delta\mathbf{m}_2)\rangle + \langle \mathbf{u}_1^\dagger \cdot \nabla_m\mathbf{L}(\delta_1\mathbf{u})\,\delta\mathbf{m}_2\rangle . \end{aligned} \tag{8.59}$$

Our focus is now on the first two terms on the right-hand side of Eq. (8.59) where the direction $\delta\mathbf{m}_2$ appears implicitly via $\delta_2\mathbf{u} = \nabla_m\mathbf{u}\,\delta\mathbf{m}_2$. Invoking the adjoint $\nabla_m\mathbf{L}^\dagger$ of $\nabla_m\mathbf{L}$ we can write

$$\begin{aligned} &\langle \nabla_u\nabla_u\chi_1(\delta_1\mathbf{u}, \delta_2\mathbf{u})\rangle + \langle \mathbf{u}_1^\dagger \cdot \nabla_m\mathbf{L}(\delta_2\mathbf{u})\,\delta\mathbf{m}_1\rangle \\ &= \langle \delta_2\mathbf{u} \cdot \nabla_u\nabla_u\chi_1^\dagger(\delta_1\mathbf{u})\rangle + \langle \delta_2\mathbf{u} \cdot \nabla_m\mathbf{L}^\dagger(\mathbf{u}_1^\dagger)\,\delta\mathbf{m}_1\rangle \\ &= \langle \delta_2\mathbf{u} \cdot [\nabla_u\nabla_u\chi_1^\dagger(\delta_1\mathbf{u}) + \nabla_m\mathbf{L}^\dagger(\mathbf{u}_1^\dagger)\,\delta\mathbf{m}_1]\rangle . \end{aligned} \tag{8.60}$$

In Eq. (8.60) we have recognised that $\nabla_u\nabla_u\chi_1(\delta_1\mathbf{u}, \delta_2\mathbf{u})$ can be interpreted as a linear operator, $\nabla_u\nabla_u\chi_1(\delta_1\mathbf{u})$, acting on $\delta_2\mathbf{u}$, i.e. $\nabla_u\nabla_u\chi_1(\delta_1\mathbf{u}, \delta_2\mathbf{u}) =$

$\nabla_u \nabla_u \chi_1(\delta_1 \mathbf{u})\, \delta_2 \mathbf{u}$. At this point of the development we define the *secondary adjoint field*, $\mathbf{u}_2^\dagger$, as the solution of the *secondary adjoint equation*

$$\mathbf{L}^\dagger(\mathbf{u}_2^\dagger) := -[\nabla_u \nabla_u \chi_1^\dagger(\delta_1 \mathbf{u}) + \nabla_m \mathbf{L}^\dagger(\mathbf{u}_1^\dagger)\, \delta\mathbf{m}_1]\,. \tag{8.61}$$

Substituting (8.61) into (8.60) then yields

$$\begin{aligned}\langle \nabla_u \nabla_u \chi_1(\delta_1 \mathbf{u}, \delta_2 \mathbf{u})\rangle &+ \langle \mathbf{u}_1^\dagger \cdot \nabla_m \mathbf{L}(\delta_2 \mathbf{u})\, \delta\mathbf{m}_1\rangle \\ &= -\langle \delta_2 \mathbf{u} \cdot \mathbf{L}^\dagger(\mathbf{u}_2^\dagger)\rangle = -\langle \mathbf{u}_2^\dagger \cdot \mathbf{L}(\delta_2 \mathbf{u})\rangle\,.\end{aligned} \tag{8.62}$$

To replace $\mathbf{L}(\delta_2 \mathbf{u})$ we note that the differentiation of the forward problem, $\mathbf{L}(\mathbf{u}, \mathbf{m}) = \mathbf{f}$, with respect to $\mathbf{m}$ in the direction $\delta\mathbf{m}_2$ gives

$$\nabla_m \mathbf{L}(\mathbf{u})\, \delta\mathbf{m}_2 + \mathbf{L}(\delta_2 \mathbf{u}) = \mathbf{0}\,. \tag{8.63}$$

Inserting (8.63) into (8.62) leaves us with the following expression:

$$\langle \nabla_u \nabla_u \chi_1(\delta_1 \mathbf{u}, \delta_2 \mathbf{u})\rangle + \langle \mathbf{u}_1^\dagger \cdot \nabla_m \mathbf{L}(\delta_2 \mathbf{u})\, \delta\mathbf{m}_1\rangle = \langle \mathbf{u}_2^\dagger \cdot \nabla_m \mathbf{L}(\mathbf{u})\, \delta\mathbf{m}_2\rangle\,, \tag{8.64}$$

where the second differentiation direction, $\delta\mathbf{m}_2$, appears explicitly. We can now assemble $H_\chi\, \delta\mathbf{m}_1$:

$$\begin{aligned}H_\chi \delta\mathbf{m}_1 = H_\chi(\delta\mathbf{m}_1, \circ) =& \langle \mathbf{u}_2^\dagger \cdot \nabla_m \mathbf{L}(\mathbf{u})\, \circ\rangle \\ &+ \langle \mathbf{u}_1^\dagger \cdot \nabla_m \mathbf{L}(\delta_1 \mathbf{u})\, \circ\rangle \\ &+ \langle \mathbf{u}_1^\dagger \cdot \nabla_m \nabla_m \mathbf{L}(\mathbf{u})(\delta\mathbf{m}_1)\, \circ\rangle\,.\end{aligned} \tag{8.65}$$

The symbol $\circ$ represents a place holder for a second differentiation direction $\delta\mathbf{m}_2$.

8.4.2.1 Hessian Kernels

Equation (8.65) suggests a representation of $H_\chi\,(\delta\mathbf{m}_1, \delta\mathbf{m}_2)$ in terms of volumetric integral kernels that are reminiscent of the Fréchet kernels introduced in Eqs. (8.18) and (8.19):

$$H_\chi\,(\delta\mathbf{m}_1, \delta\mathbf{m}_2) = \int_G \left(K^a_{\delta m_1} + K^b_{\delta m_1} + K^c_{\delta m_1}\right) \delta\mathbf{m}_2\, \mathrm{d}^3\mathbf{x}\,, \tag{8.66}$$

with the *Hessian kernels* defined by

$$K^a_{\delta m_1} := \int_T \mathbf{u}_2^\dagger \cdot \nabla_m \mathbf{L}(\mathbf{u})\, \mathrm{d}t\,, \tag{8.67}$$

$$K^b_{\delta m_1} := \int_T \mathbf{u}_1^\dagger \cdot \nabla_m \mathbf{L}(\delta_1 \mathbf{u})\, \mathrm{d}t\,, \tag{8.68}$$

$$K^c_{\delta m_1} := \int_T \mathbf{u}_1^\dagger \cdot \nabla_m \nabla_m \mathbf{L}(\mathbf{u})(\delta\mathbf{m}_1)\, \mathrm{d}t\,. \tag{8.69}$$

Unlike Fréchet kernels, the Hessian kernels depend on the model perturbation, $\delta\mathbf{m}_1$, either explicitly as in $K^c_{\delta m_1}$ or implicity via $\delta_1\mathbf{u}$ and $\mathbf{u}_2^\dagger$. Section 9.3 provides physical interpretations of the Hessian kernels and explicit formulas for a selection of seismologically relevant Earth model parameters.

8.4.2.2 Translation to the Discretised Model Space

In a discretised model space the components $m_i(\mathbf{x})$ of the space-continuous model $\mathbf{m}(\mathbf{x})$ are expressed in terms of a linear combination of a finite number of basis functions, $b_j(\mathbf{x})$, as already stated in Eqs. (8.20) and (8.21).

With the help of the Hessian operator, H_χ, we can then compute the components of the Hessian matrix, $\mathbf{H}_\chi$, that we used extensively in the chapter on non-linear minimisation (Chap. 7). The component $(\mathbf{H}_\chi)_{ij,kl}$ of the Hessian matrix $\mathbf{H}_\chi$ is the second derivative of χ with respect to the discrete model parameters, μ_{ij} and μ_{kl}, that is

$$(\mathbf{H}_\chi)_{ij,kl} = \frac{\partial^2 \chi}{\partial \mu_{ij}\, \partial \mu_{kl}} \,. \tag{8.70}$$

Making use of Eq. (8.22) we find

$$(\mathbf{H}_\chi)_{ij,kl} = \frac{\partial}{\partial \mu_{kl}} \left(\frac{\partial \chi}{\partial \mu_{ij}} \right) = \frac{\partial}{\partial \mu_{kl}} (\nabla_{m_i} \chi\, b_j) = \nabla_{m_k} \nabla_{m_i} \chi\, (b_j, b_l)\,. \tag{8.71}$$

The component $(\mathbf{H}_\chi)_{ij,kl}$ of the Hessian matrix is equal to the Hessian operator $\nabla_{m_k} \nabla_{m_i} \chi$ applied to the basis functions b_j and b_l.

8.4.2.3 Summary of the Adjoint Method for Second Derivatives

We now summarise the most relevant equations for the computation of $H_\chi \delta\mathbf{m}_1$:

Forward problem:

$$\mathbf{L}(\mathbf{u}) = \mathbf{f}\,.$$

Primary adjoint equation:

$$\mathbf{L}^\dagger(\mathbf{u}_1^\dagger) = -\nabla_u \chi_1^\dagger\,.$$

Scattered wave field:

$$\delta_1\mathbf{u} = \nabla_m \mathbf{u}\, \delta\mathbf{m}_1\,.$$

Secondary adjoint equation:

$$\mathbf{L}^\dagger(\mathbf{u}_2^\dagger) = -\nabla_u \nabla_u \chi_1^\dagger \, \delta_1 \mathbf{u} - \nabla_m \mathbf{L}^\dagger(\mathbf{u}_1^\dagger)\, \delta\mathbf{m}_1 \,,$$

Hessian applied to model perturbation:

$$H_\chi \, \delta\mathbf{m}_1 = \langle \mathbf{u}_2^\dagger \cdot \nabla_m \mathbf{L}(\mathbf{u}) \circ \rangle + \langle \mathbf{u}_1^\dagger \cdot \nabla_m \mathbf{L}(\delta_1 \mathbf{u}) \circ \rangle + \langle \mathbf{u}_1^\dagger \cdot \nabla_m \nabla_m \mathbf{L}(\mathbf{u})(\delta\mathbf{m}_1) \circ \rangle \,.$$

Equation (8.65) is of general validity for operators that are linear in $\mathbf{u}$. One may attach intuitive physical meaning to each of the summands that compose H_χ. This will be the topic of Sect. 9.3.

8.5 Application to the Elastic Wave Equation

8.5.1 Derivation of the Adjoint Equations

The previous development of the adjoint method provides a simple and very general recipe that can be applied to a large class of objective functionals, χ, and physical theories, $\mathbf{L}(\mathbf{u}, \mathbf{m}) = \mathbf{f}$. In the context of seismic tomography we are interested in the special case where $\mathbf{L}$ is the elastic wave equation and where $\mathbf{m}$ represents the structural properties of the Earth.

In our development we closely follow Tarantola (1988) and Fichtner et al. (2006): The elastic wave equation, as introduced in Sect. 2.2, is given by

$$\rho(\mathbf{x})\ddot{\mathbf{u}}(\mathbf{x}, t) - \nabla \cdot \boldsymbol{\sigma}(\mathbf{x}, t) = \mathbf{f}(\mathbf{x}, t) \,, \quad \mathbf{x} \in G \subset \mathbb{R}^3 \,, \quad t \in [t_0, \infty) \subset \mathbb{R} \,. \tag{8.72}$$

Equation (8.72) relates the elastic displacement field $\mathbf{u}$ in the Earth $G \subset \mathbb{R}^3$ to the mass density ρ, the stress tensor $\boldsymbol{\sigma}$ and an external force density $\mathbf{f}$. Under the assumption of a linear visco-elastic rheology, the stress tensor $\boldsymbol{\sigma}$ is related to the displacement gradient $\nabla\mathbf{u}$ via the constitutive relation

$$\boldsymbol{\sigma}(\mathbf{x}, t) = \int_{\tau=t_0}^{\infty} \dot{\mathbf{C}}(\mathbf{x}, t-\tau) : \nabla\mathbf{u}(\mathbf{x}, \tau)\, d\tau. \tag{8.73}$$

Inserting Eq. (8.73) into Eq. (8.72) allows us to express the wave equation in terms of an operator $\mathbf{L}$:

$$\mathbf{L}(\mathbf{u}, \rho, \mathbf{C}) = \mathbf{f} \,, \tag{8.74}$$

$$\mathbf{L}(\mathbf{u}, \rho, \mathbf{C}) = \rho(\mathbf{x})\, \ddot{\mathbf{u}}(\mathbf{x}, t) - \nabla \cdot \int_{\tau=t_0}^{t} \dot{\mathbf{C}}(\mathbf{x}, t-\tau) : \nabla\mathbf{u}(\mathbf{x}, \tau)\, d\tau \,. \tag{8.75}$$

The elastic wave operator, **L**, is accompanied by the initial and boundary conditions

$$\mathbf{u}|_{t\le t_0} = \dot{\mathbf{u}}|_{t\le t_0} = \mathbf{0}\,, \quad \mathbf{n}\cdot\boldsymbol{\sigma}|_{\mathbf{x}\in\partial G} = \mathbf{0}\,. \tag{8.76}$$

To compute the derivative of an objective functional, χ, with respect to ρ and **C** we need to find the adjoint operator $\nabla_u \mathbf{L}^\dagger$ of $\nabla_u \mathbf{L}$, as defined in Eq. (8.13):

$$\langle \mathbf{u}\cdot\nabla_u\mathbf{L}^\dagger\,\mathbf{u}^\dagger\rangle = \langle \mathbf{u}^\dagger\cdot\nabla_u\mathbf{L}\,\mathbf{u}\rangle\,. \tag{8.77}$$

The explicit version of $\langle \mathbf{u}^\dagger\cdot\nabla_u\mathbf{L}\,\mathbf{u}\rangle$ is

$$\begin{aligned}\langle \mathbf{u}^\dagger\cdot\nabla_u\mathbf{L}\,\mathbf{u}\rangle &= \int_T\int_G \mathbf{u}^\dagger\cdot\nabla_u\mathbf{L}\,\mathbf{u}\,\mathrm{d}t\,\mathrm{d}^3\mathbf{x} = \int_T\int_G \rho\,\mathbf{u}^\dagger\cdot\ddot{\mathbf{u}}\,\mathrm{d}t\,\mathrm{d}^3\mathbf{x}\\ &\quad - \int_T\int_G \mathbf{u}^\dagger\cdot\left[\nabla\cdot\int_{\tau=t_0}^{t}\dot{\mathbf{C}}(t-\tau):\nabla\mathbf{u}(\tau)\,\mathrm{d}\tau\right]\mathrm{d}t\,\mathrm{d}^3\mathbf{x}\,,\end{aligned} \tag{8.78}$$

where we have omitted all dependencies on **x** in the interest of clarity. Our goal is to isolate **u** in Eq. (8.78) so that it is not differentiated. We start with the first term on the right-hand side, i.e. with

$$\langle \mathbf{u}^\dagger\cdot\rho\ddot{\mathbf{u}}\rangle = \int_T\int_G \rho\,\mathbf{u}^\dagger\cdot\ddot{\mathbf{u}}\,\mathrm{d}^3\mathbf{x}\,\mathrm{d}t\,. \tag{8.79}$$

Repeated integration by parts yields

$$\begin{aligned}\int_T\int_G \rho\,\mathbf{u}^\dagger\cdot\ddot{\mathbf{u}}\,\mathrm{d}^3\mathbf{x}\,\mathrm{d}t &= \int_T\int_G \rho\,\mathbf{u}\cdot\ddot{\mathbf{u}}^\dagger\,\mathrm{d}^3\mathbf{x}\,\mathrm{d}t\\ &\quad + \int_G \rho\,\dot{\mathbf{u}}\cdot\mathbf{u}^\dagger\,\mathrm{d}^3\mathbf{x}\Big|_{t=t_1} - \int_G \rho\,\mathbf{u}\cdot\dot{\mathbf{u}}^\dagger\,\mathrm{d}^3\mathbf{x}\Big|_{t=t_1}\,,\end{aligned} \tag{8.80}$$

where we already used the initial conditions $\mathbf{u}|_{t\le t_0} = \dot{\mathbf{u}}|_{t\le t_0} = \mathbf{0}$. By imposing the *terminal conditions*

$$\mathbf{u}^\dagger|_{t\ge t_1} = \dot{\mathbf{u}}^\dagger|_{t\ge t_1} = \mathbf{0}\,, \tag{8.81}$$

we can obtain the first ingredient of the adjoint operator:

$$\langle \mathbf{u}^\dagger\cdot\rho\ddot{\mathbf{u}}\rangle = \langle \mathbf{u}\cdot\rho\ddot{\mathbf{u}}^\dagger\rangle\,. \tag{8.82}$$

We now turn to the spatial differentiation of **u**. For this, we consider the expression

$$\Upsilon := \langle \mathbf{u}^\dagger\cdot(\nabla\cdot\boldsymbol{\sigma})\rangle = \int_T\int_G \mathbf{u}^\dagger\cdot\left[\nabla\cdot\int_{\tau=t_0}^{t}\dot{\mathbf{C}}(t-\tau):\nabla\mathbf{u}(\tau)\,\mathrm{d}\tau\right]\mathrm{d}^3\mathbf{x}\,\mathrm{d}t\,. \tag{8.83}$$

Upon invoking the symmetries of the elastic tensor, **C**, we find the relation

$$\mathbf{u}^\dagger \cdot [\nabla \cdot (\dot{\mathbf{C}} : \nabla \mathbf{u})] = \nabla \cdot (\mathbf{u}^\dagger \cdot \dot{\mathbf{C}} : \nabla \mathbf{u}) - \nabla \cdot (\mathbf{u} \cdot \dot{\mathbf{C}} : \nabla \mathbf{u}^\dagger) + \mathbf{u} \cdot [\nabla \cdot (\dot{\mathbf{C}} : \nabla \mathbf{u}^\dagger)] \,, \quad (8.84)$$

which we can use to transform Eq. (8.83):

$$\begin{aligned} \Upsilon = & \int_G \int_{t=t_0}^{t_1} \int_{\tau=t_0}^{t} \nabla \cdot [\mathbf{u}^\dagger(t) \cdot \dot{\mathbf{C}}(t-\tau) : \nabla \mathbf{u}(\tau)] \, \mathrm{d}\tau \, \mathrm{d}t \, \mathrm{d}^3\mathbf{x} \\ & - \int_G \int_{t=t_0}^{t_1} \int_{\tau=t_0}^{t} \nabla \cdot [\mathbf{u}(\tau) \cdot \dot{\mathbf{C}}(t-\tau) : \nabla \mathbf{u}^\dagger(t)] \, \mathrm{d}\tau \, \mathrm{d}t \, \mathrm{d}^3\mathbf{x} \\ & + \int_G \int_{t=t_0}^{t_1} \int_{\tau=t_0}^{t} \mathbf{u}(\tau) \cdot [\nabla \cdot (\dot{\mathbf{C}}(t-\tau) : \nabla \mathbf{u}^\dagger(\tau))] \, \mathrm{d}\tau \, \mathrm{d}t \, \mathrm{d}^3\mathbf{x} \,. \end{aligned} \quad (8.85)$$

With the help of Gauss' theorem and the identity

$$\int_{\tau=t_0}^{t} \int_{t=t_0}^{t_1} \mathrm{d}t \, \mathrm{d}\tau = \int_{t=\tau}^{t_1} \int_{\tau=t_0}^{t_1} \mathrm{d}t \, \mathrm{d}\tau \quad (8.86)$$

we write Υ in a form that allows us to eliminate two of the contributing integrands:

$$\begin{aligned} \Upsilon = & \int_{\partial G} \int_{t=t_0}^{t_1} \mathbf{u}^\dagger(t) \cdot \left[\int_{\tau=t_0}^{t} \dot{\mathbf{C}}(t-\tau) : \nabla \mathbf{u}(\tau) \, \mathrm{d}\tau \right] \cdot \mathbf{n} \, \mathrm{d}t \, \mathrm{d}^2\mathbf{x} \\ & - \int_{\partial G} \int_{\tau=t_0}^{t_1} \mathbf{u}(\tau) \cdot \left[\int_{t=\tau}^{t_1} \dot{\mathbf{C}}(t-\tau) : \nabla \mathbf{u}^\dagger(t) \, \mathrm{d}t \right] \cdot \mathbf{n} \, \mathrm{d}\tau \, \mathrm{d}^2\mathbf{x} \\ & + \int_G \int_{\tau=t_0}^{t_1} \mathbf{u}(\tau) \cdot \left[\nabla \cdot \int_{t=\tau}^{t_1} \dot{\mathbf{C}}(t-\tau) : \nabla \mathbf{u}^\dagger(t) \, \mathrm{d}t \right] \mathrm{d}\tau \, \mathrm{d}^3\mathbf{x} \,. \end{aligned} \quad (8.87)$$

The first term in Eq. (8.87) is equal to zero because the expression in square brackets is the stress tensor $\boldsymbol{\sigma}$, and $\boldsymbol{\sigma} \cdot \mathbf{n} = \mathbf{0}$ on ∂G as a consequence of the free surface boundary condition. Since $\mathbf{u}^\dagger$ is so far only constrained by the terminal conditions (8.81) we are free to impose a boundary condition that forces the second term to zero:

$$\mathbf{n} \cdot \boldsymbol{\sigma}^\dagger|_{\mathbf{x} \in \partial G} = \mathbf{0} \,, \quad (8.88)$$

with the *adjoint stress tensor* $\boldsymbol{\sigma}^\dagger$ defined by

$$\boldsymbol{\sigma}^\dagger(t) = \int_{\tau=t}^{t_1} \dot{\mathbf{C}}(\tau - t) : \nabla \mathbf{u}^\dagger(\tau) \, \mathrm{d}\tau \,. \quad (8.89)$$

The third integrand in Eq. (8.87) is already of the form that we require. It thus remains to assemble the complete adjoint $\nabla_u \mathbf{L}^\dagger$: Combining Eqs. (8.78), (8.82) (8.87), (8.88) and (8.89) yields

$$\langle \mathbf{u} \cdot \nabla_u \mathbf{L}^\dagger \mathbf{u}^\dagger \rangle = \langle \mathbf{u} \cdot \rho \ddot{\mathbf{u}}^\dagger \rangle - \Upsilon = \langle \mathbf{u} \cdot (\rho \, \ddot{\mathbf{u}}^\dagger - \nabla \cdot \boldsymbol{\sigma}^\dagger) \rangle \,. \tag{8.90}$$

It follows that the adjoint operator $\nabla_u \mathbf{L}^\dagger$ is given by

$$\nabla_u \mathbf{L}^\dagger \mathbf{u}^\dagger = \rho \, \ddot{\mathbf{u}}^\dagger - \nabla \cdot \boldsymbol{\sigma}^\dagger \,. \tag{8.91}$$

Thus, to compute the derivative of the objective functional, $\nabla_m \chi$, one needs to solve the adjoint equation

$$\rho \ddot{\mathbf{u}}^\dagger - \nabla \cdot \boldsymbol{\sigma}^\dagger = -\nabla_u \chi_1^\dagger \,, \tag{8.92}$$

subject to the terminal and boundary conditions

$$\mathbf{u}^\dagger|_{t \ge t_1} = \dot{\mathbf{u}}^\dagger|_{t \ge t_1} = \mathbf{0} \,, \quad \mathbf{n} \cdot \boldsymbol{\sigma}^\dagger|_{\mathbf{x} \in \partial G} = \mathbf{0} \,. \tag{8.93}$$

In non-dissipative media the elastic wave operator $\mathbf{L}$ is *self-adjoint*, meaning that $\mathbf{L} = \mathbf{L}^\dagger$. Since the adjoint equation (8.92) is still of the wave equation type, it can be solved numerically using the same methods as for the solution of the regular wave equation.

The obvious numerical difficulty in solving the adjoint equation is the occurrence of the terminal conditions (8.93) that require that the adjoint field be zero at time $t = t_1$ when the observation ends. In practice, this condition can only be met by solving the adjoint equation backwards in time, that is by reversing the time axis from $t_0 \to t_1$ to $t_1 \to t_0$. The terminal conditions then act as zero initial conditions, at least in the numerical simulation.

Time reversal appears in numerous applications including reverse time migration (e.g. Baysal et al., 1983) and the time-reversal imaging of seismic sources (e.g. Larmat et al., 2006; Sect. 9.2). Most of these are closely related to the adjoint method.

The adjoint sources are fully specified by the objective functional, that may be used, for instance, to measure the misfit between observed and synthetic seismograms. In this context the adjoint method is often classified as a *back projection technique* where the waveform residuals are propagated backwards in time towards the location from where they originated. The adjoint operator is sometimes referred to as *back projection operator*.

Equation (8.89) reveals that the adjoint stress tensor $\boldsymbol{\sigma}^\dagger$ at time t depends on future strain from t to t_1. This results in a growth of elastic energy when the wave field propagates in the regular time direction from t_0 to t_1. In reversed time, however, the elastic energy decays, so that numerical instabilities do not occur (e.g. Tarantola, 1988).

In our analysis we did not consider perturbations in the topography of the Earth's internal discontinuities. For a derivation of Fréchet kernels with respect to topographical variations the reader is referred to Liu & Tromp (2008).

8.5.2 Practical Implementation

To compute the derivatives of χ via the time integral from Eq. (8.16), the regular wave field – propagating forward in time – and the adjoint wave field – propagating backwards in time – must be available simultaneously. In a medium without dissipation and without absorbing boundaries, one can solve this problem as follows: First, the regular wave equation is solved, and its final state, that is $\mathbf{u}(\mathbf{x}, t_1)$ and $\dot{\mathbf{u}}(\mathbf{x}, t_1)$, is stored. Following the computation of the adjoint source, the adjoint equation is solved backwards in time. Then, as $\mathbf{u}^\dagger$ propagates from t_1 towards the initial time t_0, the regular wave field is propagated simultaneously in reverse time, starting from the final stage that has been stored before. Since the regular and the adjoint fields are known at coincident time steps, the time integral can be solved on the fly, that is, in the course of the adjoint simulation.

This strategy works because the earlier states of the regular wave field $\mathbf{u}$ can be perfectly reconstructed from the final state at $t = t_1$ when the wave equation is invariant with respect to a sign change of the time variable, which is equivalent to the perfect conservation of elastic energy.

In the presence of dissipation or absorbing boundaries, however, elastic energy is lost during the propagation of the regular wave field, so that its earlier states *cannot* be reconstructed from the final state. It follows that we are forced to store $\mathbf{u}$ at intermediate time steps. The stored regular wave field is then loaded during the adjoint simulation and used to approximate the time integral.

In the extreme case, $\mathbf{u}$ is stored in every single time step of the forward simulation. Since the resulting amount of data can easily exceed conventional capacities, *checkpointing algorithms* have been developed (e.g. Griewank & Walther, 2000; Charpentier, 2001). The idea is to store the regular wave field at a smaller number of time steps, called checkpoints, and to solve the forward problem from there until the current time of the adjoint calculation is reached. The storage requirements are thus reduced at the expense of significantly increased computation time.

A less time-consuming alternative to checkpointing is *data compression*, in the sense that a reduced version of the regular wave field is stored. This can often be done in a natural way within the numerical modelling code. In the spectral-element method, for instance (see Chap. 4), the dynamic fields are represented in terms of Nth-order Lagrange polynomials. Since neighbouring elements share grid points, the storage requirements are proportional to N^3. This can be reduced by storing lower order polynomial representations of the forward dynamic fields that are then re-converted to Nth-degree polynomials during the adjoint calculation. A reduction of the polynomial degree from 6 to 4 results in a compression ratio of 3.4 with no significant deterioration of the sensitivity kernels.

As it is very often the case, the numerical costs can be reduced significantly with the help of physical insight. Following a detailed discussion of the anatomy of Fréchet kernels in Chap. 12, we will present a time integration scheme that limits the storage requirements to a more acceptable level (Sect. 12.4).

Chapter 9
First and Second Derivatives with Respect to Structural and Source Parameters

This chapter provides specific examples for the first and second derivatives of an objective functional with respect to selected structural and source parameters. We discuss the general characteristics of Fréchet kernels in the context of single scattering from within a first-order influence zone. With the help of the adjoint method we then derive explicit formulas for Fréchet kernels in isotropic, radially anisotropic and visco-elastic media. First derivatives with respect to source parameters are shown to be intimately related to the time-reverse imaging of seismic rupture processes. A strong focus is on the interpretation of the Hessian especially in terms of resolution and trade-off kernels. We demonstrate the relation between Hessian kernels and several forms of second-order scattering. We give a recipe for the computation of second derivatives, followed by a collection of Hessian kernels for a variety of structural parameters.

9.1 First Derivatives with Respect to Selected Structural Parameters

The most general expression for the derivative of an objective functional $\chi(\mathbf{m})$ in the direction $\delta\mathbf{m}$ is given by Eq. (8.16) that we repeat here for convenience:

$$\nabla_m \chi\, \delta\mathbf{m} = \langle \mathbf{u}^\dagger \cdot \nabla_m \mathbf{L}\, \delta\mathbf{m} \rangle\,. \tag{9.1}$$

Substituting the wave operator defined in Eq. (8.75) yields the explicit formula

$$\nabla_m \chi\, \delta\mathbf{m} = \int_T \int_G \mathbf{u}^\dagger(t) \cdot \left[\delta\rho\, \ddot{\mathbf{u}}(t) - \nabla \cdot \int_{\tau=t_0}^{t} \delta\dot{\mathbf{C}}(t-\tau) : \nabla\mathbf{u}(\tau)\, \mathrm{d}\tau \right] \mathrm{d}t\, \mathrm{d}^3\mathbf{x}\,, \tag{9.2}$$

with $\delta\mathbf{m} = (\delta\rho, \delta\mathbf{C})$. Again, we omitted spatial dependencies in the interest of readability. Integrating by parts provides a more symmetric and more useful version of Eq. (9.2):

A. Fichtner, *Full Seismic Waveform Modelling and Inversion*, Advances in Geophysical and Environmental Mechanics and Mathematics,
DOI 10.1007/978-3-642-15807-0_9, © Springer-Verlag Berlin Heidelberg 2011

$$\nabla_m \chi\, \delta \mathbf{m} = -\int_T \int_G \delta\rho\, \dot{\mathbf{u}}^\dagger(t) \cdot \dot{\mathbf{u}}(t)\, \mathrm{d}t\, \mathrm{d}^3\mathbf{x} + \int_T \int_G \left[\int_{\tau=t_0}^{t} \nabla \mathbf{u}^\dagger(t) : \delta \dot{\mathbf{C}}(t-\tau) : \nabla \mathbf{u}(\tau)\, \mathrm{d}\tau \right] \mathrm{d}t\, \mathrm{d}^3\mathbf{x}\,. \tag{9.3}$$

The Fréchet kernels associated with (9.3) are

$$K_\rho = -\int_T \dot{\mathbf{u}}^\dagger(t) \cdot \dot{\mathbf{u}}(t)\, \mathrm{d}t \tag{9.4}$$

and

$$\mathbf{K}_C(\tau) = \int_T \nabla \mathbf{u}^\dagger(t) \otimes \nabla \mathbf{u}(t+\tau)\, \mathrm{d}t\,, \tag{9.5}$$

where $\otimes$ denotes the tensor or dyadic product. Both kernels are non-zero only within the *primary influence zone* where the regular and adjoint wave fields interact at some time between t_0 and t_1. The primary influence zone, illustrated in Fig. 9.1, is the region where a model perturbation, $\delta\mathbf{m}$, causes the regular wave field, $\mathbf{u}$, to generate

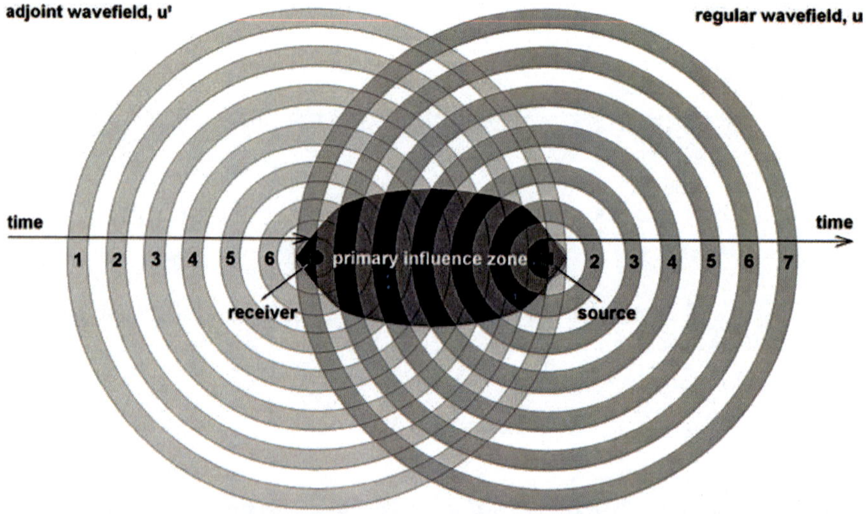

Fig. 9.1 Illustration of the primary influence zone where the regular wave field, $\mathbf{u}$, interacts with the adjoint wave field, $\mathbf{u}^\dagger$. Numbers are used to mark the regular and adjoint wavefronts at successive points in time. As time goes on, the regular wave field propagates away from the source while the adjoint wave field collapses into the receiver. In numerical simulations the adjoint equations are solved backwards in time in order to satisfy the terminal conditions. On the reverse time axis, the adjoint field propagates away from the receiver, starting at the final observation time. The primary influence zone marks the region where a model perturbation, $\delta\mathbf{m}$, generates a first-order scattered wave field that affects the measurement at the receiver. Perturbations located outside the primary influence zone have no first-order effect on the measurement

a first-order or single scattered wave that affects the measurement at the receiver. A perturbation located outside the primary influence zone has no first-order effect on the measurement.

For most seismic phases, the primary influence zone is a roughly cigar-shaped region connecting the source and the receiver. Its precise geometry depends on many factors including the frequency content, the length of the considered time window, the type of measurement and the reference Earth model, **m**. Specific examples for common seismic phases and measurements can be found in Chap. 12.

In what follows we derive the Fréchet kernels for several widely used rheologies, i.e. for special choices of $\mathbf{C}(t)$.

9.1.1 Perfectly Elastic and Isotropic Medium

The defining property of a perfectly elastic medium is that present stresses do not depend on past deformation. This is expressed mathematically by a time dependence of $\mathbf{C}$ and $\delta\mathbf{C}$ that has the form of a unit-step function:

$$\mathbf{C}(\mathbf{x}, t) = \mathbf{C}(\mathbf{x})\, H(t)\,, \qquad \delta\mathbf{C}(\mathbf{x}, t) = \delta\mathbf{C}(\mathbf{x})\, H(t)\,. \tag{9.6}$$

Upon inserting (9.6) into Eq. (9.3) we obtain a simplified expression for $\nabla_m \chi$:

$$\begin{aligned} \nabla_m \chi\, \delta\mathbf{m} = &- \int_T \int_G \delta\rho\, \dot{\mathbf{u}}^\dagger(t) \cdot \dot{\mathbf{u}}(t)\, dt\, d^3\mathbf{x} \\ &+ \int_T \int_G \nabla \mathbf{u}^\dagger(t) : \delta\mathbf{C} : \nabla \mathbf{u}(t)\, dt\, d^3\mathbf{x}\,. \end{aligned} \tag{9.7}$$

In an isotropic medium the components of $\mathbf{C}$ are given by (e.g. Aki & Richards, 2002)

$$C_{ijkl} = \lambda \delta_{ij}\delta_{kl} + \mu \delta_{ik}\delta_{jl} + \mu \delta_{il}\delta_{jk}\,. \tag{9.8}$$

The symbols λ and μ denote the Lamé parameters. It follows that the complete derivative of χ is composed of three terms:

$$\nabla_m \chi\, \delta\mathbf{m} = \nabla_\rho \chi\, \delta\rho + \nabla_\lambda \chi\, \delta\lambda + \nabla_\mu \chi\, \delta\mu\,, \tag{9.9}$$

with

$$\nabla_\rho \chi\, \delta\rho = -\int_T \int_G \delta\rho\, \dot{\mathbf{u}}^\dagger \cdot \dot{\mathbf{u}}\, dt\, d^3\mathbf{x}\,, \tag{9.10a}$$

$$\nabla_\lambda \chi\, \delta\lambda = \int_T \int_G \delta\lambda\, (\nabla \cdot \mathbf{u})(\nabla \cdot \mathbf{u}^\dagger)\, dt\, d^3\mathbf{x}\,, \tag{9.10b}$$

$$\nabla_\mu \chi \delta\mu = \int_T \int_G \delta\mu\, [(\nabla \mathbf{u}^\dagger) : (\nabla \mathbf{u}) + (\nabla \mathbf{u}^\dagger) : (\nabla \mathbf{u})^T]\, dt\, d^3\mathbf{x}\,. \tag{9.10c}$$

We obtain the corresponding sensitivity kernels by simply dropping the spatial integration:

$$K_\rho^0 = -\int_T \dot{\mathbf{u}}^\dagger \cdot \dot{\mathbf{u}}\, dt\,, \tag{9.11a}$$

$$K_\lambda^0 = \int_T (\nabla \cdot \mathbf{u})(\nabla \cdot \mathbf{u}^\dagger)\, dt\,, \tag{9.11b}$$

$$K_\mu^0 = \int_T [(\nabla \mathbf{u}^\dagger) : (\nabla \mathbf{u}) + (\nabla \mathbf{u}^\dagger) : (\nabla \mathbf{u})^T]\, dt\,. \tag{9.11c}$$

The superscript ^{0} symbolises that the Fréchet kernels correspond to the fundamental parameterisation $\mathbf{m} = (\rho, \lambda, \mu)$. Eqs. (9.11a), (9.11b) and (9.11c) confirm the intuitive expectation that K_λ^0 vanishes for pure shear waves, while both shear and compressional waves contribute to K_μ^0. Depending on personal preference and numerical convenience, one may rewrite some of the above equations in terms of the regular strain tensor $\boldsymbol{\varepsilon} = \frac{1}{2}[(\nabla \mathbf{u}) + (\nabla \mathbf{u})^T]$ and the adjoint strain tensor $\boldsymbol{\varepsilon}^\dagger = \frac{1}{2}[(\nabla \mathbf{u}^\dagger) + (\nabla \mathbf{u}^\dagger)^T]$. For this we can use the relations

$$(\nabla \cdot \mathbf{u})(\nabla \cdot \mathbf{u}^\dagger) = (\operatorname{tr} \boldsymbol{\varepsilon})\, (\operatorname{tr} \boldsymbol{\varepsilon}^\dagger) \tag{9.12}$$

and

$$\begin{aligned}
&(\nabla \mathbf{u}^\dagger) : (\nabla \mathbf{u}) + (\nabla \mathbf{u}^\dagger) : (\nabla \mathbf{u})^T = 2(\nabla \mathbf{u}^\dagger) : \boldsymbol{\varepsilon} \\
=&2(\nabla \mathbf{u}^\dagger) : \boldsymbol{\varepsilon} - (\nabla \mathbf{u}^\dagger)^T : \boldsymbol{\varepsilon} + (\nabla \mathbf{u}^\dagger)^T : \boldsymbol{\varepsilon} \\
=&[(\nabla \mathbf{u}^\dagger) : \boldsymbol{\varepsilon} + (\nabla \mathbf{u}^\dagger)^T : \boldsymbol{\varepsilon}] + [(\nabla \mathbf{u}^\dagger) : \boldsymbol{\varepsilon} - (\nabla \mathbf{u}^\dagger)^T : \boldsymbol{\varepsilon}] = 2\boldsymbol{\varepsilon}^\dagger : \boldsymbol{\varepsilon}\,.
\end{aligned} \tag{9.13}$$

The last equality holds because the dot product of a symmetric tensor and an anti-symmetric tensor vanishes.

Based on Eqs. (9.11a), (9.11b) and (9.11c) we easily derive formulas for the Fréchet kernels with respect to other parameters, such as the bulk modulus, $\kappa = \lambda + 2\mu/3$, the S wave speed, $v_S = \sqrt{\mu/\rho}$ or the P wave speed, $v_P = \sqrt{(\lambda + 2\mu)/\rho}$. Defining ρ, κ and μ as independent parameters, we find the following set of kernels:

$$K_\rho = K_\rho^0\,, \tag{9.14a}$$
$$K_\kappa = K_\lambda^0\,, \tag{9.14b}$$
$$K_\mu = K_\mu^0 - \frac{2}{3} K_\lambda^0\,. \tag{9.14c}$$

With ρ, v_P and v_S as independent parameters we have

$$K_\rho = K_\rho^0 + (v_P^2 - 2v_S^2)\, K_\lambda^0 + v_S^2\, K_\mu^0\,, \tag{9.15a}$$
$$K_{v_S} = 2\rho v_S\, K_\mu^0 - 4\rho v_S\, K_\lambda^0\,, \tag{9.15b}$$
$$K_{v_P} = 2\rho v_P\, K_\lambda^0\,. \tag{9.15c}$$

A comparison between Eqs. (9.14a) and (9.15a) makes it particularly obvious that the sensitivity with respect to density, ρ, depends strongly on the choice of free parameters.

9.1.2 Perfectly Elastic Medium with Radial Anisotropy

Anisotropy is the dependence of the elastic tensor on the orientation of the coordinate system. Its most direct seismological expressions are the splitting of shear waves and the dependence of seismic velocities on the propagation and polarisation directions. Polarisation anisotropy is particularly pronounced in the Earth's uppermost mantle, and it leads to the readily observable *Love–Rayleigh discrepancy*: The propagation speeds of Love and Rayleigh waves travelling in the same direction can rarely be explained with an isotropic model. Pure polarisation anisotropy occurs in elastic media with radial symmetry axis that are fully described by five independent elastic tensor components. They can be summarised in two 3×3 matrices (e.g. Babuska & Cara, 1991):

$$\begin{pmatrix} C_{rrrr} & C_{rr\phi\phi} & C_{rr\theta\theta} \\ C_{\phi\phi rr} & C_{\phi\phi\phi\phi} & C_{\phi\phi\theta\theta} \\ C_{\theta\theta rr} & C_{\theta\theta\phi\phi} & C_{\theta\theta\theta\theta} \end{pmatrix} = \begin{pmatrix} \lambda+2\mu & \lambda+c & \lambda+c \\ \lambda+c & \lambda+2\mu+a & \lambda+a \\ \lambda+c & \lambda+a & \lambda+2\mu+a \end{pmatrix} \tag{9.16a}$$

$$\begin{pmatrix} C_{\phi\theta\phi\theta} & C_{\phi\theta r\theta} & C_{\phi\theta r\phi} \\ C_{r\theta\phi\theta} & C_{r\theta r\theta} & C_{r\theta r\phi} \\ C_{r\phi\phi\theta} & C_{r\phi r\theta} & C_{r\phi r\phi} \end{pmatrix} = \begin{pmatrix} \mu & 0 & 0 \\ 0 & \mu+b & 0 \\ 0 & 0 & \mu+b \end{pmatrix}. \tag{9.16b}$$

All components of $\mathbf{C}$ that do not appear in Eq. (9.16) are equal to zero. Many of the examples given in the following chapters will be based on this type of anisotropy. For $a = b = c = 0$ we retrieve the elastic tensor of an isotropic medium. In homogeneous media we can give intuitive interpretations of λ, μ, a, b and c: The velocity of a horizontally propagating plane S wave is

$$v_{\mathrm{SV}} = \sqrt{\frac{\mu + b}{\rho}} \quad \text{or} \quad v_{\mathrm{SH}} = \sqrt{\frac{\mu}{\rho}} \tag{9.17}$$

depending on whether it is vertically (v_{SV}) or horizontally (v_{SH}) polarised. For vertically and horizontally propagating P waves we find the propagation speeds v_{PV} and v_{PH}, given by

$$v_{\mathrm{PV}} = \sqrt{\frac{\lambda + 2\mu}{\rho}} \quad \text{and} \quad v_{\mathrm{PH}} = \sqrt{\frac{\lambda + 2\mu + a}{\rho}}\,, \tag{9.18}$$

respectively. The parameter c can only be determined from P waves that do not travel in exactly radial or horizontal directions. It is commonly absorbed in a new parameter, η, defined by (e.g. Dziewonski & Anderson, 1981)

$$\eta := \frac{\lambda + c}{\lambda + a}\,. \tag{9.19}$$

After some tedious but straightforward algebraic manipulations we find the following set of Fréchet kernels with respect to ρ, λ, μ, a, b and c:

$$K_\rho^0 = -\int_T \dot{\mathbf{u}}^\dagger \cdot \dot{\mathbf{u}}\,\mathrm{d}t\,, \tag{9.20a}$$

$$K_\lambda^0 = \int_T (\mathrm{tr}\,\boldsymbol{\varepsilon}^\dagger)(\mathrm{tr}\,\boldsymbol{\varepsilon})\,\mathrm{d}t\,, \tag{9.20b}$$

$$K_\mu^0 = 2\int_T \boldsymbol{\varepsilon}^\dagger : \boldsymbol{\varepsilon}\,\mathrm{d}t\,, \tag{9.20c}$$

$$K_a^0 = \int_T (\varepsilon_{\phi\phi}^\dagger + \varepsilon_{\theta\theta}^\dagger)(\varepsilon_{\phi\phi} + \varepsilon_{\theta\theta})\,\mathrm{d}t\,, \tag{9.20d}$$

$$K_b^0 = 4\int_T (\varepsilon_{r\theta}^\dagger \varepsilon_{r\theta} + \varepsilon_{r\phi}^\dagger \varepsilon_{r\phi})\,\mathrm{d}t\,, \tag{9.20e}$$

$$K_c^0 = \int_T [\varepsilon_{rr}^\dagger(\varepsilon_{\phi\phi} + \varepsilon_{\theta\theta}) + \varepsilon_{rr}(\varepsilon_{\phi\phi}^\dagger + \varepsilon_{\theta\theta}^\dagger)]\,\mathrm{d}t\,. \tag{9.20f}$$

Fréchet kernels for the seismologically more relevant parameters v_{SV}, v_{SH}, v_{PV}, v_{PH} and η can now be expressed as linear combinations of the kernels in Eqs. (9.20a), (9.20b), (9.20c), (9.20d), (9.20e) and (9.20f):

$$\begin{aligned} K_\rho &= K_\rho^0 + v_{\text{SH}}^2 K_\mu^0 + (v_{\text{PV}}^2 - 2v_{\text{SH}}^2) K_\lambda^0 + (v_{\text{PH}}^2 - v_{\text{PV}}^2) K_a^0 \\ &\quad + (v_{\text{SV}}^2 - v_{\text{SH}}^2) K_b^0 + \left[2(1-\eta) v_{\text{SH}}^2 + \eta v_{\text{PH}}^2 - v_{\text{PV}}^2\right] K_c^0 \\ &= K_\rho^0 + \rho^{-1} (\mu\, K_\mu^0 + \lambda\, K_\lambda^0 + a\, K_a^0 + b\, K_b^0 + c\, K_c^0)\,, \end{aligned} \tag{9.21a}$$

$$K_{v_{\text{SH}}} = 2\rho v_{\text{SH}} [K_\mu^0 - 2K_\lambda^0 - K_b^0 + 2(1-\eta)\, K_c^0]\,, \tag{9.21b}$$

$$K_{v_{\text{SV}}} = 2\rho v_{\text{SV}} K_b^0\,, \tag{9.21c}$$

$$K_{v_{\text{PH}}} = 2\rho v_{\text{PH}} (K_a^0 + \eta\, K_c^0)\,, \tag{9.21d}$$

$$K_{v_{\text{PV}}} = 2\rho v_{\text{PV}} (K_\lambda^0 - K_a^0 - K_c^0)\,, \tag{9.21e}$$

$$K_\eta = \rho\, (v_{\text{PH}}^2 - 2v_{\text{SH}}^2)\, K_c^0 = (\lambda + a)\, K_c^0\,. \tag{9.21f}$$

Equations (9.21a), (9.21b), (9.21c), (9.21d), (9.21e) and (9.21f) can be simplified in the case of an isotropic reference medium where $v_{\text{SH}} = v_{\text{SV}} = v_{\text{S}}$, $v_{\text{PH}} = v_{\text{PV}} = v_{\text{P}}$ and $\eta = 1$:

$$K_\rho = K_\rho^0 + v_{\text{S}}^2 K_\mu^0 + (v_{\text{P}}^2 - 2v_{\text{S}}^2) K_\lambda^0\,, \tag{9.22a}$$

$$K_{v_{\text{SH}}} = 2\rho v_{\text{S}} (K_\mu^0 - 2K_\lambda^0 - K_b^0)\,, \tag{9.22b}$$

$$K_{v_{\text{SV}}} = 2\rho v_{\text{S}} K_b^0\,, \tag{9.22c}$$

$$K_{v_{\text{PH}}} = 2\rho v_{\text{P}} (K_a^0 + K_c^0)\,, \tag{9.22d}$$

$$K_{v_{\text{PV}}} = 2\rho v_{\text{P}} (K_\lambda^0 - K_a^0 - K_c^0)\,, \tag{9.22e}$$

$$K_\eta = \rho\, (v_{\text{P}}^2 - 2v_{\text{S}}^2)\, K_c^0\,. \tag{9.22f}$$

The comparison of Eqs. (9.15a), (9.15b) and (9.15c) and (9.22a), (9.22b), (9.22c), (9.22d), (9.22e) and (9.22f) yields two interesting relations between the Fréchet kernels for v_{PV}, v_{PH}, v_{SV}, v_{SH}, v_{P} and v_{S} in the case of an isotropic background medium:

$$K_{v_{\text{P}}} = K_{v_{\text{PH}}} + K_{v_{\text{PV}}}\,, \tag{9.23}$$

$$K_{v_{\text{S}}} = K_{v_{\text{SH}}} + K_{v_{\text{SV}}}\,. \tag{9.24}$$

Depending on the data used, the kernels $K_{v_{\text{P}}}$ and $K_{v_{\text{S}}}$ can be small, in the sense that the data are hardly affected by changes in v_{P} or v_{S}. The above relations, however, imply that nearly vanishing Fréchet kernels with respect to v_{P} and v_{S} may in fact be composed of significantly non-zero Fréchet kernels with respect to v_{PV}, v_{PH}, v_{SV} and v_{SH}. Certain P and S wave sensitivities can become alive when the model is allowed to be anisotropic. An example of this phenomenon was found by Dziewonski & Anderson (1981) who noted that the P wave sensitivity of fundamental-mode Rayleigh waves with periods around 120 s is restricted to the uppermost 100 km. Sensitivities with respect to the PH and PV wave speeds, however, are large down to 400 km depth but opposite in sign.

9.1.3 Isotropic Visco-Elastic Medium: Q_μ and Q_κ

The conversion of elastic into thermal energy in the Earth is largely controlled by the spatial distribution of the quality factors Q_μ and Q_κ associated with the shear modulus, μ, and the bulk modulus, $\kappa = \lambda + \frac{2}{3}\mu$. To derive expressions for the Fréchet derivatives of χ with respect to Q_μ and Q_κ we shall make frequent use of the results obtained in Chap. 5 on the mathematical description of visco-elastic dissipation. Restricting our attention to isotropic media, the time-dependent elastic tensor components are given by

$$C_{ijkl}(t) = \left[\kappa(t) - \frac{2}{3}\mu(t)\right] \delta_{ij}\delta_{kl} + \mu(t)\,\delta_{ik}\delta_{jl} + \mu(t)\,\delta_{il}\delta_{jk}\,. \tag{9.25}$$

For $\kappa(t)$ and $\mu(t)$ we choose a superposition of N standard linear solids, as proposed in Eq. (5.5):

$$\mu(t) = \mu_r \left[1 + \frac{1}{Q_\mu K} \sum_{n=1}^{N} \mathrm{e}^{-t/\tau_n}\right] H(t)\,, \tag{9.26}$$

$$\kappa(t) = \kappa_r \left[1 + \frac{1}{Q_\kappa K} \sum_{n=1}^{N} \mathrm{e}^{-t/\tau_n}\right] H(t)\,. \tag{9.27}$$

This rheology assumes that Q_μ and Q_κ are nearly constant within the frequency band of interest, $[\omega_1, \omega_2]$, which is determined by the set of relaxation times, τ_n, with $n = 1, \ldots, N$. The parameter K is constant for $\omega \in [\omega_1, \omega_2]$, and it is practically independent of the quality factors Q_μ and Q_κ. Based on Eq. (9.3) we find that the Fréchet derivative of χ with respect to Q_μ is then given by

$$\nabla_{Q_\mu}\chi\,\delta Q_\mu = \int_T \int_G \int_{\tau=t_0}^{t} \delta\dot{\mu}(t-\tau) \left[2\,\boldsymbol{\varepsilon}^\dagger(t) : \boldsymbol{\varepsilon}(\tau) - \frac{2}{3}\mathrm{tr}\,\boldsymbol{\varepsilon}^\dagger(t)\,\mathrm{tr}\,\boldsymbol{\varepsilon}(\tau)\right] d\tau\, dt\, d^3\mathbf{x}\,, \tag{9.28}$$

where

$$\delta\mu(t) = \nabla_{Q_\mu}\mu\,\delta Q_\mu = -\frac{\mu_r \delta Q_\mu}{Q_\mu^2 K} \left(\sum_{n=1}^{N} \mathrm{e}^{-t/\tau_n}\right) H(t) \tag{9.29}$$

is the infinitesimal change of the shear modulus μ caused by an infinitesimal change in Q_μ. Substituting (9.29) into (9.28) results in a numerically inconvenient double integral with respect to time. Fortunately, with the help of the memory variables defined in Eq. (5.7), we can eliminate the time convolutions from (9.28):

$$\int_{\tau=t_0}^{t} \delta\dot{\mu}(t-\tau)\,\boldsymbol{\varepsilon}(\tau)\,\mathrm{d}\tau = \int_{\tau=t_0}^{t} \delta\mu(t-\tau)\,\dot{\boldsymbol{\varepsilon}}(\tau)\,\mathrm{d}\tau = \mu_r \frac{\delta Q_\mu}{Q_\mu^2}\frac{N}{K}\sum_{n=1}^{N} \tau_n \mathbf{M}_n(t)\,. \tag{9.30}$$

The symbol $\mathbf{M}_n$ denotes the tensor of memory variables defined by

$$\mathbf{M}_n(t) := -\frac{1}{N\,\tau_n}\int_{\tau=t_0}^{t} \mathrm{e}^{-(t-\tau)/\tau_n}\,H(t-\tau)\,\boldsymbol{\varepsilon}(\tau)\,\mathrm{d}\tau\,. \tag{9.31}$$

An equation similar to (9.30) holds for the term involving $\mathrm{tr}\,\boldsymbol{\varepsilon}$ in Eq. (9.28). In terms of memory variables, the Fréchet derivative $\nabla_{Q_\mu}\chi\,\delta Q_\mu$ is now given by

$$\nabla_{Q_\mu}\chi\,\delta Q_\mu = \sum_{n=1}^{N}\int_T\int_G \mu_r \frac{\delta Q_\mu}{Q_\mu^2}\frac{N}{K}\tau_n\left[2\mathbf{M}_n : \boldsymbol{\varepsilon}^\dagger - \frac{2}{3}\mathrm{tr}\,\mathbf{M}_n\,\mathrm{tr}\,\boldsymbol{\varepsilon}^\dagger\right]\mathrm{d}t\,\mathrm{d}^3\mathbf{x}\,. \tag{9.32}$$

Omitting the spatial integration yields the Fréchet kernel K_{Q_μ}:

$$K_{Q_\mu} = \frac{\mu_r}{Q_\mu^2}\frac{N}{K}\sum_{n=1}^{N}\tau_n\int_T\left[2\mathbf{M}_n : \boldsymbol{\varepsilon}^\dagger - \frac{2}{3}\mathrm{tr}\,\mathbf{M}_n\,\mathrm{tr}\,\boldsymbol{\varepsilon}^\dagger\right]\mathrm{d}t\,. \tag{9.33}$$

Following the same procedure, we find the Fréchet kernel for Q_κ:

$$K_{Q_\kappa} = \frac{\kappa_r}{Q_\kappa^2}\frac{N}{K}\sum_{n=1}^{N}\tau_n\int_T \mathrm{tr}\,\mathbf{M}_n\,\mathrm{tr}\,\boldsymbol{\varepsilon}^\dagger\,\mathrm{d}t\,. \tag{9.34}$$

The kernels K_{Q_μ} and K_{Q_κ} depend on Q_μ^{-2} and Q_κ^{-2}, respectively. This property is disadvantageous in the context of a tomographic inversion where the initial Q model may be poorly constrained. In fact, the kernels can lead to significant updates of the initial Q model only in those regions where Q_μ and Q_κ are very small. In regions where the initial Q values are large, the kernels are effectively zero. The most straightforward solution is to consider the inverse parameters Q_μ^{-1} and Q_κ^{-1}.

The corresponding kernels,

$$K_{Q_\mu^{-1}} = -Q_\mu^2 K_{Q_\mu}\,,\quad K_{Q_\kappa^{-1}} = -Q_\kappa^2 K_{Q_\kappa}\,, \tag{9.35}$$

are effectively independent of the initial Q model. We note that kernels for elastic parameters and density also depend on the initial model. The effect is, however, not as severe as in the case of Q which can vary by several orders of magnitude inside the Earth.

9.2 First Derivatives with Respect to Selected Source Parameters

9.2.1 Distributed Sources and the Relation to Time-Reversal Imaging

In Sect. 8.3 we derived an expression for the derivative of the objective functional, χ, with respect to a general source, $\mathbf{f}$ (Eq. 8.33):

$$\nabla_f \chi \, \delta\mathbf{f} = -\langle \mathbf{u}^\dagger \cdot \delta\mathbf{f} \rangle \, .$$

The negative adjoint field, $-\mathbf{u}^\dagger$, acts as a time-dependent sensitivity kernel, much similar to the time-independent kernels for structural parameters that we considered in the previous paragraphs. In an iterative finite-source inversion, $-\mathbf{u}^\dagger$ therefore indicates how to change the current source model in order to better fit the data, that is, to reduce the misfit, χ.

The *time-reversal imaging of seismic sources* is a special case of this very general scenario, where observed waveforms are re-injected at the receiver positions and then propagated backwards in time (e.g. Fink, 1999; Larmat et al., 2006, 2009; Lokmer et al., 2009). The wave field propagating in reverse time plays the role of the adjoint field, $\mathbf{u}^\dagger$. Under favourable circumstances, $\mathbf{u}^\dagger$ collapses into the source region, thus providing information on the location and time evolution of the seismic source.

The original justification for the time-reversal imaging is the invariance of the elastic wave equation with respect to a sign change of the time variable from $+t$ to $-t$, at least in the case of negligible dissipation. When the re-injected seismograms bear sufficiently complete information on the original seismic wave field, the back-propagated field, $\mathbf{u}^\dagger$, indeed converges in the vicinity of the source region.

The adjoint method suggests an alternative interpretation of the time-reversal technique in the context of an iterative finite-source inversion (Kawakatsu & Montagner, 2008): Starting from the zero initial source model, $\mathbf{f} = \mathbf{0}$, the negative adjoint field, $-\mathbf{u}^\dagger$, corresponds to the direction of steepest descent towards a source model that better explains the data. Time-reversal imaging is, in this sense, equivalent to the first iteration in a non-linear finite-source inversion.

The time-reversal imaging concept is closely related to waveform inversion for the Earth structure. Sub-wavelength heterogeneities in the Earth act as first-order scatterers, i.e. as secondary sources that generate a scattered wave field. Back-propagating the scattered field then corresponds to the imaging of the scatterers. A regional-scale example where the back-propagation of scattered surface waves was used for the imaging of lateral heterogeneities may be found in Stich et al. (2009).

9.2.2 Moment Tensor Point Source

Most tectonic earthquakes observed at long periods and in the far field can be described by a moment tensor point source

$$\mathbf{f}(\mathbf{x},t) = -\nabla \cdot [\mathbf{M}\,\delta(\mathbf{x}-\mathbf{x}^s)\,H(t-t^s)]\,. \tag{9.36}$$

The symbols $\mathbf{M}$, $\mathbf{x}^s$ and t^s denote the moment tensor, the source location and the source origin time, respectively. We can compute the Fréchet derivatives of the objective functional χ with respect to the source parameters using again Eq. (8.33). Inserting the perturbations $\delta\mathbf{f}$ corresponding to the perturbations $\delta\mathbf{M}$, $\delta\mathbf{x}^s$ and δt^s yields a set of Fréchet derivatives with respect to the parameters describing the moment tensor point source:

$$\nabla_M \chi\,\delta\mathbf{M} = -\int_T \delta\mathbf{M} : \boldsymbol{\varepsilon}^\dagger(\mathbf{x}^s,t)\,\mathrm{d}t\,, \tag{9.37}$$

$$\nabla_{x^s} \chi\,\delta\mathbf{x}^s = -\delta\mathbf{x}^s \cdot \nabla \int_T \mathbf{M} : \boldsymbol{\varepsilon}^\dagger(\mathbf{x}^s,t)\,\mathrm{d}t\,, \tag{9.38}$$

$$\nabla_{t^s} \chi\,\delta t^s = \delta t^s\,\mathbf{M} : \boldsymbol{\varepsilon}^\dagger(\mathbf{x}^s,t^s)\,. \tag{9.39}$$

Equations (9.37), (9.38) and (9.39) only depend on the moment tensor and on the adjoint strain field $\boldsymbol{\varepsilon}^\dagger$ at the source position $\mathbf{x}^s$.

9.3 Second Derivatives with Respect to Selected Structural Parameters

As we have seen in Sect. 9.1, Fréchet kernels can be interpreted straightforwardly in terms of first-order scattering from within a primary influence zone. The interpretation of second derivatives is more involved. Therefore, we first attach physical meaning to the Hessian, before delving into the derivation of explicit formulas for a variety of seismologically relevant rheologies.

9.3.1 Physical Interpretation and Structure of the Hessian

9.3.1.1 Resolution and Trade-Off (RETRO) Kernels

As discussed already in Sect. 8.4.1.1, the Hessian evaluated at the optimal model, $H_\chi(\tilde{\mathbf{m}})$, is the carrier of the covariance information. It can be used to quantify the resolution of and trade-offs between model parameters. Large diagonal elements of $H_\chi(\tilde{\mathbf{m}})$ indicate that the corresponding model parameters are well resolved and vice versa. The size of the off-diagonal elements is a measure of the independence of model parameters. Zero off-diagonal elements in one column (or row) of the Hessian imply that the corresponding parameter can be constrained independently. Large off-diagonal elements, however, indicate a strong dependence between model parameters.

In this sense, the sum of the Hessian kernels introduced in Eqs. (8.66), (8.67), (8.68) and (8.69)

$$K_{\delta m_1} := K^a_{\delta m_1} + K^b_{\delta m_1} + K^c_{\delta m_1}\,, \tag{9.40}$$

can be interpreted as the *resolution and trade-off (RETRO) kernel* of $\delta\mathbf{m}_1$. We expect $K_{\delta m_1}$ to be large in the vicinity of $\delta\mathbf{m}_1$ itself and in those regions of the Earth model where structure trades off with $\delta\mathbf{m}_1$. An exemplary RETRO kernel is presented in Sect. 12.3.

9.3.1.2 First- and Second-Order Scattering

To understand the structure of the Hessian, $H_\chi = \nabla_m \nabla_m \chi$, and the physical meaning of its components, we consider the specific example of a pure perturbation in density and the objective functional

$$\chi(\mathbf{m}) = \frac{1}{2}\int_T \int_G [\mathbf{u}(\mathbf{m};\mathbf{x},t) - \mathbf{u}^0(\mathbf{x},t)]^2\, \delta(\mathbf{x}-\mathbf{x}^{\mathrm{r}})\, dt\, d^3\mathbf{x} \tag{9.41}$$

that we already introduced in Eq. (8.5). The choice of this deliberately simplistic and illustrative example facilitates the following discussion without affecting the generality of the results. In Sect. 8.4 we found that the Hessian applied to the model perturbations $\delta\mathbf{m}_1$ and $\delta\mathbf{m}_2$ can be expressed in terms of the first derivatives of the forward field $\mathbf{u}$ and the primary adjoint field $\mathbf{u}_1^\dagger$, which is the solution of the primary adjoint equation $\nabla_u \mathbf{L}^\dagger\, \mathbf{u}_1^\dagger = \mathbf{L}^\dagger(\mathbf{u}_1^\dagger) = -\nabla_u \chi_1^\dagger$:

$$\begin{aligned} H_\chi(\delta\mathbf{m}_1, \delta\mathbf{m}_2) &= \langle \nabla_u \nabla_u \chi_1(\delta_1\mathbf{u}, \delta_2\mathbf{u})\rangle + \langle \mathbf{u}_1^\dagger \cdot \nabla_m \nabla_m \mathbf{L}(\mathbf{u})(\delta\mathbf{m}_1, \delta\mathbf{m}_2)\rangle \\ &\quad + \langle \mathbf{u}_1^\dagger \cdot \nabla_m \mathbf{L}(\delta_2\mathbf{u})\,\delta\mathbf{m}_1\rangle + \langle \mathbf{u}_1^\dagger \cdot \nabla_m \mathbf{L}(\delta_1\mathbf{u})\,\delta\mathbf{m}_2\rangle\,. \end{aligned} \tag{9.42}$$

The first derivatives

$$\delta_1\mathbf{u} = \nabla_m \mathbf{u}\,\delta\mathbf{m}_1 \approx \mathbf{u}(\mathbf{m}+\delta\mathbf{m}_1) - \mathbf{u}(\mathbf{m}) \tag{9.43}$$

and

$$\delta_2\mathbf{u} = \nabla_m \mathbf{u}\,\delta\mathbf{m}_2 \approx \mathbf{u}(\mathbf{m}+\delta\mathbf{m}_2) - \mathbf{u}(\mathbf{m})\,, \tag{9.44}$$

are the first-order scattered wave fields that are excited when the forward wave field in the reference medium, $\mathbf{u}(\mathbf{m})$, impinges on the perturbations $\delta\mathbf{m}_1$ and $\delta\mathbf{m}_2$, respectively.

Interaction of First-Order Scattered Fields

Our interpretation of the different contributions to the Hessian, H_χ, starts with the first term on the right-hand side of Eq. (9.42). With the objective functional defined in (9.41) we find that $\langle \nabla_u \nabla_u \chi_1(\delta_1\mathbf{u}, \delta_2\mathbf{u})\rangle$ is the time-integrated product of the scattered fields $\delta_1\mathbf{u}$ and $\delta_2\mathbf{u}$ as seen at the receiver:

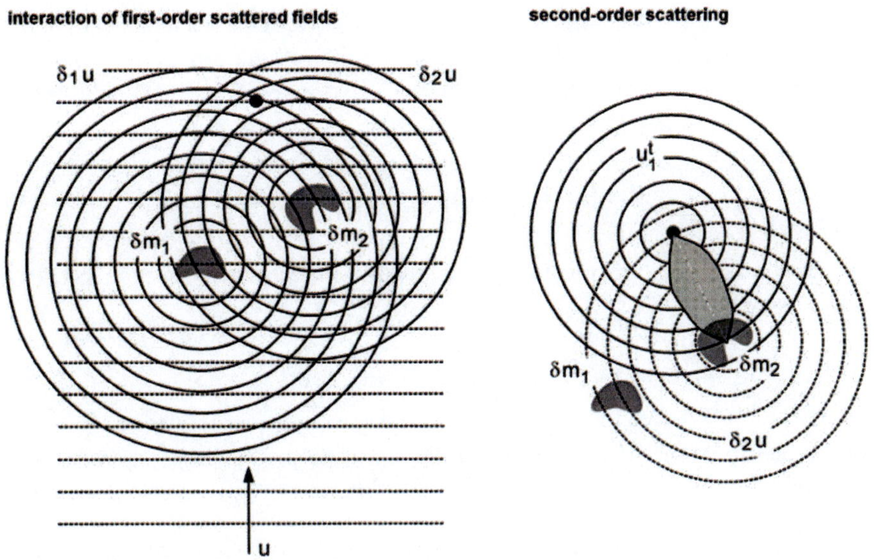

Fig. 9.2 Schematic illustration of the contributions to the Hessian as given in Eq. (9.42). *Left*: Scattering of the forward wave field, $\mathbf{u}$, at the perturbations $\delta\mathbf{m}_1$ and $\delta\mathbf{m}_2$ generates the first-order scattered waves $\delta_1\mathbf{u}$ and $\delta_2\mathbf{u}$, respectively. The contribution $\langle\nabla_u\nabla_u\chi_1(\delta_1\mathbf{u},\delta_2\mathbf{u})\rangle$ measures the interaction of the scattered wave fields $\delta_1\mathbf{u}$ and $\delta_2\mathbf{u}$ at the receiver position, indicated by a filled circle (•). It is largest when the traveltimes from the perturbations to the source are nearly equal. *Right*: The contribution $\langle\mathbf{u}_1^\dagger\cdot\nabla_m\mathbf{L}(\delta_2\mathbf{u})\,\delta\mathbf{m}_1\rangle$ measures the interaction between the primary adjoint field $\mathbf{u}_1^\dagger$ that emanates from the receiver and the scattered wave field $\delta_2\mathbf{u}$ that is excited as $\mathbf{u}$ impinges on the perturbation $\delta\mathbf{m}_2$. The associated Hessian kernel, $K_{\delta\rho_2}$, defined in Eq. (9.47), is non-zero within a secondary influence zone, indicated by the shaded region between $\delta\mathbf{m}_2$ and the receiver. Within the secondary influence zone, a model perturbation causes the scattered field $\delta_2\mathbf{u}$ to generate another scattered field that affects the measurement at the receiver

$$\langle\nabla_u\nabla_u\chi_1(\delta_1\mathbf{u},\delta_2\mathbf{u})\rangle=\int_T\delta_1\mathbf{u}(\mathbf{x}^r,t)\cdot\delta_2\mathbf{u}(\mathbf{x}^r,t)\,dt\,. \tag{9.45}$$

This contribution to the Hessian, illustrated in the left panel of Fig. 9.2, measures the interaction of the scattered wave fields at the receiver position, $\mathbf{x}^r$. It is large only when the traveltimes from $\delta\mathbf{m}_1$ and $\delta\mathbf{m}_2$ to the receiver are nearly equal, so that both scattered wave fields, $\delta_1\mathbf{u}$ and $\delta_2\mathbf{u}$, arrive within the measurement interval $T=[t_0,t_1]$. This implies that $\nabla_u\nabla_u\chi_1$ tends to be dominated by its diagonal elements.

Second-Order Scattering

We now consider the term $\langle\mathbf{u}_1^\dagger\cdot\nabla_m\mathbf{L}(\delta_2\mathbf{u})\,\delta\mathbf{m}_1\rangle$. Inserting the wave equation operator and a pure density variation, $\delta\rho_1$, gives

$$\langle\mathbf{u}_1^\dagger\cdot\nabla_m\mathbf{L}(\delta_2\mathbf{u})\,\delta\mathbf{m}_1\rangle=-\int_T\int_G\delta\rho_1(\mathbf{x})\,\dot{\mathbf{u}}_1^\dagger(\mathbf{x},t)\cdot\delta_2\dot{\mathbf{u}}(\mathbf{x},t)\,dt\,d^3\mathbf{x}\,. \tag{9.46}$$

The integral in Eq. (9.46) measures the interaction between the scattered wave field, $\delta_2\mathbf{u}$, and the primary adjoint field, $\mathbf{u}_1^\dagger$, at the position of the density perturbation, $\delta\rho_1$. Omitting the spatial integration defines a Hessian kernel that is similar to the ones introduced in Eqs. (8.67), (8.68) and (8.69)

$$K_{\delta\rho_2}(\mathbf{x}) := -\int_T \dot{\mathbf{u}}_1^\dagger(\mathbf{x},t)\cdot\delta_2\dot{\mathbf{u}}(\mathbf{x},t)\,dt\,. \tag{9.47}$$

The structure of $K_{\delta\rho_2}$ is similar to the structure of the Fréchet kernels from Sect. 9.1. In fact, we found

$$K_\rho^0 = -\int_T \dot{\mathbf{u}}_1^\dagger(\mathbf{x},t)\cdot\dot{\mathbf{u}}(\mathbf{x},t)\,dt\,, \tag{9.48}$$

for the density kernel in a perfectly elastic and isotropic medium, parameterised in terms of ρ, λ and μ (Eq. 9.11a). The Fréchet kernel K_ρ^0 localises a *primary influence zone* where a secondary wave field scattered off a density perturbation affects the objective functional, χ, at the receiver position, $\mathbf{x}^\mathrm{r}$. Similarly, the Hessian kernel $K_{\delta\rho_2}$ localises a *secondary influence zone* where a density perturbation, $\delta\rho_1$, can cause the scattered field, $\delta_2\mathbf{u}$, to generate a secondary scattered field that is able to affect the observation at the receiver. This is schematically shown in the right panel of Fig. 9.2. The Hessian therefore accounts for *second-order scattering*, that is the scattering of scattered waves. An interpretation similar to the one for $\langle\mathbf{u}_1^\dagger\cdot\nabla_m\mathbf{L}(\delta_2\mathbf{u})\,\delta\mathbf{m}_1\rangle$ holds, of course, for the contribution $\langle\mathbf{u}_1^\dagger\cdot\nabla_m\mathbf{L}(\delta_1\mathbf{u})\,\delta\mathbf{m}_2\rangle$ that we alternatively expressed in terms of the Hessian kernel $K_{\delta m_1}^b = K_{\delta\rho_1}^b$:

$$\langle\mathbf{u}_1^\dagger\cdot\nabla_m\mathbf{L}(\delta_1\mathbf{u})\,\delta\mathbf{m}_2\rangle = \int_G K_{\delta m_1}^b\,\delta\mathbf{m}_2\,d^3\mathbf{x}\,, \tag{9.49}$$

with

$$K_{\delta\rho_1}^b = -\int_T \dot{\mathbf{u}}_1^\dagger(\mathbf{x},t)\cdot\delta_1\dot{\mathbf{u}}(\mathbf{x},t)\,dt\,. \tag{9.50}$$

Combination Term

Finally, the contribution $\langle\mathbf{u}_1^\dagger\cdot\nabla_m\nabla_m\mathbf{L}(\mathbf{u})(\delta\mathbf{m}_1,\delta\mathbf{m}_2)\rangle$ to the Hessian from Eq. (9.42) is similar to the expressions for first derivatives from Eq. (8.16). The only difference is that $\nabla_m\nabla_m(\mathbf{u})\mathbf{L}(\delta\mathbf{m}_1,\delta\mathbf{m}_2)$ appears instead of $\nabla_m\mathbf{L}\,\delta\mathbf{m}$. This suggests to interpret $\langle\mathbf{u}_1^\dagger\cdot\nabla_m\nabla_m\mathbf{L}(\mathbf{u})(\delta\mathbf{m}_1,\delta\mathbf{m}_2)\rangle$ in terms of first-order scattering of the regular field, $\mathbf{u}$, caused by a *combination* of the perturbations $\delta\mathbf{m}_1$ and $\delta\mathbf{m}_2$. Note that this is different from the scenarios that we had before, where scattering of a wave field (regular or already scattered) was due to either $\delta\mathbf{m}_1$ or $\delta\mathbf{m}_2$. Whether the *combination term* is zero or not depends strongly on the parameterisation. For instance, $\mathbf{L}$ is linear with respect to the model parameters in a medium parame-

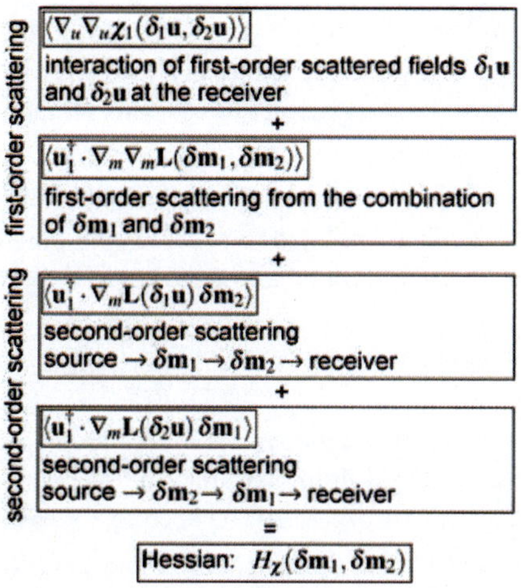

Fig. 9.3 Physical interpretation of the contributions to the Hessian, as given in Eq. (9.42). The second derivatives are affected by both first-order and second-order scattering effects

terised in terms of density and the elastic parameters; and therefore $\nabla_m \nabla_m \mathbf{L} = \mathbf{0}$. Parameterisations in terms of seismic velocities generate quadratic dependencies and $\nabla_m \nabla_m \mathbf{L} \neq \mathbf{0}$. In Sects. 9.3.4, 9.3.5 and 9.3.6 we study the combination term $\langle \mathbf{u}_1^\dagger \cdot \nabla_m \nabla_m \mathbf{L}(\mathbf{u})(\delta \mathbf{m}_1, \delta \mathbf{m}_2) \rangle$ for several seismologically relevant rheologies. Therefore, it will become evident that $\langle \mathbf{u}_1^\dagger \cdot \nabla_m \nabla_m \mathbf{L}(\mathbf{u})(\delta \mathbf{m}_1, \delta \mathbf{m}_2) \rangle$ accounts for the first-order scattering that results from the combination of the two perturbations $\delta \mathbf{m}_1$ and $\delta \mathbf{m}_2$.

Figure 9.3 summarises the role played by the different contributions to the Hessian. They are due to either first-order or second-order scattering.

Interpretation of Hessian Kernels

As discussed already in Sect. 8.4, it is often desirable to compute $H_\chi\, \delta \mathbf{m}_1$ instead of $H_\chi(\delta \mathbf{m}_1, \delta \mathbf{m}_2)$. For this we derived an alternative version of Eq. (9.42) that allows us to isolate the second direction, $\delta \mathbf{m}_2$ (Eq. 8.65):

$$\begin{aligned} H_\chi(\delta \mathbf{m}_1, \delta \mathbf{m}_2) &= \langle \mathbf{u}_2^\dagger \cdot \nabla_m \mathbf{L}(\mathbf{u})\, \delta \mathbf{m}_2 \rangle + \langle \mathbf{u}_1^\dagger \cdot \nabla_m \mathbf{L}(\delta_1 \mathbf{u})\, \delta \mathbf{m}_2 \rangle \\ &\quad + \langle \mathbf{u}_1^\dagger \cdot \nabla_m \nabla_m \mathbf{L}(\mathbf{u})(\delta \mathbf{m}_1)\, \delta \mathbf{m}_2 \rangle \\ &= \int_G \left(K^a_{\delta m_1} + K^b_{\delta m_1} + K^c_{\delta m_1} \right) \delta \mathbf{m}_2\, d^3\mathbf{x}\,, \end{aligned} \tag{9.51}$$

with the Hessian kernels, $K^a_{\delta m_1}$, $K^b_{\delta m_1}$ and $K^c_{\delta m_1}$ defined in Eqs. (8.67), (8.68) and (8.69). The secondary adjoint field, $\mathbf{u}_2^\dagger$, is defined as the solution of the secondary adjoint equation

$$\mathbf{L}^\dagger(\mathbf{u}_2^\dagger) = -\nabla_u \nabla_u \chi_1^\dagger \, \delta_1 \mathbf{u} - \nabla_m \mathbf{L}^\dagger(\mathbf{u}_1^\dagger) \, \delta \mathbf{m}_1 \, . \tag{9.52}$$

We are already familiar with the second and third terms on the right-hand side of Eq. (9.51). In order to attach physical meaning to the contribution $\langle \mathbf{u}_2^\dagger \cdot \nabla_m \mathbf{L}(\mathbf{u}) \, \delta \mathbf{m}_2 \rangle$ and the associated Hessian kernel $K^a_{\delta m_1}$, we first take a closer look at the sources of $\mathbf{u}_2^\dagger$: Inserting the misfit functional defined in (9.41) and a pure density perturbation, $\delta\rho_1$, transforms the secondary adjoint equation (9.52) to

$$\mathbf{L}^\dagger(\mathbf{u}_2^\dagger) = -\delta_1 \mathbf{u}(\mathbf{x}, t) \, \delta(\mathbf{x} - \mathbf{x}^r) - \delta\rho_1(\mathbf{x}) \, \ddot{\mathbf{u}}_1^\dagger(\mathbf{x}, t) \, . \tag{9.53}$$

The term $\delta_1 \mathbf{u}(\mathbf{x}, t) \, \delta(\mathbf{x} - \mathbf{x}^r)$ represents a point source localised at the receiver position, $\mathbf{x}^r$, that radiates the scattered field, $\delta_1 \mathbf{u}$, back into the medium. This is shown schematically in the left panel of Fig. 9.4. The second source term, $\delta\rho_1(\mathbf{x}) \, \ddot{\mathbf{u}}_1^\dagger(\mathbf{x}, t)$, acts when the primary adjoint field, $\mathbf{u}_1^\dagger$, reaches the model perturbation $\delta \mathbf{m}_1$. It radiates a wave field emanating from $\delta \mathbf{m}_1$ into the medium (Fig. 9.4, right). Now we consider the specific form of $\langle \mathbf{u}_2^\dagger \cdot \nabla_m \mathbf{L}(\mathbf{u}) \, \delta \mathbf{m}_2 \rangle$:

$$\langle \mathbf{u}_2^\dagger \cdot \nabla_m \mathbf{L}(\mathbf{u}) \, \delta \mathbf{m}_2 \rangle = \int_T \int_G \delta\rho_2 \, \mathbf{u}_2^\dagger \cdot \ddot{\mathbf{u}} \, \mathrm{d}t \, \mathrm{d}^3\mathbf{x} = \int_G K^a_{\delta\rho_1} \, \delta\rho_2 \, \mathrm{d}^3\mathbf{x} \, . \tag{9.54}$$

Since $\mathbf{u}_2^\dagger$ has two sources, its interaction with the forward field, $\mathbf{u}$, generates two secondary influence zones that define the support of the kernel $K^a_{\delta m_1}$. These are shown as shaded areas in Fig. 9.4. The secondary influence zone associated with the point source, $\delta_1 \mathbf{u}(\mathbf{x}, t) \, \delta(\mathbf{x} - \mathbf{x}^r)$, localises the region where a model perturbation generates a scattered wave that affects the measurement of $\delta_1 \mathbf{u}$ at the receiver. Similarly, the secondary influence zone corresponding to the adjoint source $\delta\rho_1(\mathbf{x}) \, \ddot{\mathbf{u}}_1^\dagger(\mathbf{x}, t)$ defines the region where model perturbations generate scattered waves that are then scattered again from $\delta \mathbf{m}_1$ and finally reach the receiver such that they affect the measurement. This is where the effect of second-order scattering on the Hessian becomes most explicit.

9.3.2 Practical Resolution of the Secondary Adjoint Equation

The secondary adjoint equation (9.52) is inconvenient from a numerical point of view because it involves the source term, $-\nabla_m \mathbf{L}^\dagger(\mathbf{u}_1^\dagger) \, \delta \mathbf{m}_1$, that becomes active when the primary adjoint field, $\mathbf{u}_1^\dagger$, interacts with the model perturbation $\delta \mathbf{m}_1$. The proper implementation of this spatially distributed source may require substantial modifications of pre-existing codes. To avoid this complication, we propose an alternative to the direct solution of the secondary adjoint equation. For this, we consider the primary adjoint equation

$$\mathbf{L}^\dagger(\mathbf{u}_1^\dagger) = -\nabla_u \chi_1^\dagger \, . \tag{9.55}$$

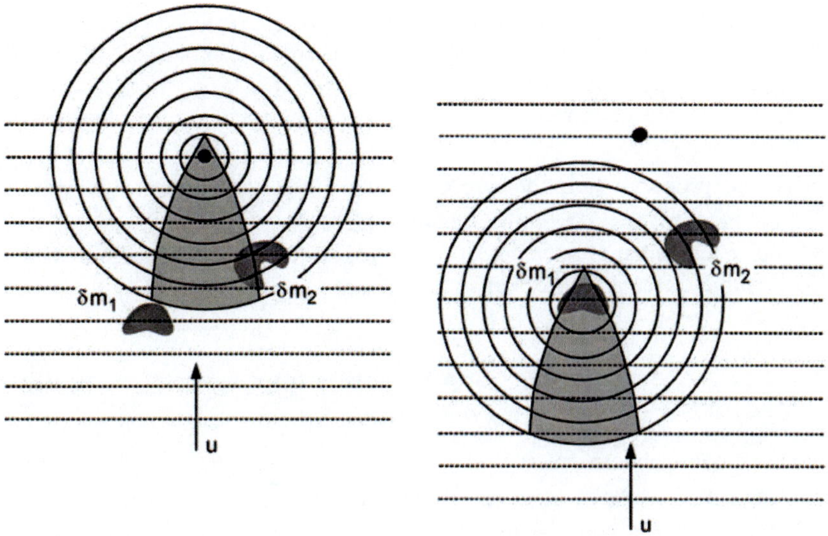

Fig. 9.4 Schematic illustration of the contribution $\langle \mathbf{u}_2^\dagger \cdot \nabla_m \mathbf{L}(\mathbf{u})\, \delta\mathbf{m}_2 \rangle = \int_G K_{\delta m_1}^a \, \delta\mathbf{m}_2 \, d^3\mathbf{x}$ to the alternative form of the Hessian given in Eq. (9.51). The secondary adjoint field, $\mathbf{u}_2^\dagger$, emanates from two sources: a point source localised at the receiver position, (•), and a distributed source located at the perturbation $\delta\mathbf{m}_1$. These two components are shown in the form of solid circles in the *left* and the *right panels*, respectively. *Left*: The first component of the secondary adjoint field, $\mathbf{u}_2^\dagger$, propagates the recorded scattered $\delta_1\mathbf{u}$ back into the medium. There it interacts with the forward field, $\mathbf{u}$. This interaction defines a secondary influence zone, shown as shaded area, where model perturbations generate scattered fields that affect the measurement of $\delta_1\mathbf{u}$. *Right*: The second component of $\mathbf{u}_2^\dagger$ is excited when the primary adjoint wave field, $\mathbf{u}_1^\dagger$, impinges on the model perturbation $\delta\mathbf{m}_1$. A secondary influence zone results from the interaction of this component with the forward field. It localises the region where secondary scattering from some model perturbation via $\delta\mathbf{m}_1$ can affect the measurement at the receiver

Differentiating with respect to $\mathbf{m}$ in the direction $\delta\mathbf{m}_1$ and rearranging terms gives

$$\mathbf{L}^\dagger(\delta_1\mathbf{u}_1^\dagger) = -\nabla_u\nabla_u\chi_1^\dagger\,\delta_1\mathbf{u} - \nabla_m\mathbf{L}^\dagger(\mathbf{u}_1^\dagger)\,\delta\mathbf{m}_1\,. \tag{9.56}$$

Equation (9.56) demonstrates that the first-order scattered adjoint field, $\delta_1\mathbf{u}_1^\dagger = \nabla_m\mathbf{u}_1^\dagger\,\delta\mathbf{m}_1$, solves the secondary adjoint equation (9.52). This observation suggests a recipe for the computation of $H_\chi\,\delta\mathbf{m}_1$ that we summarise in the following paragraph.

9.3.3 Hessian Recipe

To practically compute the Hessian applied to a model perturbation, $H_\chi\,\delta\mathbf{m}_1$, we proceed as follows:

1. Solve the forward problem, $\mathbf{L}(\mathbf{u}) = \mathbf{f}$, and store the regular wave field, $\mathbf{u}$, at sufficiently many time steps to allow for the accurate computation of the time integrals involved in $H_\chi\, \delta\mathbf{m}_1$.
2. Evaluate the objective functional, $\chi(\mathbf{u}) = \langle \chi_1(\mathbf{u}) \rangle$, and from it compute the adjoint source $-\nabla_u \chi_1^\dagger$.
3. Solve the primary adjoint equation, $\mathbf{L}^\dagger(\mathbf{u}_1^\dagger) = -\nabla_u \chi_1^\dagger$, and store the primary adjoint field, $\mathbf{u}_1^\dagger$, at sufficiently many time steps.
4. Compute the scattered wave field, $\delta_1\mathbf{u} = \nabla_m \mathbf{u}\, \delta\mathbf{m}_1$, via either the adjoint method or the finite-difference approximation

$$\delta_1\mathbf{u} \approx \frac{1}{\varepsilon}[\mathbf{u}(\mathbf{m} + \varepsilon\delta\mathbf{m}_1) - \mathbf{u}(\mathbf{m})]\,, \tag{9.57}$$

with a suitably chosen $\varepsilon > 0$.
5. Compute the scattered primary adjoint field, $\delta_1\mathbf{u}_1^\dagger$, using a finite-difference approximation similar to the one in (9.57):

$$\delta_1\mathbf{u}_1^\dagger \approx \frac{1}{\varepsilon}[\mathbf{u}_1^\dagger(\mathbf{m} + \varepsilon\delta\mathbf{m}_1) - \mathbf{u}_1^\dagger(\mathbf{m})]\,.$$

From (9.56) we infer that $\delta_1\mathbf{u}_1^\dagger$ solves the secondary adjoint equation (9.52), i.e. $\mathbf{u}_2^\dagger = \delta_1\mathbf{u}_1^\dagger$.
6. Compute the Hessian kernels

$$\begin{aligned}
K_{\delta m_1}^a &= \int_T \mathbf{u}_2^\dagger \cdot \nabla_m \mathbf{L}(\mathbf{u})\, \mathrm{d}t\,, \\
K_{\delta m_1}^b &= \int_T \mathbf{u}_1^\dagger \cdot \nabla_m \mathbf{L}(\delta_1\mathbf{u})\, \mathrm{d}t\,, \\
K_{\delta m_1}^c &= \int_T \mathbf{u}_1^\dagger \cdot \nabla_m \nabla_m \mathbf{L}(\mathbf{u})(\delta\mathbf{m}_1)\, \mathrm{d}t\,.
\end{aligned}$$

7. Assemble $H_\chi\, \delta\mathbf{m}_1$ according to

$$H_\chi\, \delta\mathbf{m}_1 = \int_G \left(K_{\delta m_1}^a + K_{\delta m_1}^b + K_{\delta m_1}^c\right) \circ \mathrm{d}^3\mathbf{x}\,,$$

where $\circ$ is a place holder for a second model perturbation, $\delta\mathbf{m}_2$.

In the following paragraphs we study the Hessian kernels for several rheologies that are of special seismological interest.

9.3.4 Perfectly Elastic and Isotropic Medium

The components of the Hessian, $H_\chi(\delta\mathbf{m}_1, \delta\mathbf{m}_2)$, can be represented as the sum over three terms involving the regular wave field, $\mathbf{u}$, the primary adjoint wave field, $\mathbf{u}_1^\dagger$, and their scattered versions, $\delta_1\mathbf{u}$ and $\delta_1\mathbf{u}_1^\dagger = \mathbf{u}_2^\dagger$ (Eq. 9.51). The first two contributions,

$$\langle \mathbf{u}_2^\dagger \cdot \nabla_m \mathbf{L}(\mathbf{u})\, \delta\mathbf{m}_2 \rangle = \int_G K_{\delta m_1}^a\, \delta\mathbf{m}_2\, \mathrm{d}^3\mathbf{x}$$

and

$$\langle \mathbf{u}_1^\dagger \cdot \nabla_m \mathbf{L}(\delta_1\mathbf{u})\, \delta\mathbf{m}_2 \rangle = \int_G K_{\delta m_1}^b\, \delta\mathbf{m}_2\, \mathrm{d}^3\mathbf{x}\,,$$

are of a form that we already encountered in the computation of first derivatives and Fréchet kernels. In fact, for $\nabla_m \chi_1\, \delta\mathbf{m}$, we already found (see Eq. 8.16)

$$\nabla_m \chi_1\, \delta\mathbf{m} = \langle \mathbf{u}_1^\dagger \cdot \nabla_m \mathbf{L}(\mathbf{u})\, \delta\mathbf{m} \rangle = \int_G K_m\, \delta\mathbf{m}\, \mathrm{d}^3\mathbf{x}\,.$$

We may therefore use the formulas derived in Sect. 9.1 for the Fréchet kernels to compute the first two contributions to the Hessian. In what follows, we focus on the combination term $\langle \mathbf{u}_1^\dagger \cdot \nabla_m \nabla_m \mathbf{L}(\mathbf{u})(\delta\mathbf{m}_1)\, \delta\mathbf{m}_2 \rangle$.

As a preparatory step towards explicit expressions for special rheologies, we substitute the elastic wave operator (8.75) for $\mathbf{L}$:

$$\begin{aligned}
\langle \mathbf{u}_1^\dagger \cdot \nabla_m \nabla_m \mathbf{L}(\delta\mathbf{m}_1)\, \delta\mathbf{m}_2 \rangle &= \int_G K_{\delta m_1}^c\, \delta\mathbf{m}_2\, \mathrm{d}^3\mathbf{x} \\
&= \int_T \int_G \nabla_m \nabla_m \rho\, (\delta\mathbf{m}_1, \delta\mathbf{m}_2)\, \mathbf{u}_1^\dagger(t) \cdot \ddot{\mathbf{u}}(t)\, \mathrm{d}t\, \mathrm{d}^3\mathbf{x} \\
&- \int_T \int_G \mathbf{u}_1^\dagger(t) \cdot \left[\nabla \cdot \int_{\tau=t_0}^{t} \nabla_m \nabla_m \dot{\mathbf{C}}(t-\tau)(\delta\mathbf{m}_1, \delta\mathbf{m}_2) : \nabla\mathbf{u}(\tau) \right] \mathrm{d}\tau\, \mathrm{d}t\, \mathrm{d}^3\mathbf{x}\,.
\end{aligned} \tag{9.58}$$

In Eq. (9.58) we did not specify a particular parameterisation. This is accounted for by the notations $\nabla_m \nabla_m \rho$ and $\nabla_m \nabla_m \dot{\mathbf{C}}$ that allow ρ and $\mathbf{C}$ to depend on model parameters, $\mathbf{m}$, other than density and the elastic coefficients themselves. Integrating Eq. (9.58) by parts and writing the resulting expression in terms of the regular and primary adjoint strain fields, $\boldsymbol{\varepsilon}$ and $\boldsymbol{\varepsilon}_1^\dagger$, give

$$\int_G K^c_{\delta m_1}\,\delta\mathbf{m}_2\,\mathrm{d}^3\mathbf{x} = -\int_T\int_G \nabla_m\nabla_m\rho\,(\delta\mathbf{m}_1,\delta\mathbf{m}_2)\,\dot{\mathbf{u}}_1^\dagger(t)\cdot\dot{\mathbf{u}}(t)\,\mathrm{d}t\,\mathrm{d}^3\mathbf{x}$$
$$+\int_T\int_G\int_{\tau=t_0}^{t} \boldsymbol{\varepsilon}_1^\dagger(t) : \left[\nabla_m\nabla_m\dot{\mathbf{C}}(t-\tau)\,(\delta\mathbf{m}_1,\delta\mathbf{m}_2)\right] : \boldsymbol{\varepsilon}(\tau)\,\mathrm{d}\tau\,\mathrm{d}t\,\mathrm{d}^3\mathbf{x}\,. \quad (9.59)$$

We now insert the perfectly elastic and isotropic rheology described by Eqs. (9.6) and (9.8):

$$\int_G K^c_{\delta m_1}\,\delta\mathbf{m}_2\,\mathrm{d}^3\mathbf{x} = -\int_T\int_G \nabla_m\nabla_m\rho\,(\delta\mathbf{m}_1,\delta\mathbf{m}_2)\,\dot{\mathbf{u}}_1^\dagger\cdot\dot{\mathbf{u}}\,\mathrm{d}t\,\mathrm{d}^3\mathbf{x}$$
$$+\int_T\int_G \nabla_m\nabla_m\lambda\,(\delta\mathbf{m}_1,\delta\mathbf{m}_2)\,\mathrm{tr}\,\boldsymbol{\varepsilon}_1^\dagger\,\mathrm{tr}\,\boldsymbol{\varepsilon}\;\mathrm{d}t\,\mathrm{d}^3\mathbf{x}$$
$$+\int_T\int_G \nabla_m\nabla_m\mu\,(\delta\mathbf{m}_1,\delta\mathbf{m}_2)\,\boldsymbol{\varepsilon}_1^\dagger : \boldsymbol{\varepsilon}\,\mathrm{d}t\,\mathrm{d}^3\mathbf{x}\,. \quad (9.60)$$

When the isotropic medium is parameterised in terms of ρ, λ and μ or, alternatively, ρ, κ and μ, we find

$$K^c_{\delta m_1} = 0\,, \quad (9.61)$$

meaning that the combination term is identically zero. This result depends critically on the choice of the free parameters. Changing, for instance, the set of free parameters to ρ, the P wave speed, $v_{\mathrm{P}} = \sqrt{(\lambda+2\mu)/\rho}$, and the S wave speed, $v_{\mathrm{S}} = \sqrt{\mu/\rho}$, results in

$$\nabla_m\nabla_m\rho\,(\delta\mathbf{m}_1,\delta\mathbf{m}_2) = 0\,,$$
$$\nabla_m\nabla_m\mathbf{C}(\delta\mathbf{m}_1,\delta\mathbf{m}_2) = \delta\mathbf{m}_1\cdot\begin{pmatrix} 0 & 2v_{\mathrm{P}}\delta_{ij}\delta_{kl} & 2v_{\mathrm{S}}\gamma_{ijkl} \\ 2v_{\mathrm{P}}\delta_{ij}\delta_{kl} & 2\rho\delta_{ij}\delta_{kl} & 0 \\ 2v_{\mathrm{S}}\gamma_{ijkl} & 0 & 2\rho\gamma_{ijkl}\end{pmatrix}\cdot\delta\mathbf{m}_2\,, \quad (9.62)$$

with the model vector

$$\delta\mathbf{m}_i = (\delta\rho_i\,,\,\delta v_{\mathrm{P},i}\,,\,\delta v_{\mathrm{S},i})\,,\quad i=1,2\,, \quad (9.63)$$

and the auxiliary variable γ_{ijkl} defined by

$$\gamma_{ijkl} := \delta_{ik}\delta_{jl} + \delta_{il}\delta_{jk} - 2\delta_{ij}\delta_{kl}\,. \quad (9.64)$$

The combination term can now be written in a very convenient form

$$\int_G K^c_{\delta m_1}\,\delta\mathbf{m}_2\,\mathrm{d}^3\mathbf{x} = \int_G \delta\mathbf{m}_1\cdot\mathbf{J}\cdot\delta\mathbf{m}_2\,\mathrm{d}^3\mathbf{x}\,, \quad (9.65)$$

where we defined the symmetric 3×3 matrix $\mathbf{J}$ through

$$\mathbf{J} := \int_T \begin{pmatrix} 0 & 2v_\mathrm{P}\,\mathrm{tr}\,\boldsymbol{\varepsilon}_1^\dagger\,\mathrm{tr}\,\boldsymbol{\varepsilon} & 4v_\mathrm{S}\,(\boldsymbol{\varepsilon}_1^\dagger : \boldsymbol{\varepsilon} - \mathrm{tr}\,\boldsymbol{\varepsilon}_1^\dagger\,\mathrm{tr}\,\boldsymbol{\varepsilon}) \\ 2v_\mathrm{P}\mathrm{tr}\,\boldsymbol{\varepsilon}_1^\dagger\,\mathrm{tr}\,\boldsymbol{\varepsilon} & 2\rho\,\mathrm{tr}\,\boldsymbol{\varepsilon}_1^\dagger\,\mathrm{tr}\,\boldsymbol{\varepsilon} & 0 \\ 4v_\mathrm{S}\,(\boldsymbol{\varepsilon}_1^\dagger : \boldsymbol{\varepsilon} - \mathrm{tr}\,\boldsymbol{\varepsilon}_1^\dagger\,\mathrm{tr}\,\boldsymbol{\varepsilon}) & 0 & 4\rho\,(\boldsymbol{\varepsilon}_1^\dagger : \boldsymbol{\varepsilon} - \mathrm{tr}\,\boldsymbol{\varepsilon}_1^\dagger\,\mathrm{tr}\,\boldsymbol{\varepsilon}) \end{pmatrix} dt\,. \tag{9.66}$$

From Eq. (9.65) we immediately infer that the Hessian kernel $H_{\delta m_1}^c$ is explicitly given by

$$K_{\delta m_1}^c = \mathbf{J}\cdot\delta\mathbf{m}_1\,. \tag{9.67}$$

It is interesting to note that the P and S wave speed Fréchet kernels, K_{v_P} and K_{v_S}, from Sect. 9.1.1 reappear in the combination term. In fact, we may write $\mathbf{J}$ in the following form:

$$\mathbf{J} = \begin{pmatrix} 0 & \rho^{-1}K_{v_\mathrm{P}} & \rho^{-1}K_{v_\mathrm{S}} \\ \rho^{-1}K_{v_\mathrm{P}} & v_\mathrm{P}^{-1}K_{v_\mathrm{P}} & 0 \\ \rho^{-1}K_{v_\mathrm{S}} & 0 & v_\mathrm{S}^{-1}K_{v_\mathrm{S}} \end{pmatrix} \tag{9.68}$$

The Fréchet kernels for the isotropic medium can thus be recycled for the computation of the combination term, $\int_G K_{\delta m_1}^c\,\delta\mathbf{m}_2\,d^3\mathbf{x}$, that contributes to the Hessian.

Equation (9.65) is very general, and it will reappear in the following sections. Only the Fréchet kernels composing the matrix $\mathbf{J}$ depend on the rheology.

The comparison of Eqs. (9.61) and (9.65) makes it particularly obvious that the structure of the Hessian in general, and the size of its diagonal elements in particular, depends critically on the choice of the free parameters. In the context of tomographic problems the parameterisation should be chosen such that resolution is maximised. This requires the diagonal elements of H_χ to be as large as possible and the off-diagonal elements as small as possible. The Hessian therefore has the potential of playing a key role in future experimental design problems that attempt to optimise the resolution in tomographic inversions.

Another remarkable aspect of Eqs. (9.66) and (9.68) is the explicit dependence of the off-diagonal elements on ρ, v_P and v_S. Trade-offs between different classes of parameters may therefore depend significantly on the background model.

9.3.5 Perfectly Elastic Medium with Radial Anisotropy

To derive the equivalents of Eqs. (9.65) and (9.68) for an elastic medium with radial anisotropy, we start from the non-dissipative version of (9.59):

$$\int_G K_{\delta m_1}^c\,\delta\mathbf{m}_2\,d^3\mathbf{x} = -\int_T\int_G \nabla_m\nabla_m\rho\,(\delta\mathbf{m}_1,\delta\mathbf{m}_2)\,\dot{\mathbf{u}}_1^\dagger\cdot\dot{\mathbf{u}}\,dt\,d^3\mathbf{x}$$
$$+\int_T\int_G \boldsymbol{\varepsilon}_1^\dagger : [\nabla_m\nabla_m\mathbf{C}\,(\delta\mathbf{m}_1,\delta\mathbf{m}_2)] : \boldsymbol{\varepsilon}\,dt\,d^3\mathbf{x}\,. \tag{9.69}$$

As in the case of the isotropic medium, the second derivatives $\nabla_m \nabla_m \rho$ and $\nabla_m \nabla_m \mathbf{C}$ are equal to zero when the medium is parameterised in terms of the density, ρ, and any linear combination of the elastic coefficients, C_{ijkl}. We therefore consider the practically more relevant case where a model, $\mathbf{m}$, is specified by the density ρ, the wave speeds v_{PV}, v_{PH}, v_{SV} and v_{SH}, and the parameter η (see Sect. 9.1.2):

$$\mathbf{m} = (\rho, v_{\mathrm{PV}}, v_{\mathrm{PH}}, v_{\mathrm{SV}}, v_{\mathrm{SH}}, \eta)\,. \tag{9.70}$$

The second derivative $\nabla_m \nabla_m \rho$ is still zero because ρ is a free parameter. Inserting the anisotropic rheology specified in Eqs. (9.16) allows us to rewrite the second integrand on the right-hand side of (9.69):

$$\begin{aligned}\boldsymbol{\varepsilon}_1^\dagger : [\nabla_m \nabla_m \mathbf{C}\,(\delta\mathbf{m}_1, \delta\mathbf{m}_2)] : \boldsymbol{\varepsilon} = \nabla_m \nabla_m [&\rho v_{\mathrm{PV}}^2 \varepsilon_{1,rr}^\dagger \varepsilon_{rr} + \eta\rho(v_{\mathrm{PH}}^2 - 2v_{\mathrm{SH}}^2)\, \varepsilon_{1,rr}^\dagger \varepsilon_{\phi\phi} \\ &+ \eta\rho(v_{\mathrm{PH}}^2 - 2v_{\mathrm{SH}}^2)\, \varepsilon_{1,rr}^\dagger \varepsilon_{\theta\theta} + \eta\rho(v_{\mathrm{PH}}^2 - 2v_{\mathrm{SH}}^2)\, \varepsilon_{1,\phi\phi}^\dagger \varepsilon_{rr} + \rho v_{\mathrm{PH}}^2 \varepsilon_{1,\phi\phi}^\dagger \varepsilon_{\phi\phi} \\ &+ \rho(v_{\mathrm{PH}}^2 - 2v_{\mathrm{SH}}^2)\, \varepsilon_{1,\phi\phi}^\dagger \varepsilon_{\theta\theta} + \eta\rho(v_{\mathrm{PH}}^2 - 2v_{\mathrm{SH}}^2)\, \varepsilon_{1,\theta\theta} \varepsilon_{rr} + \rho(v_{\mathrm{PH}}^2 - 2v_{\mathrm{SH}}^2)\, \varepsilon_{1,\theta\theta}^\dagger \varepsilon_{\phi\phi} \\ &+ \rho v_{\mathrm{PH}}^2 \varepsilon_{1,\theta\theta}^\dagger \varepsilon_{\theta\theta} + 4\rho v_{\mathrm{SH}}^2 \varepsilon_{1,r\phi}^\dagger \varepsilon_{r\phi} + 4\rho v_{\mathrm{SV}}^2 \varepsilon_{r\theta}^\dagger \varepsilon_{r\theta} + 4\rho v_{\mathrm{SH}}^2 \varepsilon_{1,\phi\theta}^\dagger \varepsilon_{\phi\theta}](\delta\mathbf{m}_1, \delta\mathbf{m}_2)\,.\end{aligned} \tag{9.71}$$

Tedious but straightforward algebra then leads to

$$\int_G K_{\delta m_1}^c\, \delta\mathbf{m}_2\, \mathrm{d}^3\mathbf{x} = \int_G \delta\mathbf{m}_1 \cdot \mathbf{J} \cdot \delta\mathbf{m}_2\, \mathrm{d}^3\mathbf{x} \tag{9.72}$$

in analogy to Eq. (9.65). The components of the symmetric 6×6 matrix

$$\mathbf{J} = \begin{pmatrix} J_{\rho\rho} & J_{\rho v_{\mathrm{PV}}} & J_{\rho v_{\mathrm{PH}}} & J_{\rho v_{\mathrm{SV}}} & J_{\rho v_{\mathrm{SH}}} & J_{\rho\eta} \\ J_{v_{\mathrm{PV}}\rho} & J_{v_{\mathrm{PV}} v_{\mathrm{PV}}} & J_{v_{\mathrm{PV}} v_{\mathrm{PH}}} & J_{v_{\mathrm{PV}} v_{\mathrm{SV}}} & J_{v_{\mathrm{PV}} v_{\mathrm{SH}}} & J_{v_{\mathrm{PV}}\eta} \\ J_{v_{\mathrm{PH}}\rho} & J_{v_{\mathrm{PH}} v_{\mathrm{PV}}} & J_{v_{\mathrm{PH}} v_{\mathrm{PH}}} & J_{v_{\mathrm{PH}} v_{\mathrm{SV}}} & J_{v_{\mathrm{PH}} v_{\mathrm{SH}}} & J_{v_{\mathrm{PH}}\eta} \\ J_{v_{\mathrm{SV}}\rho} & J_{v_{\mathrm{SV}} v_{\mathrm{PV}}} & J_{v_{\mathrm{SV}} v_{\mathrm{PH}}} & J_{v_{\mathrm{SV}} v_{\mathrm{SV}}} & J_{v_{\mathrm{SV}} v_{\mathrm{SH}}} & J_{v_{\mathrm{SV}}\eta} \\ J_{v_{\mathrm{SH}}\rho} & J_{v_{\mathrm{SH}} v_{\mathrm{PV}}} & J_{v_{\mathrm{SH}} v_{\mathrm{PH}}} & J_{v_{\mathrm{SH}} v_{\mathrm{SV}}} & J_{v_{\mathrm{SH}} v_{\mathrm{SH}}} & J_{v_{\mathrm{SH}}\eta} \\ J_{\eta\rho} & J_{\eta v_{\mathrm{PV}}} & J_{\eta v_{\mathrm{PH}}} & J_{\eta v_{\mathrm{SV}}} & J_{\eta v_{\mathrm{SH}}} & J_{\eta\eta} \end{pmatrix} \tag{9.73}$$

can again be expressed in terms of the Fréchet kernels for the radially anisotropic medium, introduced in Eqs. (9.20) and (9.22):

$$J_{\rho\rho} = 0\,, \tag{9.74a}$$

$$J_{\eta\eta} = 0\,, \tag{9.74b}$$

$$J_{v_{PV}v_{PV}} = 2\rho\left(K_\lambda^0 - K_c^0 - K_a^0\right) = v_P^{-1} K_{v_{PV}}\,, \tag{9.74c}$$

$$J_{v_{PH}v_{PH}} = 2\rho\left(K_a^0 + K_c^0\right) = v_P^{-1} K_{v_{PH}}, \tag{9.74d}$$

$$J_{v_{SV}v_{SV}} = 2\rho\, K_b^0 = v_S^{-1} K_{v_{SV}}\,, \tag{9.74e}$$

$$J_{v_{SH}v_{SH}} = 2\rho\left(K_\mu^0 - 2K_\lambda^0 - K_b^0\right) = v_S^{-1} K_{v_{SH}}\,, \tag{9.74f}$$

$$J_{v_{PV}\rho} = J_{\rho v_{PV}} = 2v_P\left(K_\lambda^0 - K_c^0 - K_a^0\right) = \rho^{-1} K_{v_{PV}}\,, \tag{9.74g}$$

$$J_{v_{PH}\rho} = J_{\rho v_{PH}} = 2v_P\left(K_a^0 + K_c^0\right) = \rho^{-1} K_{v_{PH}}\,, \tag{9.74h}$$

$$J_{v_{SV}\rho} = J_{\rho v_{SV}} = 2v_S\, K_b^0 = \rho^{-1} K_{v_{SV}}, \tag{9.74i}$$

$$J_{v_{SH}\rho} = J_{\rho v_{SH}} = 2v_S\left(K_\mu^0 - 2K_\lambda^0 - K_b^0\right) = \rho^{-1} K_{v_{SH}}\,, \tag{9.74j}$$

$$J_{\eta\rho} = J_{\rho\eta} = \left(v_P^2 - 2v_S^2\right) K_c^0 = \rho^{-1} K_\eta, \tag{9.74k}$$

$$J_{v_{PV}\eta} = J_{\eta v_{PV}} = 0\,, \tag{9.74l}$$

$$J_{v_{PH}\eta} = J_{\eta v_{PH}} = 2\rho v_P\, K_c^0 = 2v_P\left(v_P^2 - 2v_S^2\right)^{-1} K_\eta\,, \tag{9.74m}$$

$$J_{v_{SV}\eta} = J_{\eta v_{SV}} = 0\,, \tag{9.74n}$$

$$J_{v_{SH}\eta} = J_{\eta v_{SH}} = -4\rho v_S\, K_c^0 = -4v_S\left(v_P^2 - 2v_S^2\right)^{-1} K_\eta\,, \tag{9.74o}$$

$$J_{v_{PH}v_{PV}} = J_{v_{PV}v_{PH}} = 0\,, \tag{9.74p}$$

$$J_{v_{SV}v_{PV}} = J_{v_{PV}v_{SV}} = 0\,, \tag{9.74q}$$

$$J_{v_{SH}v_{PV}} = J_{v_{PV}v_{SH}} = 0\,, \tag{9.74r}$$

$$J_{v_{SV}v_{PH}} = J_{v_{PH}v_{SV}} = 0\,, \tag{9.74s}$$

$$J_{v_{SH}v_{PH}} = J_{v_{PH}v_{SH}} = 0\,, \tag{9.74t}$$

$$J_{v_{SH}v_{SV}} = J_{v_{SV}v_{SH}} = 0\,. \tag{9.74u}$$

In the interest of readability, we assumed an isotropic background medium, where $v_{PV} = v_{PH} = v_P$, $v_{SV} = v_{SH} = v_S$ and $\eta = 1$.

9.3.6 Isotropic Visco-Elastic Medium

We consider the combination term, $\int_G K_{\delta m_1}^c\, \delta\mathbf{m}_2\, d^3\mathbf{x}$, for an isotropic and visco-elastic medium parameterised in terms of the shear modulus, μ, the bulk modulus, κ, and the inverse quality factors Q_μ^{-1} and Q_κ^{-1}. The choice for Q_μ^{-1} and Q_κ^{-1} instead of Q_μ and Q_κ is motivated by the independence of the corresponding Fréchet kernels, $K_{Q_\mu^{-1}}$ and $K_{Q_\kappa^{-1}}$ from the background Q model, that we already discussed in

Sect. 9.1.3. The components of the isotropic elastic tensor are

$$C_{ijkl}(t) = \left[\kappa(t) - \frac{2}{3}\mu(t)\right] \delta_{ij}\delta_{kl} + \mu(t)\,\delta_{ik}\delta_{jl} + \mu(t)\,\delta_{il}\delta_{jk}\,, \tag{9.75}$$

with the time-dependent elastic coefficients defined as superpositions of N standard linear solids

$$\mu(t) = \mu_\mathrm{r}\left[1 + \frac{Q_\mu^{-1}}{K}\sum_{n=1}^{N} \mathrm{e}^{-t/\tau_n}\right] H(t)\,, \tag{9.76}$$

$$\kappa(t) = \kappa_\mathrm{r}\left[1 + \frac{Q_\kappa^{-1}}{K}\sum_{n=1}^{N} \mathrm{e}^{-t/\tau_n}\right] H(t) \tag{9.77}$$

that we already analysed in Chap. 5. Equations (9.76) and (9.77) are valid under the assumption that Q_μ^{-1} and Q_κ^{-1} are small compared to 1 and approximately constant over the frequency range of interest, $[\omega_1, \omega_2]$. This frequency range is determined by the relaxation times τ_n, with $n = 1, \ldots, N$. The parameter K is constant within the frequency band $[\omega_1, \omega_2]$, and it is the same for both κ and μ.

As free parameters we consider the inverse quality factors Q_μ^{-1} and Q_κ^{-1} and their corresponding relaxed moduli μ_r and κ_r. Following the procedure from the previous paragraphs, we arrive at a familiar expression for the combination term

$$\int_G K^c_{\delta m_1}\,\delta\mathbf{m}_2\,\mathrm{d}^3\mathbf{x} = \int_G \delta\mathbf{m}_1 \cdot \mathbf{J} \cdot \delta\mathbf{m}_2\,\mathrm{d}^3\mathbf{x}\,, \tag{9.78}$$

with the symmetric 4×4 matrix,

$$\mathbf{J} = \begin{pmatrix} 0 & J_{\mu_\mathrm{r}\,Q_\mu^{-1}} & 0 & 0 \\ J_{\mu_\mathrm{r}\,Q_\mu^{-1}} & 0 & 0 & 0 \\ 0 & 0 & 0 & J_{\kappa_\mathrm{r}\,Q_\kappa^{-1}} \\ 0 & 0 & J_{\kappa_\mathrm{r}\,Q_\kappa^{-1}} & 0 \end{pmatrix} \tag{9.79}$$

and the model perturbations

$$\delta\mathbf{m}_i = (\delta\mu_{\mathrm{r},i}, \delta Q^{-1}_{\mu,i}, \delta\kappa^{-1}_{\mathrm{r},i}, \delta Q^{-1}_{\kappa,i})\,, \quad i = 1, 2\,. \tag{9.80}$$

The non-zero components of $\mathbf{J}$ are

$$J_{\mu_\mathrm{r}\,Q_\mu^{-1}} = \int_T \int_{\tau=t_0}^{t} \left[2\boldsymbol{\varepsilon}_1^\dagger(t) : \boldsymbol{\varepsilon}(\tau) - \frac{2}{3}\,\mathrm{tr}\,\boldsymbol{\varepsilon}_1^\dagger(t)\,\mathrm{tr}\,\boldsymbol{\varepsilon}(\tau)\right] \dot{F}(t-\tau)\,\mathrm{d}\tau\,\mathrm{d}t\,, \tag{9.81}$$

$$J_{\kappa_\mathrm{r}\,Q_\kappa^{-1}} = \int_T \int_{\tau=t_0}^{t} \mathrm{tr}\,\boldsymbol{\varepsilon}_1^\dagger(t)\,\mathrm{tr}\,\boldsymbol{\varepsilon}(\tau)\,\dot{F}(t-\tau)\,\mathrm{d}\tau\,\mathrm{d}t\,. \tag{9.82}$$

For convenience we defined the auxiliary function

$$F(t) := \frac{1}{K}\left[\sum_{n=1}^{N} \mathrm{e}^{-t/\tau_n}\right] H(t)\,. \tag{9.83}$$

To facilitate the numerical computation of (9.81) and (9.82), we express the time convolution in terms of the memory variables defined in (5.7). For this, we note that the following identity holds:

$$\int_{\tau=t_0}^{t} \boldsymbol{\varepsilon}(\tau)\,\dot{F}(t-\tau)\,\mathrm{d}\tau = \int_{\tau=t_0}^{t} F(t-\tau)\,\dot{\boldsymbol{\varepsilon}}(\tau)\,\mathrm{d}\tau = -\frac{N}{K}\sum_{n=1}^{N} \tau_n \mathbf{M}_n(t)\,. \tag{9.84}$$

Introducing (9.84) into (9.81) and (9.82), and comparing the result with the Fréchet kernels $K_{Q_\mu^{-1}}$ and $K_{Q_\kappa^{-1}}$ from Eqs. (9.33), (9.34) and (9.35), allows us to express $\mathbf{J}$ in a condensed and numerically convenient form:

$$\mathbf{J} = \begin{pmatrix} 0 & \mu_\mathrm{r}^{-1} K_{Q_\mu^{-1}} & 0 & 0 \\ \mu_\mathrm{r}^{-1} K_{Q_\mu^{-1}} & 0 & 0 & 0 \\ 0 & 0 & 0 & \kappa_\mathrm{r}^{-1} K_{Q_\kappa^{-1}} \\ 0 & 0 & \kappa_\mathrm{r}^{-1} K_{Q_\kappa^{-1}} & 0 \end{pmatrix}. \tag{9.85}$$

Just as the Fréchet kernels $K_{Q_\mu^{-1}}$ and $K_{Q_\kappa^{-1}}$, the Hessian kernel, $K^c_{\delta m_1}$, is independent of the background Q model. This is a very desirable property given that Q varies by several orders of magnitude inside the Earth. It ensures that the resolution of Q_μ^{-1} and Q_κ^{-1} is nearly independent of the initial model in a tomographic inversion.

Chapter 10
The Frequency-Domain Discrete Adjoint Method

The discrete adjoint method is a special case of the more general continuous adjoint method, applied to physical systems that are governed by algebraic equations. Either these systems can be inherently discrete or they can be the result of a discretisation process.

In this chapter we introduce the discrete adjoint method applied to a discretised frequency-domain wave equation, as it is commonly used for 2D full waveform inversion (see Sect. 2.5.2 and Chap. 14). We put special emphasis on close relations to the continuous adjoint method and on the computation of second derivatives.

10.1 General Formulation

The time-domain continuous adjoint method presented in Chap. 8 is very general because it is independent of the numerical methods used to solve the optimisation problem in practice. As an alternative to the continuous adjoint method, one may derive the adjoint equations based on the already discretised forward problem. This approach is particularly attractive when the wave equation is solved in the frequency domain.

To avoid duplications, our discussion of the discrete adjoint method will be brief compared to the chapters on the continuous adjoint method. In fact, most discrete equations have a direct correspondence in the continuous world.

To derive the frequency-domain discrete adjoint method, we consider the generic form of the space-discretised wave equation in the frequency domain, presented already in Sect. 2.5.2:

$$-\,\omega^2 \mathbf{M}\cdot\bar{\mathbf{u}}(\omega) + \mathbf{K}\cdot\bar{\mathbf{u}}(\omega) = \bar{\mathbf{f}}(\omega)\,, \tag{10.1}$$

where $\mathbf{M}$, $\mathbf{K}$ and $\bar{\mathbf{u}}$ represent the mass matrix, the stiffness matrix and a discrete version of the elastic displacement field. The variables ω and $\bar{\mathbf{f}}$ are the angular frequency and the discrete external force, respectively. Defining the *impedance matrix* $\mathbf{L}(\omega) := -\omega^2\mathbf{M}+\mathbf{K}$, we can rewrite Eq. (10.1) in the form of a simple matrix-vector equation:

A. Fichtner, *Full Seismic Waveform Modelling and Inversion*, Advances in Geophysical and Environmental Mechanics and Mathematics,
DOI 10.1007/978-3-642-15807-0_10,

$$\mathbf{L} \cdot \bar{\mathbf{u}} = \mathbf{f}. \tag{10.2}$$

The matrix $\mathbf{L}$ now contains all the structural information of the Earth model, i.e. $\mathbf{L} = \mathbf{L}(\mathbf{m})$. Furthermore, we assume that the model space $\mathfrak{M}$ is finite dimensional, so that any model $\mathbf{m} \in \mathfrak{M}$ can be written in the form of a vector with $n < \infty$ components:

$$\mathbf{m} = (m_1, m_2, \ldots, m_n). \tag{10.3}$$

In the following paragraphs we slightly generalise the frequency-domain discrete adjoint method as described, for example, by Pratt et al. (1998) and Pratt (1999).

We are interested in the partial derivatives of the objective functional $\chi(\mathbf{m}) = \chi[\bar{\mathbf{u}}(\mathbf{m})]$ with respect to the model parameters m_i:

$$\frac{\partial \chi}{\partial m_i} = \nabla_u \chi \cdot \frac{\partial \bar{\mathbf{u}}}{\partial m_i}. \tag{10.4}$$

To eliminate $\partial \bar{\mathbf{u}} / \partial m_i$ from Eq. (10.4) we differentiate the discrete wave Eq. (10.2) with respect to m_i:

$$\frac{\partial \mathbf{L}}{\partial m_i} \cdot \bar{\mathbf{u}} + \mathbf{L} \cdot \frac{\partial \bar{\mathbf{u}}}{\partial m_i} = \mathbf{0} \quad \Rightarrow \quad \frac{\partial \bar{\mathbf{u}}}{\partial m_i} = -\mathbf{L}^{-1} \cdot \frac{\partial \mathbf{L}}{\partial m_i} \cdot \bar{\mathbf{u}}. \tag{10.5}$$

The appearance of the inverse $\mathbf{L}^{-1}$ in Eq. (10.5) is purely symbolic. It does not have to be computed in practice. Substituting Eq. (10.5) into Eq. (10.4) yields

$$\frac{\partial \chi}{\partial m_i} = -\nabla_u \chi \cdot \left(\mathbf{L}^{-1} \cdot \frac{\partial \mathbf{L}}{\partial m_i} \right) \cdot \bar{\mathbf{u}} = -\bar{\mathbf{u}} \cdot \left[\frac{\partial \mathbf{L}^{\mathrm{T}}}{\partial m_i} \cdot (\mathbf{L}^{-1})^{\mathrm{T}} \right] \cdot \nabla_u \chi. \tag{10.6}$$

We now define the *discrete adjoint wave field* $\bar{\mathbf{u}}^\dagger$ as the solution of the *adjoint equation*

$$\mathbf{L}^{\mathrm{T}} \cdot \bar{\mathbf{u}}^\dagger = -\nabla_u \chi. \tag{10.7}$$

Clearly, Eq. (10.7) corresponds to the adjoint equation $\nabla_u \mathbf{L}^\dagger \bar{\mathbf{u}}^\dagger = -\nabla_u \chi_1^\dagger$ that we found in the continuous case (Eq. 8.15). With the help of the discrete adjoint field $\bar{\mathbf{u}}^\dagger$ we can now obtain a simple expression for $\partial \chi / \partial m_i$:

$$\frac{\partial \chi}{\partial m_i} = \bar{\mathbf{u}}^\dagger \cdot \frac{\partial \mathbf{L}}{\partial m_i} \cdot \bar{\mathbf{u}}. \tag{10.8}$$

As in the continuous case, the computation of the partial derivatives of χ reduces to the solution of the adjoint equation (10.7) with an adjoint source determined by the objective functional. The continuous counterpart of Eq. (10.8) is $\nabla_m \chi \, \delta\mathbf{m} = \langle \bar{\mathbf{u}}^\dagger \cdot \nabla_m \mathbf{L} \, \delta\mathbf{m} \rangle$ (Eq. 8.16).

10.2 Second Derivatives

Following the recipe from the previous section, we can derive an expression for the second derivative of χ in terms of the adjoint field $\bar{\mathbf{u}}^\dagger$, as defined in Eq. (10.7). Differentiating χ first with respect to m_i and then with respect to m_j gives

$$H_{ij} := \frac{\partial^2 \chi}{\partial m_i \partial m_j} = \bar{H}_{ij} + \nabla_u \chi \cdot \frac{\partial^2 \bar{\mathbf{u}}}{\partial m_i \partial m_j}, \tag{10.9}$$

where the components of the *approximate Hessian* are defined as

$$\bar{H}_{ij} := \frac{\partial \bar{\mathbf{u}}}{\partial m_i} \cdot (\nabla_u \nabla_u \chi) \cdot \frac{\partial \bar{\mathbf{u}}}{\partial m_j}. \tag{10.10}$$

In contrast to the *full Hessian*, H_{ij}, the approximate Hessian merely involves first derivatives which makes its practical computation via the standard adjoint method comparatively straightforward. The Gauss–Newton and Levenberg–Marquardt methods of non-linear optimisation therefore use the approximate Hessian as a substitute of the full Hessian (see Sect. 7.3.3). The second term in Eq. (10.9) is often assumed to be too expensive to compute. Its neglect can be justified when the forward problem is nearly linear and when the misfit is small (Tarantola, 1987). To eliminate the second derivative $\frac{\partial^2 \bar{\mathbf{u}}}{\partial m_i \partial m_j}$ from Eq. (10.9), we differentiate the forward problem (10.2) twice:

$$\frac{\partial^2 \mathbf{L}}{\partial m_i \partial m_j} \cdot \bar{\mathbf{u}} + \frac{\partial \mathbf{L}}{\partial m_i} \cdot \frac{\partial \bar{\mathbf{u}}}{\partial m_j} + \frac{\partial \mathbf{L}}{\partial m_j} \cdot \frac{\partial \bar{\mathbf{u}}}{\partial m_i} + \mathbf{L} \cdot \frac{\partial^2 \bar{\mathbf{u}}}{\partial m_i \partial m_j} = \mathbf{0}. \tag{10.11}$$

A rearrangement of Eq. (10.11) provides an explicit expression for $\frac{\partial^2 \bar{\mathbf{u}}}{\partial m_i \partial m_j}$,

$$\frac{\partial^2 \bar{\mathbf{u}}}{\partial m_i \partial m_j} = -\mathbf{L}^{-1} \cdot \left(\frac{\partial^2 \mathbf{L}}{\partial m_i \partial m_j} \cdot \bar{\mathbf{u}} + \frac{\partial \mathbf{L}}{\partial m_i} \cdot \frac{\partial \bar{\mathbf{u}}}{\partial m_j} + \frac{\partial \mathbf{L}}{\partial m_j} \cdot \frac{\partial \bar{\mathbf{u}}}{\partial m_i} \right), \tag{10.12}$$

that we substitute into Eq. (10.9):

$$\begin{aligned} H_{ij} &= \bar{H}_{ij} - \nabla_u \chi \cdot \left[\mathbf{L}^{-1} \cdot \left(\frac{\partial^2 \mathbf{L}}{\partial m_i \partial m_j} \cdot \bar{\mathbf{u}} + \frac{\partial \mathbf{L}}{\partial m_i} \cdot \frac{\partial \bar{\mathbf{u}}}{\partial m_j} + \frac{\partial \mathbf{L}}{\partial m_j} \cdot \frac{\partial \bar{\mathbf{u}}}{\partial m_i} \right) \right] \\ &= \bar{H}_{ij} - \left[\left(\frac{\partial^2 \mathbf{L}}{\partial m_i \partial m_j} \cdot \bar{\mathbf{u}} + \frac{\partial \mathbf{L}}{\partial m_i} \cdot \frac{\partial \bar{\mathbf{u}}}{\partial m_j} + \frac{\partial \mathbf{L}}{\partial m_j} \cdot \frac{\partial \bar{\mathbf{u}}}{\partial m_i} \right)^{\mathrm{T}} \cdot (\mathbf{L}^{-1})^{\mathrm{T}} \right] \cdot \nabla_u \chi\,. \end{aligned} \tag{10.13}$$

Again defining the adjoint field $\bar{\mathbf{u}}^\dagger$ as the solution of

$$\mathbf{L}^{\mathrm{T}} \cdot \bar{\mathbf{u}}^\dagger = -\nabla_u \chi \tag{10.14}$$

yields the desired formula for H_{ij} that is free of the explicit inverse of $\mathbf{L}$ and that does not contain second derivatives of $\bar{\mathbf{u}}$:

$$H_{ij} = \bar{H}_{ij} + \bar{\mathbf{u}}^{\dagger} \cdot \left(\frac{\partial^2 \mathbf{L}}{\partial m_i \partial m_j} \cdot \bar{\mathbf{u}} + \frac{\partial \mathbf{L}}{\partial m_i} \cdot \frac{\partial \bar{\mathbf{u}}}{\partial m_j} + \frac{\partial \mathbf{L}}{\partial m_j} \cdot \frac{\partial \bar{\mathbf{u}}}{\partial m_i} \right) . \tag{10.15}$$

Equation (10.15) is the discrete analogue of the Hessian for time- and space-continuous problems (Eq. 8.59). The structure of the Hessian and the physical interpretation of its constituents are identical to the continuous case, treated in Sect. 9.3. For a detailed discussion of the discrete Hessian, the reader is referred to Pratt et al. (1998).

Chapter 11
Misfit Functionals and Adjoint Sources

The early developments of full waveform inversion for 2D acoustic problems (e.g. Tarantola, 1984; Gauthier et al., 1986) were almost immediately followed by the recognition that the choice of a suitable misfit functional is crucial for a successful application to real data. In one of the first large-scale full waveform inversions, Crase et al. (1990) proposed a series of robust misfit measures that are comparatively insensitive to seismic noise. Their study was extended by Brossier et al. (2010). The L_2 distance between observed and synthetic seismograms is efficient for the detection of sharp material contrasts, but the recovery of long-wavelength Earth structure requires misfit functionals that explicitly extract phase information. In their pioneering work, Luo & Schuster (1991) therefore proposed to measure the cross-correlation time shift between observed and synthetic waveforms (Sect. 11.3). This idea was extended by van Leeuwen & Mulder (2010). An alternative approach based on the measurement of time–frequency misfits was proposed by Fichtner et al. (2008) (Sect. 11.5). Closely related is the quantification of waveform differences based on the instantaneous phase and envelope (Bozdağ & Trampert, 2010).

In the following sections we present a selection of misfit functionals and derive the corresponding adjoint sources. This is intended to offer both physical insight and concrete solutions. Colourful examples for the resulting Fréchet kernels are shown in Chap. 12. To avoid clutter, we consider measurements made at a single station. The generalisation to measurements at multiple stations is straightforward.

As a preparatory step, we introduce the concept of the *adjoint Greens function* that will play a key role in our subsequent developments. The adjoint Greens function, $\mathbf{g}_i^\dagger(\boldsymbol{\xi}, \tau; \mathbf{x}, t)$, is defined as the solution of the adjoint equation (8.17) with a source that acts in i-direction at the position $\mathbf{x} = \boldsymbol{\xi}$ and at time $t = \tau$:

$$\mathbf{L}^\dagger[\mathbf{g}_i^\dagger(\boldsymbol{\xi}, \tau; \mathbf{x}, t)] := \mathbf{e}_i\, \delta(\mathbf{x} - \boldsymbol{\xi})\, \delta(t - \tau)\,. \tag{11.1}$$

The symbol $\mathbf{e}_i$ denotes the unit vector in i-direction. The importance of the definition (11.1) is related to the fact that we can express the adjoint field, $\mathbf{u}^\dagger$, corresponding to an arbitrary adjoint source, $\mathbf{f}^\dagger(\mathbf{x}, t) = \sum_{j=1}^3 \mathbf{e}_j f_j^\dagger(\mathbf{x}, t)$, in terms of $\mathbf{g}_i^\dagger(\boldsymbol{\xi}, \tau; \mathbf{x}, t)$. To see this, we multiply Eq. (11.1) by $f_i^\dagger(\boldsymbol{\xi}, \tau)$ and sum over i from 1 to 3:

A. Fichtner, *Full Seismic Waveform Modelling and Inversion*, Advances in Geophysical and Environmental Mechanics and Mathematics,
DOI 10.1007/978-3-642-15807-0_11,

$$\sum_{i=1}^{3} f_i^\dagger(\boldsymbol{\xi},\tau)\,\mathbf{L}^\dagger[\mathbf{g}_i^\dagger(\boldsymbol{\xi},\tau;\mathbf{x},t)] = \mathbf{f}^\dagger(\boldsymbol{\xi},\tau)\,\delta(\mathbf{x}-\boldsymbol{\xi})\,\delta(t-\tau)\,. \tag{11.2}$$

The adjoint wave operator, $\mathbf{L}^\dagger$, is linear and it does not involve derivatives with respect to $\boldsymbol{\xi}$ and τ. We can therefore rewrite (11.2) in the form

$$\mathbf{L}^\dagger\left[\sum_{i=1}^{3} f_i^\dagger(\boldsymbol{\xi},\tau)\,\mathbf{g}_i^\dagger(\boldsymbol{\xi},\tau;\mathbf{x},t)\right] = \mathbf{f}^\dagger(\boldsymbol{\xi},\tau)\,\delta(\mathbf{x}-\boldsymbol{\xi})\,\delta(t-\tau)\,. \tag{11.3}$$

Integrating Eq. (11.4) over time, τ, and space, $\boldsymbol{\xi}$, yields

$$\mathbf{L}^\dagger\left[\sum_{i=1}^{3} \int_T\int_G f_i^\dagger(\boldsymbol{\xi},\tau)\,\mathbf{g}_i^\dagger(\boldsymbol{\xi},\tau;\mathbf{x},t)\,\mathrm{d}\tau\,\mathrm{d}^3\boldsymbol{\xi}\right] = \mathbf{f}^\dagger(\mathbf{x},t)\,. \tag{11.4}$$

This implies that the adjoint field

$$\mathbf{u}^\dagger(\mathbf{x},t) = \sum_{i=1}^{3} \int_T\int_G f_i^\dagger(\boldsymbol{\xi},\tau)\,\mathbf{g}_i^\dagger(\boldsymbol{\xi},\tau;\mathbf{x},t)\,\mathrm{d}\tau\,\mathrm{d}^3\boldsymbol{\xi} \tag{11.5}$$

is the solution of the adjoint equation (8.17) with the adjoint source $\mathbf{f}^\dagger(\mathbf{x},t)$. Equation (11.5) is the *representation theorem* for adjoint fields. We are now set for the study of specific misfit functionals and their adjoint sources.

11.1 Derivative of the Pure Wave Field and the Adjoint Greens Function

The adjoint method can be used as a tool to linearise the forward problem. For this we let χ be the i-component of the elastic displacement field, $\mathbf{u}$, measured at the receiver position $\mathbf{x} = \mathbf{x}^\mathrm{r}$ and at time $t = t^\mathrm{r}$:

$$\chi(\mathbf{m}) = u_i(\mathbf{m};\mathbf{x}^\mathrm{r},t^\mathrm{r})\,. \tag{11.6}$$

To derive the adjoint source corresponding to the objective functional defined in Eq. (11.6), we write χ in integral form, as proposed in Eq. (8.6)

$$\begin{aligned}\chi(\mathbf{m}) = \langle\chi_1(\mathbf{m})\rangle &= \int\limits_T\int\limits_G \chi_1(\mathbf{m})\,\mathrm{d}t\,\mathrm{d}^3\mathbf{x}\\ &= \int\limits_T\int\limits_G \mathbf{e}_i\cdot\mathbf{u}(\mathbf{m};x,t)\,\delta(\mathbf{x}-\mathbf{x}^\mathrm{r})\delta(t-t^\mathrm{r})\,\mathrm{d}t\,\mathrm{d}^3\mathbf{x}\,.\end{aligned} \tag{11.7}$$

The symbol $\mathbf{e}_i$ denotes the unit vector in i-direction. It follows from Eq. (11.7) that the integrand, χ_1, is given by

$$\chi_1(\mathbf{m}) = \mathbf{e}_i \cdot \mathbf{u}(\mathbf{m}; x, t)\, \delta(\mathbf{x} - \mathbf{x}^r)\delta(t - t^r)\,. \tag{11.8}$$

Applying the recipe from Eq. (8.15) to the above expression for χ_1 yields the adjoint source, $\mathbf{f}^\dagger$, that corresponds to the objective functional $\chi(\mathbf{m}) = u_i(\mathbf{m}; x^r, t^r)$:

$$\mathbf{f}^\dagger(\mathbf{x}, t) = -\nabla_u \chi_1 = -\mathbf{e}_i\, \delta(\mathbf{x} - \mathbf{x}^r)\, \delta(t - t^r)\,. \tag{11.9}$$

The adjoint source, $\mathbf{f}^\dagger$, is point-localised in both space and time. It acts at the receiver position, $\mathbf{x}^r$, and at the observation time, t^r. The direction of the adjoint source is opposite to the direction in which the observation was made. Substituting (11.9) into (8.17) yields the adjoint equation

$$\mathbf{L}^\dagger(\mathbf{u}^\dagger) = -\mathbf{e}_i\, \delta(\mathbf{x} - \mathbf{x}^r)\, \delta(t - t^r)\,. \tag{11.10}$$

The comparison of (11.10) with the definition (11.1) implies that the adjoint field, $\mathbf{u}^\dagger$, is equal to the negative adjoint Greens function:

$$\mathbf{u}^\dagger(\mathbf{x}, t) = -\mathbf{g}_i^\dagger(\mathbf{x}^r, t^r; \mathbf{x}, t)\,. \tag{11.11}$$

For the derivative of χ we then find

$$\nabla_m \chi\, \delta\mathbf{m} = \nabla_m u_i(\mathbf{x}^r, t^r)\, \delta\mathbf{m} = -\langle \mathbf{g}_i^\dagger(\mathbf{x}^r, t^r) \cdot \nabla_m \mathbf{L}\, \delta\mathbf{m}\rangle\,. \tag{11.12}$$

In the special case of a homogeneous and unbounded 3D medium, the adjoint Greens function consists of spherical waves with infinitesimally short wavelength (e.g. Aki & Richards, 2002). This scenario is shown schematically in Fig. 11.1. The sensitivity kernels are products of the regular wave field, $\mathbf{u}$, and the adjoint Greens function $\mathbf{g}_i^\dagger$. They are non-zero only within the primary influence zone where the adjoint field and the regular field overlap at a given time between t_0 and t^r. The width of the primary influence zone is proportional to the wavelength of the regular wave $\mathbf{u}$, and it tends to zero as the wavelength of $\mathbf{u}$ decreases.

11.2 L_2 Waveform Difference

The misfit functional classically used in full waveform inversion (e.g. Bamberger et al., 1982; Tarantola, 1984; Igel et al., 1996) is the L_2 distance between the observed seismogram, $\mathbf{u}^0$, and the synthetic seismogram, $\mathbf{u}$, at the receiver position $\mathbf{x}^r$:

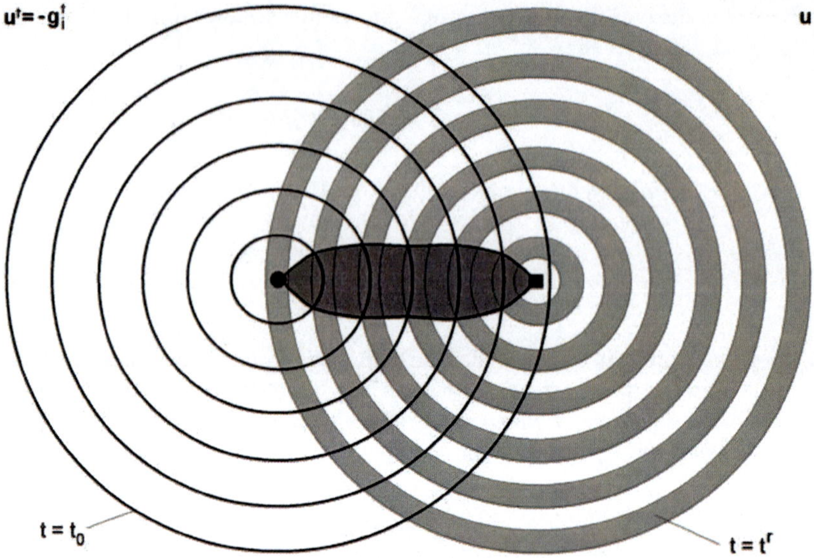

Fig. 11.1 Illustration of a sensitivity kernel that arises from the interaction of the regular wave field **u** emanating from the source (■) and the negative adjoint Greens function, $\mathbf{u}^\dagger = -\mathbf{g}_i^\dagger$ that propagates from the receiver (•) towards the source. The kernel is non-zero only in the dark shaded region where the regular and adjoint fields overlap

$$\chi(\mathbf{m}) = \frac{1}{2} \int_T [\mathbf{u}^0(\mathbf{x}^r, t) - \mathbf{u}(\mathbf{m}; \mathbf{x}^r, t)]^2 \, dt \,. \tag{11.13}$$

In the hypothetical case where χ is equal to zero, the data are explained perfectly by the Earth model. The integrand χ_1 corresponding to χ is

$$\chi_1(\mathbf{m}) = \frac{1}{2} [\mathbf{u}^0(\mathbf{x}^r, t) - \mathbf{u}(\mathbf{m}; \mathbf{x}^r, t)]^2 \, \delta(\mathbf{x} - \mathbf{x}^r) \,. \tag{11.14}$$

Invoking the recipe from Eq. (8.15) yields the adjoint source that corresponds to the misfit functional defined in (11.13):

$$\mathbf{f}^\dagger(\mathbf{x}, t) = -\nabla_u \chi_1 = [\mathbf{u}(\mathbf{m}; \mathbf{x}, t) - \mathbf{u}^0(\mathbf{x}^r, t)] \, \delta(\mathbf{x} - \mathbf{x}^r) \,. \tag{11.15}$$

The adjoint source is again point-localised at the receiver position, and its time evolution is determined by the residual time series $\mathbf{u}(t) - \mathbf{u}^0(t)$. The common expression that the adjoint method consists in *propagating the residuals back in time* largely results from Eq. (11.15).

While intuitively plausible, the L_2 distance can be problematic in practice. It is, first of all, not robust, meaning that outliers in the data can become dominant. Second, the numerical value of χ as defined in (11.13) is controlled by the large-amplitude waveforms. This means that the invaluable information contained, for instance, in the time shifts of low-amplitude P waves is almost entirely lost. The L_2 distance furthermore emphasises the non-linearity that is already inherent in the forward problem. This results in multiple local minima of the misfit functional, which is a very disadvantageous property in the context of gradient-based misfit minimisation algorithms (see Chap. 7). The existence of local minima is closely related to the incapability of the L_2 distance to detect the long-wavelength structure of the Earth. To retrieve long-wavelength features, one must adopt a multi-scale approach (see Sect. 7.4.1), where the inversion starts with the longest possible periods that are then successively decreased (e.g. Dessa et al., 2004; Bleibinhaus et al., 2007; Chap. 14). A possible alternative and complement is the combination of full waveform inversion based on the L_2 distance with classical traveltime ray tomography (e.g. Pratt & Goulty, 1991; Zhou et al., 1995; Korenaga et al., 1997; Dessa & Pascal. 2003).

Following the recognition that the L_2 distance depends very non-linearly on long-wavelength structure, various objective functionals have been designed that explicitly extract traveltime information. These are the subjects of the following paragraphs.

11.3 Cross-Correlation Time Shifts

Despite the inherent problems of the L_2 waveform distance, it remains desirable to extract as much information as possible from the difference between observed and synthetic seismograms. A milestone towards this goal was Luo & Schuster's (1991) realisation that phase information needs to be included explicitly in the objective functional. Separating the phases of seismic waveforms from their amplitudes is required to overcome the excessive non-linearity introduced by the L_2 waveform difference as defined in (11.13). Luo & Schuster's method is based on the estimation of delay times by cross-correlating data and numerically computed synthetics – a technique reminiscent of ideas expressed earlier by Dziewonski et al. (1972), Lerner-Lam & Jordan (1983) or Cara & Lévêque (1987) in the context of surface wave analysis. The cross-correlation approach was further formalised by Gee & Jordan (1992) and then directly applied to data by Zhou et al. (1995) and Chen et al. (2007). It was also used for the computation of finite-frequency delay time kernels (Dahlen et al., 2000; Tromp et al., 2005; Liu & Tromp, 2008; Sieminski et al., 2007a,b).

In our development we closely follow the concept introduced by Luo & Schuster (1991). We consider the i-component of an observed waveform, $u_i^0(\mathbf{x}^r, t)$, and the corresponding synthetic, $u_i(\mathbf{m}; \mathbf{x}^r, t)$, at a receiver position $\mathbf{x}^r$. It is implicitly assumed that a specific waveform, such as the direct P or S wave, has been isolated from both the observed and synthetic seismograms. The *cross-correlation time shift* $\mathcal{T}$ is defined as the time where the cross-correlation function

$$C(u_i^0, u_i)(\tau) := \int_T u_i^0(\mathbf{x}^r, t) u_i(\mathbf{m}; \mathbf{x}^r, t + \tau)\, dt \tag{11.16}$$

attains its global maximum. We therefore have $\mathcal{T} > 0$ when the synthetic waveform arrives later than the observed waveform and $\mathcal{T} < 0$ when the synthetic waveform arrives earlier than the observed waveform. It is assumed that both synthetic and observed waveforms have been properly filtered and isolated. The misfit functional that we now wish to minimise is

$$\chi(\mathbf{m}) = \frac{1}{2}\mathcal{T}^2(\mathbf{m})\,. \tag{11.17}$$

This is the misfit commonly used in traveltime tomography. The definition (11.16) does not provide an explicit expression for $\mathcal{T}$. We therefore need to derive a *connective function* that links u_i and u_i^0 to the secondary observable $\mathcal{T}$. For this, we note that $C(u_i^0, u_i)(\tau)$ attains a maximum for $\tau = \mathcal{T}$, and it therefore satisfies the necessary condition

$$\begin{aligned} 0 = \frac{\mathrm{d}}{\mathrm{d}\tau} C(u_i^0, u_i)(\tau)|_{\tau=\mathcal{T}} &= \int_T u_i^0(\mathbf{x}^\mathrm{r}, t)\dot{u}_i(\mathbf{m}; \mathbf{x}^\mathrm{r}, t + \mathcal{T})\,\mathrm{d}t\,, \\ &= -\int_T \dot{u}_i^0(\mathbf{x}^\mathrm{r}, t - \mathcal{T})u_i(\mathbf{m}; \mathbf{x}^\mathrm{r}, t)\,\mathrm{d}t\,. \end{aligned} \tag{11.18}$$

Equation (11.18) defines $\mathcal{T}$ implicitly, at least when there is only one maximum. That $\mathcal{T}$ corresponds indeed to the global maximum of the correlation function and not to one of several local extrema needs to be ensured by the careful observer. Invoking the implicit function differentiation yields the derivative of χ:

$$\nabla_m \chi\,\delta\mathbf{m} = \mathcal{T}\,\nabla_m \mathcal{T}(\mathbf{m})\,\delta\mathbf{m} = \frac{\mathcal{T}\int_T \dot{u}_i^0(\mathbf{x}^\mathrm{r}, t - \mathcal{T})\,\nabla_m u_i(\mathbf{m}; \mathbf{x}^\mathrm{r}, t)\,\delta\mathbf{m}\,\mathrm{d}t}{\int_T \ddot{u}_i^0(\mathbf{x}^\mathrm{r}, t - \mathcal{T})\,u_i(\mathbf{m}, \mathbf{x}^\mathrm{r}, t)\,\mathrm{d}t}\,. \tag{11.19}$$

Under the assumption that u_i^0 and u_i are purely time shifted and not otherwise distorted with respect to each other, we have $u_i^0(\mathbf{x}^\mathrm{r}, t - \mathcal{T}) = u_i(\mathbf{m}; \mathbf{x}^\mathrm{r}, t)$, and Eq. (11.19) reduces to

$$\nabla_m \chi\,\delta\mathbf{m} = -\frac{\mathcal{T}}{||\dot{u}_i||_2^2}\int_T \dot{u}_i(\mathbf{m}; \mathbf{x}^\mathrm{r}, t)\,\nabla_m u_i(\mathbf{m}; \mathbf{x}^\mathrm{r}, t)\,\delta\mathbf{m}\,\mathrm{d}t\,, \tag{11.20}$$

with

$$||\dot{u}_i||_2^2 = \int_T \dot{u}_i^2(\mathbf{m}; \mathbf{x}^\mathrm{r}, t)\,\mathrm{d}t\,. \tag{11.21}$$

For the term $\nabla_m u_i(\mathbf{m}; \mathbf{x}^\mathrm{r}, t)\,\delta\mathbf{m}$ in Eq. (11.20) we substitute the result from Eq. (11.12):

$$\nabla_m \chi \, \delta\mathbf{m} = \frac{\mathcal{T}}{||\dot{u}_i||_2^2} \int_T \dot{u}_i(\mathbf{m}; \mathbf{x}^r, t) \, \langle \mathbf{g}_i^\dagger(\mathbf{x}^r, t) \cdot \nabla_m \mathbf{L} \, \delta\mathbf{m} \rangle \, dt$$
$$= \frac{\mathcal{T}}{||\dot{u}_i||_2^2} \int_T \int_T \int_G \dot{u}_i(\mathbf{m}; \mathbf{x}^r, t) \, \mathbf{g}_i^\dagger(\mathbf{x}^r, t; \mathbf{x}, \tau) \cdot \nabla_m \mathbf{L} \, \delta\mathbf{m} \, d\tau \, dt \, d^3\mathbf{x} \,. \tag{11.22}$$

Defining the adjoint field

$$\mathbf{u}^\dagger(\mathbf{x}, \tau) := \frac{\mathcal{T}}{||\dot{u}_i||_2^2} \int_T \dot{u}_i(\mathbf{m}; \mathbf{x}^r, t) \, \mathbf{g}_i^\dagger(\mathbf{x}^r, t; \mathbf{x}, \tau) \, dt \tag{11.23}$$

allows us to write the derivative of χ in its canonical form (8.16)

$$\nabla_m \chi \, \delta\mathbf{m} = \int_T \int_G \mathbf{u}^\dagger(\mathbf{x}, t) \cdot \nabla_m \mathbf{L} \, \delta\mathbf{m} \, dt \, d^3\mathbf{x} \,. \tag{11.24}$$

Equation (11.23) has the form of the representation theorem (11.5), where the adjoint wave field, $\mathbf{u}^\dagger$, is expressed in terms of an integral over the adjoint source times the adjoint Greens function. It therefore follows that the adjoint source corresponding to χ is given by

$$\mathbf{f}^\dagger(\mathbf{x}, t) = \frac{\mathcal{T} \, \mathbf{e}_i}{||\dot{u}_i||_2^2} \dot{u}_i(\mathbf{m}; \mathbf{x}^r, t) \, \delta(\mathbf{x} - \mathbf{x}^r) \,, \tag{11.25}$$

with $\mathbf{e}_i$ denoting the unit vector in the observation direction. As a result of the pointwise measurement, the adjoint source is space-localised at the receiver position. Its time evolution is determined by the synthetic displacement velocity. The factor $||\dot{u}_i||_2^{-2}$ effectively normalises the sensitivity kernels. This ensures that their amplitude does not depend on the amplitude of the regular wave field.

The geometry of the kernels is quasi-independent of the data because the adjoint source does not contain the data, except for the factor $\mathcal{T}$. This quasi-independence rests, of course, on the assumption that the observed and synthetic waveforms are sufficiently similar to allow for the replacement of $u_i^0(\mathbf{x}^r, t - \mathcal{T})$ by $u_i(\mathbf{m}; \mathbf{x}^r, t)$, which then led to Eq. (11.20). Teleseismic body waveforms often reveal the necessary similarity between observation and synthetic, so that measurements of cross-correlation time shifts are physically meaningful. Depending on the magnitude of the events, it may become necessary to account for the source time function, using, for instance, a matched-filter approach (e.g. Sigloch & Nolet, 2006). In the case of surface waves, a sufficient waveform similarity is usually not present. It can, however, be enforced by rigorous band-pass filtering (e.g. Gee & Jordan, 1992).

Several recent applications of cross-correlation time shifts (e.g. Sigloch et al., 2008; Tape et al., 2010) were based on measurements in multiple frequency bands. This strategy increases the amount of useful information substantially, thus leading to higher spatial resolution of tomographic images.

11.4 L_2 Amplitudes

The amplitudes of seismic waveforms contain invaluable structural information that can today be exploited in tomographic inversions (e.g. Tibuleac et al., 2003; Sigloch et al., 2006, 2008). A robust measure of the amplitude of an observed waveform, u_i^0, is the L_2 norm

$$\mathcal{A}^0 := \sqrt{\int_T [u_i^0(\mathbf{x}^r, t)]^2 \, dt} \,. \tag{11.26}$$

Again, we implicitly assumed that the waveform of interest has been properly filtered and isolated. In what follows, we will refer to $\mathcal{A}$ as *the amplitude*. In analogy to (11.26) we define the amplitude of the corresponding synthetic waveform as

$$\mathcal{A}(\mathbf{m}) := \sqrt{\int_T [u_i(\mathbf{m}; \mathbf{x}^r, t)]^2 \, dt} \,. \tag{11.27}$$

The amplitude misfit that we may wish to minimise in a tomographic inversion is then

$$\chi(\mathbf{m}) = \frac{1}{2} \frac{(\mathcal{A} - \mathcal{A}^0)^2}{(\mathcal{A}^0)^2} \,. \tag{11.28}$$

Differentiating equation (11.28) with respect to the model parameters $\mathbf{m}$ gives

$$\begin{aligned} \nabla_m \chi \, \delta\mathbf{m} &= \frac{\mathcal{A} - \mathcal{A}^0}{(\mathcal{A}^0)^2} \nabla_m \mathcal{A} \, \delta\mathbf{m} \\ &= \frac{\mathcal{A} - \mathcal{A}^0}{\mathcal{A}\,(\mathcal{A}^0)^2} \int_T u_i(\mathbf{m}; \mathbf{x}^r, t) \, \nabla_m u_i(\mathbf{m}; \mathbf{x}^r, t) \, \delta\mathbf{m} \, dt \,. \end{aligned} \tag{11.29}$$

We again substitute Eq. (11.12) for $\nabla_m u_i(\mathbf{m}; \mathbf{x}^r, t)\, \delta\mathbf{m}$:

$$\begin{aligned} \nabla_m \chi \, \delta\mathbf{m} &= -\frac{\mathcal{A} - \mathcal{A}^0}{\mathcal{A}\,(\mathcal{A}^0)^2} \int_T u_i(\mathbf{m}; \mathbf{x}^r, t) \, \langle \mathbf{g}_i^\dagger(\mathbf{x}^r, t) \cdot \nabla_m \mathbf{L} \, \delta\mathbf{m} \rangle \, dt \\ &= -\frac{\mathcal{A} - \mathcal{A}^0}{\mathcal{A}\,(\mathcal{A}^0)^2} \int_T \int_T \int_G u_i(\mathbf{m}; \mathbf{x}^r, t) \, \mathbf{g}_i^\dagger(\mathbf{x}^r, t; \mathbf{x}, \tau) \cdot \nabla_m \mathbf{L} \, \delta\mathbf{m} \, d\tau \, dt \, d^3\mathbf{x} \,. \end{aligned} \tag{11.30}$$

To bring Eq. (11.30) into the canonical form $\nabla_m \chi \, \delta\mathbf{m} = \langle \mathbf{u}^\dagger \cdot \nabla_m \mathbf{L} \, \delta\mathbf{m} \rangle$ we define the adjoint field $\mathbf{u}^\dagger$ as follows:

$$\mathbf{u}^\dagger(\mathbf{x}, \tau) := -\frac{\mathcal{A} - \mathcal{A}^0}{\mathcal{A}\,(\mathcal{A}^0)^2} \int_T u_i(\mathbf{m}; \mathbf{x}^r, t)\, \mathbf{g}_i^\dagger(\mathbf{x}^r, t; \mathbf{x}, \tau)\, dt \,. \tag{11.31}$$

We infer by comparison of (11.31) with the representation theorem (11.5) that the adjoint source, $\mathbf{f}^\dagger$, is given by

$$\mathbf{f}^\dagger(\mathbf{x}, t) = -\frac{\mathcal{A} - \mathcal{A}^0}{\mathcal{A}\,(\mathcal{A}^0)^2}\, \mathbf{e}_i\, u_i(\mathbf{m}; \mathbf{x}, t)\, \delta(\mathbf{x}^r - \mathbf{x}) \,. \tag{11.32}$$

The adjoint source is, as expected, point-localised at the receiver position $\mathbf{x}^r$. Its time evolution is determined by the displacement waveform at the receiver, and the amplitude of the adjoint source is proportional to the amplitude difference $\mathcal{A} - \mathcal{A}^0$. We note that the time dependence of the adjoint source (11.32) and therefore the associated Fréchet kernels are always independent of the data. This is in contrast to the cross-correlation time shift where a quasi-independence of the data requires assumptions concerning waveform similarity.

11.5 Time-Frequency Misfits

Cross-correlation time shifts and L_2 amplitudes are robust measurements that have been applied successfully in seismic tomography. Their applicability, however, is limited to scenarios where single phases are clearly separable and where observed and synthetic waveforms are similar. In the common case where several phases interfere, both cross-correlation time shifts and L_2 amplitudes do not yield physically meaningful information. A similar effect arises when observed and synthetic waveforms are not only time shifted and amplitude scaled but distorted with respect to each other – a well-known phenomenon in the Earth where waveforms disperse due to the presence of 3D heterogeneities. Potentially useful and robust information about the Earth's structure may therefore remain unexploited.

Time–frequency misfits have recently been developed to assess the accuracy of numerical solutions to the wave equation (Kristekova et al., 2006, 2009) and for waveform tomography on continental scales (Fichtner et al., 2009, 2010; Sect. 13.3).

It is the principal objective of *time–frequency misfits* to overcome the limitations of cross-correlation time shift and L_2 amplitude measurements. Time–frequency misfits are specifically designed to fulfil the following requirements:

1. *Complete quantification of seismic waveform misfit in the frequency range of interest.* This means that observed and synthetic waveforms are identical in the hypothetical case of zero misfit. Note that the cross-correlation time shift does

not satisfy this requirement, meaning that the misfit defined in Eq. (11.17) can be zero while observed and synthetic waveforms are dissimilar.

2. *Separation of phase and amplitude information.* Each type of information constrains specific types of the Earth structure. While the phases of seismic waves largely determine seismic velocities, amplitudes mostly depend on lateral velocity gradients and the dissipation of elastic energy. Thus, for a waveform inversion to be efficient, phases and amplitudes should be separated.
3. *Relaxation of the requirements on waveform similarity needed for the measurement of pure cross-correlation time shifts.* Seismic waves disperse due to the presence of heterogeneities in the Earth, so that observed and synthetic waveforms are inherently dissimilar. Full waveform tomography therefore requires misfit measures that are physically meaningful when the compared waveforms are dissimilar to some degree.
4. *Possibility to analyse complete wave trains including body waves, surface waves and interfering phases.* Especially at short epicentral distances, numerous seismic phases do not yet have a clear separate identity, meaning that they interfere with other phases. To exploit as much information as possible, the measurement must be naturally applicable to any type of waveform, regardless of its composition of seismic phases that are independent from a ray-theoretical or normal mode perspective.

We start our development with the definition of time–frequency envelope and phase misfits. Much emphasis will be on the technical details of phase difference measurements and the required degree of waveform similarity. Subsequently, we will derive the adjoint sources corresponding to measurements of phase and envelope misfits.

11.5.1 Definition of Phase and Envelope Misfits

As in the previous sections we denote by $u_i^0(\mathbf{x}^{\mathrm{r}}, t)$ the i-component of an observed waveform recorded at the position $\mathbf{x} = \mathbf{x}^{\mathrm{r}}$. The corresponding synthetic waveform is $u_i(\mathbf{m}; \mathbf{x}^{\mathrm{r}}, t)$. For notational brevity we omit dependencies on the receiver position, $\mathbf{x}^{\mathrm{r}}$, and the Earth model, $\mathbf{m}$, wherever this is possible. We can analyse how the frequency content of the data evolves with time by computing the Fourier transform of $u_i^0(t)$ multiplied by a sliding window function $h(t - \tau)$ centred around τ:

$$\tilde{u}_i^0(t, \omega) = F_h[u_i^0](t, \omega) := \frac{1}{\sqrt{2\pi}\, ||h||_2} \int_{\mathbb{R}} u_i^0(\tau)\, h^*(\tau - t)\, \mathrm{e}^{-\mathrm{i}\omega\tau}\, \mathrm{d}\tau\,. \quad (11.33)$$

The norm $||h||_2$ of the window function h, defined as

$$||h||_2 = \sqrt{\int_{\mathbb{R}} h^2(t)\, \mathrm{d}t}\,, \quad (11.34)$$

is assumed to be non-zero. Using the complex conjugate h^* instead of h is a common convention to which we shall conform throughout this chapter. The windowed Fourier transform from Eq. (11.33) constitutes a time–frequency representation of the data. In analogy to (11.33) we define the time–frequency representation of the synthetics $u_i(t)$ as $\tilde{u}_i(t, \omega) = F_h[u_i](t, \omega)$. Both $\tilde{u}_i^0$ and $\tilde{u}_i$ can be written in exponential form

$$\tilde{u}_i^0(t, \omega) = |\tilde{u}_i^0(t, \omega)|\, \mathrm{e}^{\mathrm{i}\phi_i^0(t,\omega)}\,, \quad \tilde{u}_i(t, \omega) = |\tilde{u}_i(t, \omega)|\, \mathrm{e}^{\mathrm{i}\phi_i(t,\omega)}\,, \tag{11.35}$$

with the envelopes $|\tilde{u}_i^0(t, \omega)|$ and $|\tilde{u}_i(t, \omega)|$ and the corresponding phases $\phi_i^0(t, \omega)$ and $\phi_i(t, \omega)$. Closely following Fichtner et al. (2008), we define the envelope misfit, χ_e, and the phase misfit, χ_p, as weighted L_2 norms of the envelope difference $|\tilde{u}_i| - |\tilde{u}_i^0|$ and the phase difference $\phi_i - \phi_i^0$, respectively:

$$\chi_\mathrm{e}^2(u_i^0, u_i) := \int_{\mathbb{R}^2} W_\mathrm{e}^2(t, \omega)[|\tilde{u}_i(t, \omega)| - |\tilde{u}_i^0(t, \omega)|]^2\, \mathrm{d}t\, \mathrm{d}\omega\,, \tag{11.36}$$

$$\chi_\mathrm{p}^2(u_i^0, u_i) := \int_{\mathbb{R}^2} W_\mathrm{p}^2(t, \omega)[\phi_i(t, \omega) - \phi_i^0(t, \omega)]^2\, \mathrm{d}t\, \mathrm{d}\omega\,. \tag{11.37}$$

The symbols W_e and W_p denote positive weighting functions that we will discuss in the following sections. The envelope difference $|\tilde{u}_i^0| - |\tilde{u}_i|$ represents time- and frequency-dependent discrepancies between the amplitudes of u_i^0 and u_i. The phase difference $\Delta\phi_i = \phi_i - \phi_i^0$ can be interpreted in terms of a time shift Δt at frequency ω: $\Delta\phi_i = \omega\, \Delta t$.

In the following paragraphs we let the window h be the Gaussian

$$h(t) = (\pi\sigma^2)^{-1/4}\, \mathrm{e}^{-t^2/2\sigma^2}\,. \tag{11.38}$$

The time–frequency transform F_h is then referred to as the *Gabor transform*. The Gaussian window is real, and it is easily verified that $||h||_2 = 1$. This choice is both convenient and advantageous from a theoretical point of view because it maximises the time–frequency resolution (see, for example, Strang & Nguyen, 1996 and Appendix B). Equation (11.33) does not represent the only possible characterisation of data or synthetics in time–frequency space. Alternatively, one could employ a wavelet transform instead of a windowed Fourier transform, as suggested, for instance, by Kristekova et al. (2006).

11.5.2 Practical Implementation of Phase Difference Measurements

While envelope measurements are conceptually straightforward, phase difference measurements are more complicated. This is due to the unavoidable discontinuities in the phases ϕ_i and ϕ_i^0. For a given time t the discontinuities of ϕ_i and ϕ_i^0 generally occur at different frequencies ω. Since the phase jumps from $-\pi$ to π, or vice versa, have different locations on the frequency axis, the difference $\Delta\phi_i$ can reach values of $\pm 2\pi$, even when the signals u_i and u_i^0 are nearly identical.

Under the assumption that data and synthetics are approximately in phase we can circumvent this obstacle. For this we note that for a fixed time t the time–frequency representations $\tilde{u}_i^0$ and $\tilde{u}_i$ are the Fourier transforms of the functions $f_t^0(\tau) := u_i^0(\tau)h(\tau - t)$ and $f_t(\tau) := u_i(\tau)h(\tau - t)$, respectively. Now, let the correlation function of f_t and f_t^0, denoted by $C(f_t^0, f_t)(\tau)$, be defined through

$$C(f_t^0, f_t)(\tau) := \int_{\mathbb{R}} f_t^0(t') f_t(t' + \tau)\, \mathrm{d}t' . \tag{11.39}$$

For the regular Fourier transform of the correlation function $C(f_t^0, f_t)$ we then find

$$\begin{aligned} F[C](\omega) &= \frac{1}{\sqrt{2\pi}} \int_{\mathbb{R}} C(f_t^0, f_t)(\tau)\, \mathrm{e}^{-\mathrm{i}\omega\tau}\, \mathrm{d}\tau = \sqrt{2\pi}\, \tilde{u}_i(t,\omega)\, \tilde{u}_i^0(t,\omega)^* \\ &= \sqrt{2\pi}\, |\tilde{u}_i||\tilde{u}_i^0|\, \mathrm{e}^{\mathrm{i}(\phi_i - \phi_i^0)} = |F[C](\omega)|\, \mathrm{e}^{\mathrm{i}(\phi_i - \phi_i^0)} . \end{aligned} \tag{11.40}$$

The phase difference $\Delta\phi_i = \phi_i - \phi_i^0$ can therefore be expressed as

$$\Delta\phi_i = -\mathbf{i} \ln \left(\frac{F[C]}{|F[C]|} \right) . \tag{11.41}$$

For a given time t the phase of the Fourier-transformed correlation function C coincides with the phase difference between the time-frequency representations $\tilde{u}_i$ and $\tilde{u}_i^0$. In the regions of time–frequency space where u_i^0 and u_i are approximately in phase, Eq. (11.41) will provide phase differences that range between $-\pi$ and π. Discontinuities, as produced by directly subtracting ϕ_i from ϕ_i^0 can thus be avoided.

When the observed and synthetic waveforms are, however, significantly out of phase, discontinuities appear even when Eq. (11.41) is used for the computation of $\Delta\phi_i$. The occurrence of discontinuities indicates a level of waveform dissimilarity that needs to be tamed either by low-pass filtering $u_i(t)$ and $u_i^0(t)$ or by applying a suitable weighting function W_p to the time–frequency representations $\tilde{u}_i$ and $\tilde{u}_i^0$.

The choice of the parameter σ in the Gaussian window $h(t) = (\pi\sigma^2)^{-1/4} \mathrm{e}^{-t^2/2\sigma^2}$ influences the time–frequency representations of both data and synthetics. The technical details of the measurement process, represented by σ in our case, affect the outcome of the measurement. Optimising this outcome has always played a central role in time–frequency and spectral analysis. In classical surface wave analysis, for instance, σ is usually tuned to render measurements of group arrival times as easy as possible (Cara, 1973; Nyman & Landisman, 1977). Another example of measurement optimisation comes from multi-taper methods, which are designed to provide spectral estimates that are as free as possible from the effects of windowing functions (Thomson, 1982).

In the case of full waveform tomography the situation is slightly different from the ones encountered in classical surface wave or multi-taper analysis. The measurements, namely phase and envelope misfits, are extracted from time–frequency representations, and those time–frequency representations have, by design, a free parameter: σ. This parameter makes the subjectivity inherent in any measurement rather explicit. In principle, we cannot exclude a priori a certain value for σ as long as it results in a physically meaningful measurement and as long as we interpret the results accordingly.

We can, however, tune σ such that it produces results that are in agreement with our physical intuition and experience. In this sense, we suggest choosing σ such that the mathematically defined phase difference $\Delta\phi_i$ is interpretable in terms of the intuitive meaning of a phase difference: a time shift between two associated oscillations in the data and the synthetics. A suitable value for σ is then the dominant period of the data. Choosing σ to be several times smaller than the dominant period gives narrow Gaussian windows that cannot capture time shifts between two cycles that span many such windows. Conversely, a value for σ that is several times larger than the dominant period leads to Gaussian windows that are so wide that the resulting phase difference can no longer be associated to a specific cycle. In the case where the dominant period varies strongly with time, one may use a time- or frequency-dependent window h.

The main purpose of the weighting function W_p is to suppress phase differences in those regions of the time–frequency space where a physically meaningful measurement is not possible. Those regions include time intervals where the signal is either zero or below the noise level. Furthermore, W_p may be used to isolate time–frequency windows that are found to be particularly useful in a specific application. For instance, W_p could isolate or emphasise higher mode surface waves in order to improve the depth resolution in tomographic inversions.

For a semi-automatic data processing we suggest to construct the weighting function as follows: (1) Estimate a noise level $n(t, \omega)$. In most applications $n(t, \omega)$ will be independent of time, t, and frequency, ω. (2) Define

$$\tilde{W}_\mathrm{p}(t, \omega) := 1 - \mathrm{e}^{-|\tilde{u}^0(t,\omega)|^2/n^2(t,\omega)} \,. \tag{11.42}$$

The filter $\tilde{W}_\mathrm{p}$ is close to one where the data are significantly above the noise and close to zero elsewhere.

11.5.3 An Example

To illustrate the practical measurement of phase differences, we consider the N-S-component seismograms shown in Fig. 11.2a. The dominant period is 100 s. The body wave phases interfere strongly, thus loosing their separate identities. The phase

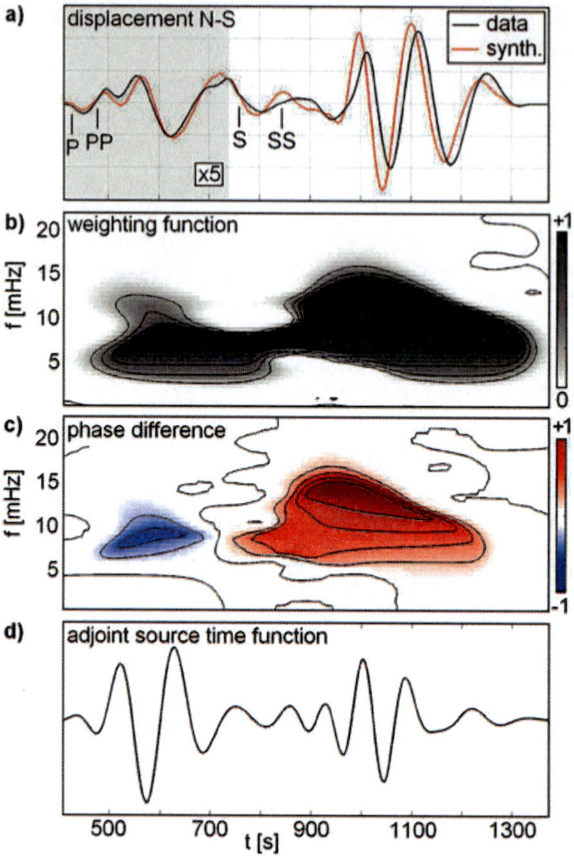

Fig. 11.2 (**a**) N-S-component waveforms at 100 s dominant period. Data: *black*, synthetics: *red*. The P wave part (*light gray shaded*) is amplified by a factor of 5 to enhance visibility. The synthetic surface wave is significantly early compared to the observed surface wave. Smaller differences exist within the body wave part. (**b**) Phase weighting function, W_{p}, as defined in Eq. (11.42). (**c**) Weighted phase difference in time–frequency space. The delay prior to 700 s maps into a negative phase difference, and the advance in the surface wave train corresponds to a positive-phase difference. (**d**) Adjoint source time function before time reversal. The body wave part has a larger amplitude despite the smaller phase difference, compared to the surface wave part. This is because the adjoint source automatically compensates for the smaller amplitude of the regular body wave, in order to make the Fréchet kernels amplitude independent

weighting function, W_{p} (Fig. 11.2b), was constructed using Eq. (11.42), with the noise level set to $n(t,\omega) = \max|\tilde{u}^0|/30$. In those parts of the time–frequency space where a meaningful measurement is possible, W_{p} attains values around $+1$. Regions where the amplitudes are too small are excluded from the computation of the phase difference, which is shown in Fig. 11.2c. The delay of the synthetics relative to the data in the earlier part of seismogram maps into a negative phase difference that is located around $t = 600\,\text{s}$ and $f = 8\,\text{mHz}$. The surface wave part of the synthetics is advanced relative to the observed surface wave. This leads to a positive phase difference, that is largest for frequencies around 15 mHz and less pronounced for lower frequencies.

Figure 11.2d displays the adjoint source time function, the computation of which will be the subject of the following paragraphs. Clearly visible are the two contributions that correspond to the positive and negative phase differences in the time–frequency plane. Note that the body wave part of the adjoint source time function has a larger amplitude than the surface wave part, despite a smaller phase difference. This is because the adjoint source automatically compensates for the smaller amplitudes of the regular waves, thus making the Fréchet kernels for phase misfit measurements amplitude independent. We already observed the same effect in the adjoint source for measurements of cross-correlation time shifts (Sect. 11.3, Eq. 11.25).

11.5.4 Adjoint Sources

11.5.4.1 A Preparatory Step for the Derivation of the Adjoint Sources for Envelope and Phase Misfits

To prepare the derivation of the adjoint sources for envelope and phase misfit measurements, we provide some useful expressions for the Fréchet derivatives $\nabla_m \tilde{u}_i \,\delta\mathbf{m}$ and $\nabla_m |\tilde{u}_i| \,\delta\mathbf{m}$ that we will use frequently in the following paragraphs: First, we differentiate the definition of $\tilde{u}_i$ with respect to the model parameters $\mathbf{m}$,

$$\nabla_m \tilde{u}_i(t,\omega)\,\delta\mathbf{m} = \frac{1}{\sqrt{2\pi}} \int_{\mathbb{R}} \nabla_m u_i(\tau)\,\delta\mathbf{m}\, h(\tau - t)\,\mathrm{e}^{-\mathrm{i}\omega\tau}\,\mathrm{d}\tau\,, \tag{11.43}$$

where we already assumed that h is a Gaussian window with $||h||_2 = 1$. Substituting Eq. (11.12) for $\nabla_m u_i(\tau)\,\delta\mathbf{m}$ gives

$$\nabla_m \tilde{u}_i(t,\omega)\,\delta\mathbf{m} = -\frac{1}{\sqrt{2\pi}} \int_T \int_G \int_{\mathbb{R}} \mathbf{g}_i^\dagger(\mathbf{x}^r,\tau;\mathbf{x},t')\cdot\nabla_m \mathbf{L}\,\delta\mathbf{m}\, h(\tau - t)\,\mathrm{e}^{-\mathrm{i}\omega\tau}\,\mathrm{d}\tau\,\mathrm{d}t'\,\mathrm{d}^3\mathbf{x}\,. \tag{11.44}$$

By defining the auxiliary adjoint field

$$\tilde{\mathbf{u}}^\dagger(\mathbf{x},t';t,\omega) := -\frac{1}{\sqrt{2\pi}} \int_{\mathbb{R}} \mathbf{g}_i^\dagger(\mathbf{x}^r,\tau;\mathbf{x},t')\, h(\tau - t)\,\mathrm{e}^{-\mathrm{i}\omega\tau}\,\mathrm{d}\tau\,, \tag{11.45}$$

we can write the derivative $\nabla_m \tilde{u}_i(t,\omega)\,\delta\mathbf{m}$ in canonical form

$$\nabla_m \tilde{u}_i(t,\omega)\,\delta\mathbf{m} = \int_T \int_G \tilde{\mathbf{u}}^\dagger(\mathbf{x},t';t,\omega)\cdot\nabla_m \mathbf{L}\,\delta\mathbf{m}\,\mathrm{d}t'\,\mathrm{d}^3\mathbf{x}\,. \tag{11.46}$$

Next, we consider the derivative $\nabla_m |\tilde{u}_i| \,\delta\mathbf{m}$ that can be expressed in terms of $\nabla_m \tilde{u}_i \,\delta\mathbf{m}$. For this we note that the following relation holds:

$$\begin{aligned}\nabla_m|\tilde{u}_i|^2\delta\mathbf{m} &= 2|\tilde{u}_i|\,\nabla_m|\tilde{u}_i|\,\delta\mathbf{m}\\ &= \nabla_m(\tilde{u}_i\tilde{u}_i^*)\,\delta\mathbf{m} = \tilde{u}_i\,\nabla_m\tilde{u}_i^*\,\delta\mathbf{m} + \tilde{u}_i^*\,\nabla\tilde{u}_i\,\delta\mathbf{m}\,.\end{aligned} \tag{11.47}$$

Solving for $\nabla_m|\tilde{u}_i|\,\delta\mathbf{m}$ then yields

$$\nabla_m|\tilde{u}_i|\,\delta\mathbf{m} = \mathfrak{Re}\left(\frac{\tilde{u}_i}{|\tilde{u}_i|}\,\nabla_m\tilde{u}_i^*\,\delta\mathbf{m}\right), \tag{11.48}$$

where $\mathfrak{Re}$ extracts the real part of the complex quantity in brackets.

11.5.4.2 Adjoint Source for the Envelope Misfit

To derive the adjoint source corresponding to measurements of the envelope misfit, we differentiate the definition of χ_e (Eq. 11.36):

$$\nabla_m\chi_e\,\delta\mathbf{m} = \frac{1}{\chi_e}\int_{\mathbb{R}^2} W_e^2\,(|\tilde{u}_i| - |\tilde{u}_i^0|)\,\nabla_m|\tilde{u}_i|\,\delta\mathbf{m}\,\mathrm{d}t\,\mathrm{d}\omega\,. \tag{11.49}$$

For $\nabla_m|\tilde{u}_i|\,\delta\mathbf{m}$ we substitute Eq. (11.48):

$$\nabla_m\chi_e\,\delta\mathbf{m} = \frac{1}{\chi_e}\,\mathfrak{Re}\int_{\mathbb{R}^2} W_e^2\,(|\tilde{u}_i| - |\tilde{u}_i^0|)\,\frac{\tilde{u}_i}{|\tilde{u}_i|}\,\nabla_m\tilde{u}_i^*\,\delta\mathbf{m}\,\mathrm{d}t\,\mathrm{d}\omega\,. \tag{11.50}$$

In the interest of a lighter notation, we define the weighted relative envelope difference $\mathcal{E}(t,\omega)$:

$$\mathcal{E}(t,\omega) := W_e^2(t,\omega)\,\frac{|\tilde{u}_i(t,\omega)| - |\tilde{u}_i^0(t,\omega)|}{|\tilde{u}_i(t,\omega)|}, \tag{11.51}$$

so that we obtain a more compact expression for $\nabla_m\chi_e\,\delta\mathbf{m}$:

$$\nabla_m\chi_e\,\delta\mathbf{m} = \frac{1}{\chi_e}\,\mathfrak{Re}\int_{\mathbb{R}^2}\mathcal{E}\,\tilde{u}_i\,\nabla_m\tilde{u}_i^*\,\delta\mathbf{m}\,\mathrm{d}t\,\mathrm{d}\omega\,. \tag{11.52}$$

With the help of the auxiliary adjoint field defined in Eq. (11.45) we can eliminate $\nabla_m\tilde{u}_i^*\,\delta\mathbf{m}$ from Eq. (11.50):

$$\nabla_m\chi_e\,\delta\mathbf{m} = \frac{1}{\chi_e}\,\mathfrak{Re}\int_{\mathbb{R}^2}\int_T\int_G \mathcal{E}(t,\omega)\,\tilde{u}_i(t,\omega)\,\tilde{\mathbf{u}}^{\dagger *}(\mathbf{x},t';t,\omega)\cdot\nabla_m\mathbf{L}\,\delta\mathbf{m}\,\mathrm{d}t'\,\mathrm{d}t\,\mathrm{d}\omega\,\mathrm{d}^3\mathbf{x}\,. \tag{11.53}$$

From (11.53) we extract the adjoint field

$$\mathbf{u}^\dagger(\mathbf{x}, t') = \frac{1}{\chi_e} \Re e \int_{\mathbb{R}^2} \mathcal{E}(t, \omega)\, \tilde{u}_i(t, \omega)\, \tilde{\mathbf{u}}^{\dagger *}(\mathbf{x}, t'; t, \omega)\, \mathrm{d}t\, \mathrm{d}\omega$$
$$= -\frac{1}{\sqrt{2\pi}\, \chi_e} \Re e \int_{\mathbb{R}^3} \mathcal{E}(t, \omega)\, \tilde{u}_i(t, \omega)\, \mathbf{g}_i^\dagger(\mathbf{x}^r, \tau; \mathbf{x}, t')\, h(\tau - t)\, e^{i\omega\tau}\, \mathrm{d}\tau\, \mathrm{d}t\, \mathrm{d}\omega\,. \tag{11.54}$$

It now follows again from the representation theorem (11.5) that the adjoint source $\mathbf{f}^\dagger$ is given by

$$\mathbf{f}^\dagger(\mathbf{x}, t) = -\frac{1}{\sqrt{2\pi}\, \chi_e} \Re e \int_{\mathbb{R}^2} \mathcal{E}(\tau, \omega)\, \tilde{u}_i(\tau, \omega)\, h(t - \tau)\, e^{i\omega t}\, \mathbf{e}_i\, \delta(\mathbf{x} - \mathbf{x}^r)\, \mathrm{d}\omega\, \mathrm{d}\tau\,. \tag{11.55}$$

We can express $\mathbf{f}^\dagger$ more conveniently in terms of the inverse time–frequency transform F_h^{-1} (see Appendix B):

$$\mathbf{f}^\dagger(\mathbf{x}, t) = -\frac{1}{\chi_e} \Re e\, F_h^{-1}\left[\mathcal{E}\, \tilde{u}_i\right](t)\, \mathbf{e}_i\, \delta(\mathbf{x} - \mathbf{x}^r)\,. \tag{11.56}$$

The time evolution of the adjoint source is determined by the inverse time–frequency transform of the synthetic waveform, $\tilde{u}_i$, times the relative envelope difference, $\mathcal{E}$.

11.5.4.3 Adjoint Source for the Phase Misfit

To derive the adjoint source corresponding to phase misfit measurements, we proceed as in the case of the envelope misfit. First, we differentiate the definition of χ_p which is given by Eq. (11.37):

$$\nabla_m \chi_p\, \delta\mathbf{m} = \frac{1}{\chi_p} \int_{\mathbb{R}^2} W_p^2\, (\phi_i - \phi_i^0)\, (\nabla_m \phi_i\, \delta\mathbf{m})\, \mathrm{d}t\, \mathrm{d}\omega\,. \tag{11.57}$$

To eliminate the term $\nabla_m \phi_i\, \delta\mathbf{m}$ we again use Eq. (11.48):

$$\nabla_m \phi_i\, \delta\mathbf{m} = -\mathrm{i}\, \nabla_m \ln\left(\frac{\tilde{u}_i}{|\tilde{u}_i|}\right) \delta\mathbf{m} = -\mathrm{i}\left(\frac{1}{\tilde{u}_i} \nabla_m \tilde{u}_i\, \delta\mathbf{m} - \frac{1}{|\tilde{u}_i|} \nabla_m |\tilde{u}_i|\, \delta\mathbf{m}\right)$$
$$= \mathrm{i}\left[\frac{1}{|\tilde{u}_i|} \Re e\left(\frac{\tilde{u}_i}{|\tilde{u}_i|} \nabla_m \tilde{u}_i^*\, \delta\mathbf{m}\right) - \frac{1}{\tilde{u}_i} \nabla_m \tilde{u}_i\, \delta\mathbf{m}\right]$$
$$= \mathrm{i}\left(\frac{\tilde{u}_i}{2|\tilde{u}_i|^2} \nabla_m \tilde{u}_i^*\, \delta\mathbf{m} - \frac{\tilde{u}_i^*}{2|\tilde{u}_i|^2} \nabla_m \tilde{u}_i\, \delta\mathbf{m}\right)$$
$$= -\Im m\left(\frac{\tilde{u}_i}{|\tilde{u}_i|^2} \nabla_m \tilde{u}_i^*\, \delta\mathbf{m}\right)\,. \tag{11.58}$$

The resulting expression for $\nabla_m \chi_p\, \delta\mathbf{m}$ is then

$$\nabla_m \chi_{\mathrm{p}}\, \delta\mathbf{m} = -\frac{1}{\chi_{\mathrm{p}}} \mathfrak{Im} \int_{\mathbb{R}^2} \mathcal{P}\, \tilde{u}_i\, \nabla_m \tilde{u}_i\, \delta\mathbf{m}\, \mathrm{d}t\, \mathrm{d}\omega\,, \tag{11.59}$$

where we defined the envelope-normalised phase difference, $\mathcal{P}$, in analogy to the relative envelope difference, $\mathcal{E}$:

$$\mathcal{P}(t,\omega) := \frac{W_{\mathrm{p}}^2(t,\omega)}{|\tilde{u}_i(t,\omega)|^2} \left[\phi_i(t,\omega) - \phi_i^0(t,\omega)\right]. \tag{11.60}$$

At this point we note the similarity between Eq. (11.59) which expresses $\nabla_m \chi_{\mathrm{p}}\, \delta\mathbf{m}$ in terms of $\mathcal{P}$ and Eq. (11.52) which expresses $\nabla_m \chi_{\mathrm{e}}\, \delta\mathbf{m}$ through $\mathcal{E}$. Thanks to this similarity we can skip the remaining part of the recipe that we used to derive the adjoint source for χ_{e}, and we directly infer that the adjoint source corresponding to the phase misfit χ_{p} is given by

$$\mathbf{f}^{\dagger}(\mathbf{x},t) = \frac{1}{\chi_{\mathrm{p}}} \mathfrak{Im}\, F_h^{-1} \left[\mathcal{P}\, \tilde{u}_i\right](t)\, \mathbf{e}_i\, \delta(\mathbf{x} - \mathbf{x}^{\mathrm{r}})\,. \tag{11.61}$$

As expected, the adjoint source is a point-localised single force pointing in the observation direction. Its time evolution is determined by the inverse time–frequency transform of $\tilde{u}_i$ times the envelope-normalised phase difference that we introduced in Eq. (11.60).

We note that the misfit functionals presented above are closely related. For instance, when the observed and synthetic waveforms are merely time shifted without being otherwise distorted relative to each other, then the cross-correlation time shift and the phase difference are nearly identical measurements. A more complete analysis of the interrelations between misfit functionals can be found in Fichtner et al. (2008).

Chapter 12
Fréchet and Hessian Kernel Gallery

Following several chapters replete with purely theoretical developments, we now delve into the description of concrete Fréchet kernels for several seismologically relevant combinations of measurements and Earth model parameters. This is, first of all, intended to advance the intuition necessary for the meaningful solution of any tomographic problem.

Throughout most of this chapter we will be concerned with measurements of cross-correlation time shifts, $\mathcal{T}$, and L_2 amplitudes, $\mathcal{A}$, as defined in Eqs. (11.16) and (11.26), respectively. This choice is motivated by the simplicity of the corresponding adjoint sources and the similarity with other time- and amplitude-like measurements that are based, for instance, on time–frequency representations (Fichtner et al., 2008) or generalised seismological data functionals (Gee & Jordan, 1992). Furthermore, we restrict our attention to single-station measurements. The Fréchet kernels for multi-station measurements are simply the superposition of the individual kernels.

As a preparatory step we recall the adjoint sources associated with the absolute Fréchet derivative $\nabla_m \mathcal{T}$ and the relative Fréchet derivative $\mathcal{A}^{-1}\,\nabla_m \mathcal{A}$.

In Sect. 11.3 we found that the adjoint source corresponding to the measurement of a cross-correlation time shift on the i-component of data and synthetics is given by

$$\mathbf{f}_{\mathcal{T}}^{\dagger} = \frac{\mathbf{e}_i}{||\dot{u}_i||_2^2}\, \dot{u}_i(\mathbf{x}^{\mathrm{r}}, t)\, \delta(\mathbf{x}^{\mathrm{r}} - \mathbf{x})\,. \tag{12.1}$$

The adjoint source, $\mathbf{f}_{\mathcal{T}}^{\dagger}$, is quasi-independent of the data because we assumed that the observed and synthetic waveforms are merely shifted in time without being otherwise distorted. This assumption is well justified for most body wave phases and for surface waves filtered to a narrow frequency band. The approximate data independence of $\mathbf{f}_{\mathcal{T}}^{\dagger}$ ensures that the resulting Fréchet kernels can be interpreted in terms of the physics of wave propagation without worrying too much about the actual data. Nevertheless, we note that the waveform similarity assumption is paradoxical to some degree because the numerical value of the cross-correlation function depends

A. Fichtner, *Full Seismic Waveform Modelling and Inversion*, Advances in Geophysical and Environmental Mechanics and Mathematics,
DOI 10.1007/978-3-642-15807-0_12, © Springer-Verlag Berlin Heidelberg 2011

on waveform distortions induced by 3D Earth structure, especially when the time shift is small.

To obtain data-independent Fréchet kernels also for amplitude measurements, we focus on the relative derivative, $\mathcal{A}^{-1}\,\nabla_m \mathcal{A}$, rather than on the absolute derivative, $\nabla_m \mathcal{A}$. The adjoint source corresponding to $\mathcal{A}^{-1}\,\nabla_m \mathcal{A}$ is

$$\mathbf{f}^{\dagger}_{\mathcal{A}} = -\frac{\mathbf{e}_i}{||u_i||_2^2}\, u_i(\mathbf{x}^{\mathrm{r}}, t)\, \delta(\mathbf{x}^{\mathrm{r}} - \mathbf{x})\,, \tag{12.2}$$

as we already found in Sect. 11.4. The resulting Fréchet kernels are genuinely data independent, irrespective of any waveform dissimilarities.

For the exemplary kernel calculations presented in the following paragraphs, we use a spectral-element method (Fichtner et al., 2009, Sect. 13.2.1) that solves the elastic wave equation in a spherical section. The Earth model is the isotropic version of PREM (Dziewonski & Anderson, 1981). For simplicity we will refer to Fréchet kernels from cross-correlation measurements as traveltime kernels and to Fréchet kernels from L_2 amplitude measurements as amplitude kernels.

Following a brief description of the computational setup, we study the anatomy of traveltime and amplitude kernels with respect to v_{P} and v_{S} for a variety of body wave phases. Section 12.2 is devoted to the sensitivity of surface waves to both isotropic and anisotropic perturbations. Hessian kernels and their use in the study of resolution and trade-offs are the subject of Sect. 12.3. Based on our mostly visual analysis, we conclude this chapter with a method for the efficient computation of Fréchet kernels (Sect. 12.4).

12.1 Body Waves

The 3D sensitivity distributions of body waves with a finite-frequency content have been analysed extensively during the past two decades. In a series of pioneering studies, Yomogida (1992), Dahlen et al. (2000) and Dahlen & Baig (2002) used the ray approximation for the computation of body wave Fréchet kernels in laterally homogeneous media. For an iterative waveform inversion, kernels need to be computed also in 3D heterogeneous media, which requires the use of fully numerical methods. Numerous examples of body wave sensitivity kernels computed with the help of finite-difference and spectral-element modelling can be found in Zhao et al. (2005), Liu & Tromp (2006, 2008) or Sieminski et al. (2007b).

For our study of body wave sensitivities we consider a geometric setup where several body wave phases are easily observable. The synthetic source is located at 400 km depth beneath western Turkey. It radiates elastic waves recorded in western Spain at an epicentral distance of 25.23°. The surface projection of the ray path for this source–receiver pair and the P wave radiation pattern are shown in Fig. 12.1. The corresponding three-component synthetic velocity seismograms in Fig. 12.2 exhibit various prominent body wave phases, including the direct P and S waves, as well as the surface-reflected phases sP and sS. The dominant period is 15 s.

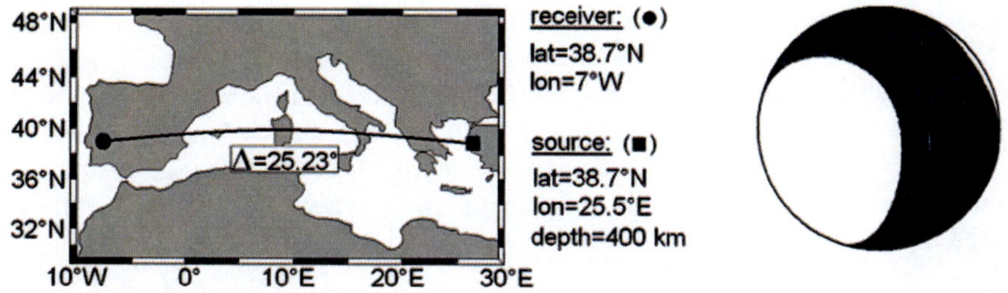

Fig. 12.1 Source–receiver geometry for a deep source beneath western Turkey and a receiver located in western Spain. The epicentral distance is 25.23°. The P wave radiation pattern is plotted to the *right*

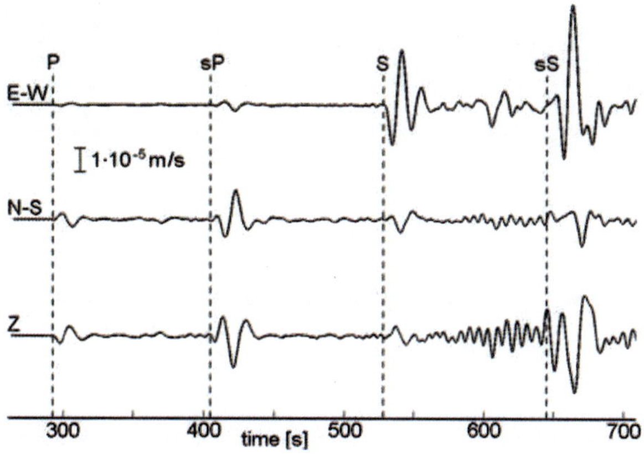

Fig. 12.2 Synthetic velocity seismograms for the source–receiver geometry shown in Fig. 12.1. The most prominent phases, P, S, sP and sS, are indicated. The dominant period is 15 s

12.1.1 Cross-Correlation Time Shifts

Cross-correlation time shifts or variants thereof are one of the most widely used measurement of finite-frequency waveform misfit (e.g. Luo & Schuster, 1991; Sigloch & Nolet, 2006; Nolet, 2008). This popularity results from the robustness of the measurement and its quasi-linear relation to the Earth structure that facilitates the solution of the tomographic inverse problem.

The critical component of cross-correlations is the isolation of a clearly identifiable waveform that has its own identity. Examples include direct P and S waves or multiply reflected body waves that do not interfere with other phases.

So far, the isolation of one specific waveform has been implicit in most of our developments. Figure 12.3 therefore illustrates in detail the construction of the adjoint source corresponding to the cross-correlation time shift measurement on the vertical component of the direct P wave displacement (Eq. 12.1). The complete vertical-component displacement seismogram, $u_z(t)$, is shown in Fig. 12.3a. It contains, besides the direct P wave, a clearly distinguishable sP phase. To construct the adjoint source time function, the P wave is isolated from the velocity seismogram, $\dot{u}_z(t)$, using a standard cosine taper, $W(t)$. The windowing produces the waveform, $W(t)\dot{u}_z(t)$, shown in Fig. 12.3c, which is then scaled by the inverse squared norm of the tapered velocity seismogram, $||W\dot{u}_z||^{-2}$. In the final step, $W(t)\dot{u}_z(t)\,||W\dot{u}_z||^{-2}$ is reversed in time to produce the adjoint source time function (Fig. 12.3e) that can be used in numerical calculations of Fréchet kernels, where the adjoint equation is solved backwards in time.

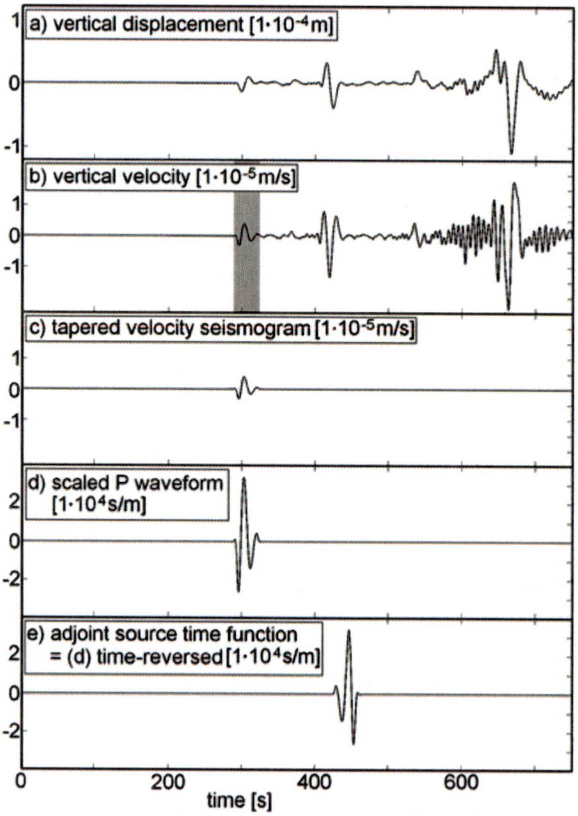

Fig. 12.3 Illustration of the different steps that lead to the construction of the adjoint source time function for a cross-correlation time shift measurement on the 15 s P waveform. **(a)** The complete vertical-component displacement seismogram, $u_z(t)$, contains clear P, sP and surface waves. **(b)** The velocity seismogram, $\dot{u}_z(t)$, is required for the computation of the adjoint source time function in Eq. (12.1). The location of the window function, $W(t)$, used to isolate the P waveform in the next step, is marked by the *gray-shaded area*. **(c)** The isolated P waveform, $W(t)\dot{u}_z(t)$, results from the application of a window function, $W(t)$, to the velocity seismogram. **(d)** Isolated P waveform scaled by the inverse squared norm of itself, $||W\dot{u}_z||^{-2}$. **(e)** Adjoint source time function, i.e. the time-reversed version of **(d)**

Cross sections through the resulting P wave speed Fréchet kernel, K_{v_P}, are presented in Fig. 12.4. The large spatial extension of the kernel is in contrast to kernels from infinite-frequency ray theory where all sensitivity is concentrated along the ray. To emphasise this difference, Fréchet kernels corresponding to waveforms with a finite-frequency content are commonly referred to as *finite-frequency kernels*. The kernel from Fig. 12.4 is the prime example of a sensitivity distribution that contradicts our intuition that is largely founded on the visualisation of seismic waves by rays. Along the geometric ray path, indicated by a dashed curve connecting source and receiver, the sensitivity is exactly equal to zero, as noted already by Yomogida (1992). It is strongest in the outer parts of the first Fresnel zone. The negative sign is expected because increasing v_P should lead to earlier-arriving P waves and a reduction in the cross-correlation time shift, as defined in Eq. (11.16). Marquering et al. (1999) jokingly described the shape of the sensitivity kernel as resembling a hollow banana parallel to the ray and a doughnut perpendicular to the ray. The expression *banana–doughnut kernel* has since then diffused into the seismological literature very efficiently. As a result of the doughnut hole, an anomaly concentrated in the outer part of the first Fresnel zone can give rise to a larger time shift than one located directly on the ray. This result was verified by Hung et al. (2000) with the help of numerical wave propagation. Hung et al. (2000) also noted that a zero time shift for perturbations along the ray path would not be present if it were measured by hand picking the onset of the arrivals. In a cross-correlation measurement, however, the complete waveform contributes to the time shift $\mathcal{T}$. The cross-correlation time shift can thus be identically zero while the difference between the P wave onset times is not. A particularly perplexing feature is the weak positive sensitivities within the second Fresnel zone. They can lead to an advance of the synthetic waveform despite the presence of a negative P wave speed anomaly.

A characteristic property shared by all Fréchet kernels – regardless of the underlying measurement – is the very large sensitivity in the vicinity of the source and the receiver. In fact, we would observe singularities directly at the source and the receiver if the numerical solution were exact. The origins of the quasi-singularities are the point source that excites the forward wave field and the point-wise measurement that implies a point-localised adjoint source.

In practice, that is in a real tomographic inversion, the effect of the quasi-singularities is less dramatic than Fig. 12.4 may suggest. This is because the Fréchet kernel is not used directly. What is needed instead are the partial derivatives of the objective functional, χ, with respect to the coefficients, m_i, of a discretised Earth model (see Eq. 8.21). The partial derivative $\partial\chi/\partial m_i$ is equal to the integral over the basis function, b_i, times the Fréchet kernel, K_m (Eq. 8.22). The integration over an extended spatial domain substantially reduces the amplitudes of the very localised extreme contributions. It can, nevertheless, be important to further suppress the quasi-singularities to prevent them from becoming dominant features in tomographic images. This can be done, for instance, by pre-conditioning the pure gradients of the misfit functional (e.g. Igel et al., 1996; Fichtner et al., 2009; Sect. 7.3.4).

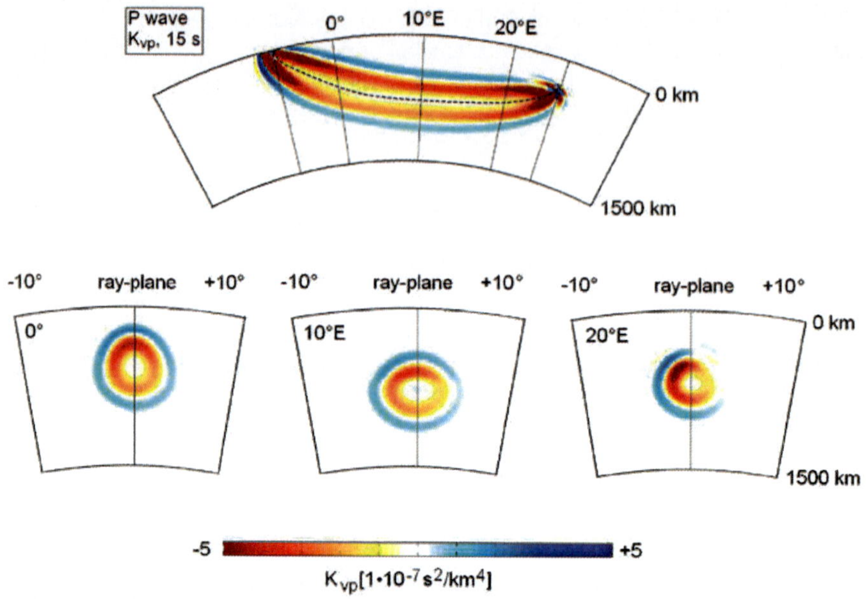

Fig. 12.4 Traveltime kernel, K_{v_P}, corresponding to a 15 s P wave from Fig. 12.2. The *dashed curve* marks the geometric ray path. *Top*: *Vertical slice* through the ray plane connecting source and receiver. Sensitivity is exactly zero along the ray path. *Bottom*: *Vertical slices* perpendicular to the ray plane at 0°, 10°E and 20°E. Owing to its characteristic shape, the traveltime kernel is often referred to as banana–doughnut kernel

Another notable feature in Fig. 12.4 is the asymmetry of the kernel relative to the geometric ray path. It is most pronounced in the slice at 20°E perpendicular to the ray plane. The asymmetry results from the asymmetric radiation pattern of the forward P wave. Regions of weak sensitivity correspond to nodal planes in the radiation pattern.

A similar asymmetry is only very weakly developed in the vicinity of the receiver where the shape of the kernel is dominated by the characteristics of the adjoint wave field. The adjoint source is a vertical single point force that generates a more isotropic radiation pattern that causes the kernel to be nearly symmetric with respect to the geometric ray.

In our next example we consider the prominent S wave arriving on the E–W component around 530 s. For the construction of the adjoint source time function we proceed as in the case of the direct P wave, the only difference being the location of the time window, $W(t)$, around the E–W-component S waveform. The resulting Fréchet kernel with respect to the S wave speed, K_{v_S}, is shown in Fig. 12.5. Its general anatomy is similar to the P wave kernel. Only the width of the kernel is reduced. This is due to both the radiation pattern and the shorter wavelength of S waves compared to the wavelength of P waves at the same period.

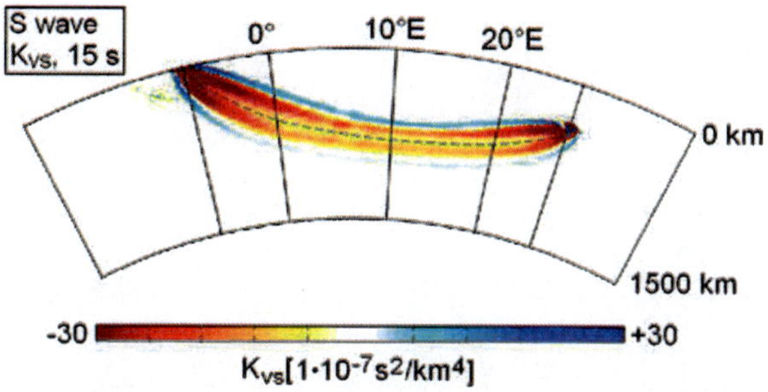

Fig. 12.5 Cross section through the traveltime Fréchet kernel, K_{v_S}, that corresponds to the E–W-component S wave with a dominant period of 15 s, as shown in Fig. 12.2

The variations of this theme are numerous and replete of interesting physics. Particularly noteworthy cases are those where an observed waveform results from P-to-S or S-to-P conversions at material interfaces. The vertical-component sP phase from Fig. 12.2, for instance, corresponds to an S wave that propagates from the source to the surface where it partly converts to a P wave that travels towards the receiver.

The two branches of sP are clearly visible in the Fréchet kernels with respect to the P wave speed, K_{v_P}, and with respect to the S wave speed, K_{v_S}. P wave speed sensitivity concentrates along the ray path that connects the surface reflection point to the receiver. Along the path segment where sP propagates in the form of an S wave, i.e. between the source and the surface reflection point, P wave speed sensitivity is small. S wave speed sensitivity, K_{v_S}, is complementary to P wave speed sensitivity, K_{v_P}. It is large between the source and the surface reflection point but nearly zero along the P wave segment of the path.

Highly oscillatory sensitivity near the source and the receiver is clearly visible especially in the P wave speed kernel, K_{v_P}. This is because the finite-frequency sP wave is spatially not as isolated as an infinite-frequency sP wave in the ray-theoretical framework. S-to-P scattered waves from the near-source region, for instance, can arrive within the sP time window and thus affect the cross-correlation measurement. From a practical point of view, highly oscillatory sensitivity is hardly relevant in the solution of a tomographic problem. The projection of the Fréchet kernels onto the basis functions (see Chap. 8 and Eq. (8.22)) usually eliminates most of the strong oscillations and emphasises the longer wavelength structure of the kernels (Fig. 12.6).

In our next example for cross-correlation measurements on body waves, we consider the prominent sS arrival on the E–W component. The corresponding Fréchet kernel, shown in Fig. 12.7, reveals a complex pattern of positive and negative sensitivity that is not as obviously associated with the geometric ray path as the kernels for P, S and sP. Yet, the large sensitivity near the reflection point indicates the implication of the surface in the generation of the sS phase.

All of the examples studied so far illustrate the delicate relationship between infinite-frequency ray theory and finite-frequency kernels. Despite its limited range

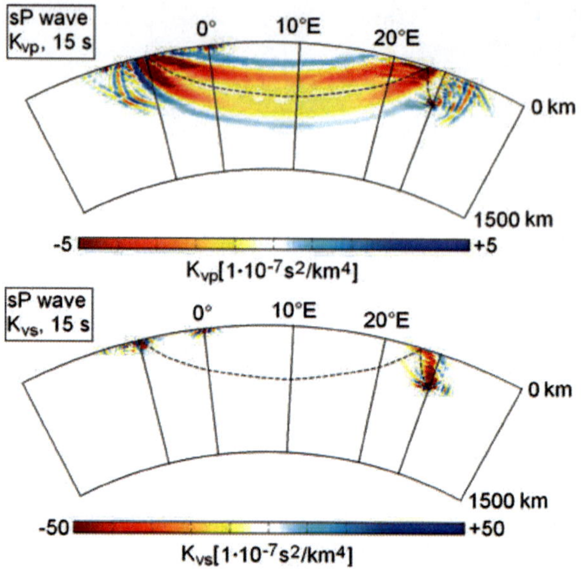

Fig. 12.6 Cross sections through traveltime Fréchet kernels for the vertical-component sP waveform with a dominant period of 15 s, as shown in Fig. 12.2. *Top*: Sensitivity with respect to the P wave speed, v_P, (K_{v_P}) is concentrated along the P branch of the ray path. *Bottom*: S wave speed sensitivity (K_{v_S}) is restricted to a region around the s branch of the ray path that represents the S wave propagating from the source to the reflection point at the surface. Highly oscillatory sensitivity appears near the receiver, but is hardly relevant from a practical point of view

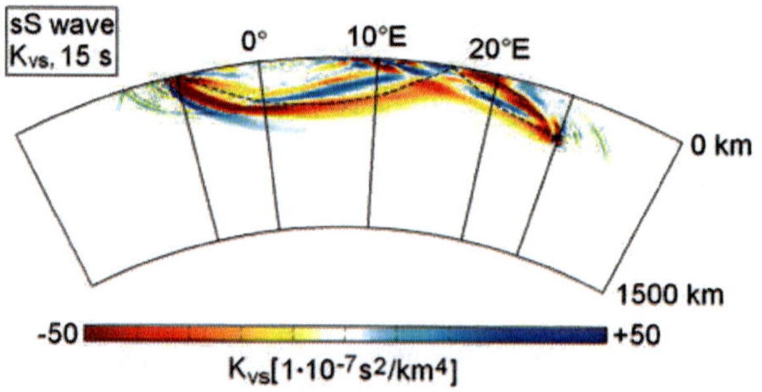

Fig. 12.7 Cross section through the traveltime Fréchet kernel, K_{v_S}, for the 15 s sS wave observed on the E–W-component in Fig. 12.2

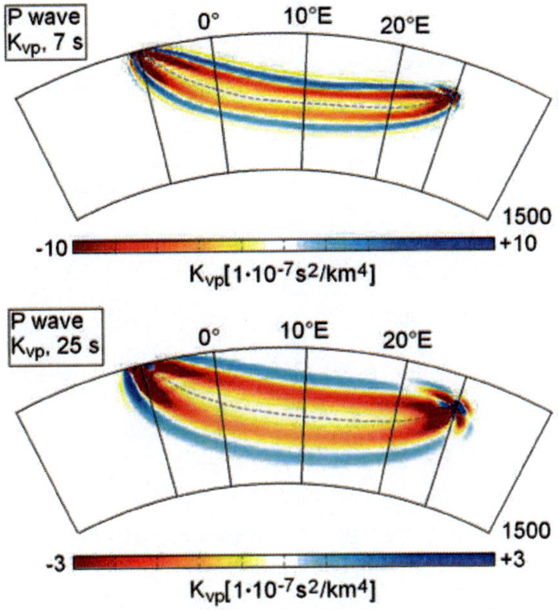

Fig. 12.8 Illustration of the frequency dependence of P wave speed Fréchet kernels, $K_{v\mathrm{p}}$, for cross-correlation measurements. *Top*: The dominant period is 7 s, which results in a comparatively slim kernel. *Bottom*: For a dominant period of 25 s the kernel extends far from the geometric ray path and the doughnut hole is particularly pronounced

of validity, ray theory is an indispensable aid in the interpretation of Fréchet kernels for measurements on isolated seismic phases. Regions of non-zero sensitivity mostly follow the infinite-frequency ray path. However, the fine structure of the kernels is strongly affected by the frequency content, the source radiation pattern, interference effects and the shape of the window function used to isolate a waveform.

As demonstrated by Dahlen et al. (2000), finite-frequency traveltime kernels collapse into infinitely thin rays as the dominant period, T_d, tends to zero. Figure 12.8 illustrates this trend. It shows traveltime kernels for the direct P wave with dominant periods of $T_d \sim 7$ s (top) and $T_d \sim 25$ s (bottom). The width of the first Fresnel zone is proportional to $\sqrt{T_d}$ which explains why the 25 s kernel is nearly twice as wide as the 7 s kernel.

12.1.2 L_2 Amplitudes

The procedure for the computation of Fréchet kernels for L_2 amplitude measurements is similar to the one outlined in the previous paragraph where we considered measurements of cross-correlation time shifts. The only notable difference lies in the time evolution of the adjoint sources: displacement velocity, $\dot{u}_i(t)$, for the time shift

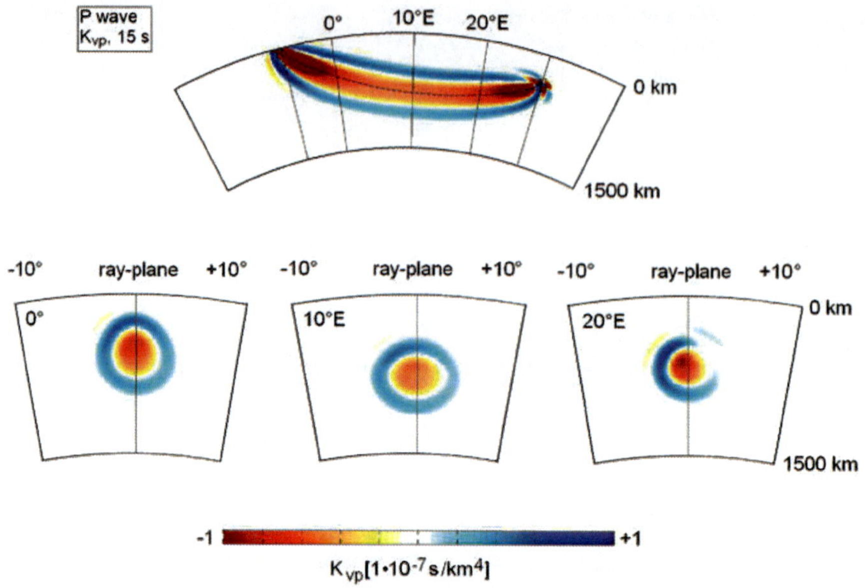

Fig. 12.9 Amplitude kernel, K_{v_P}, corresponding to a 15 s P wave. The geometric ray path is marked by the *dashed curve*. *Top*: *Vertical slice* through the ray plane connecting source and receiver. *Bottom*: *Vertical slices* perpendicular to the ray plane at 0°, 10°E and 20°E

(Eq. 12.1) and pure displacement, $u_i(t)$, for the amplitude measurement (Eq. 12.2). This apparently minor distinction causes the amplitude kernel for the 15 s P wave, shown in Fig. 12.9, to differ substantially from the traveltime kernel in Fig. 12.4.

Amplitude kernels are, in contrast to traveltime kernels, not hollow but filled, meaning that the maximum sensitivity is located along the geometric ray path. Dahlen & Baig (2002) noted that the positive contributions in the second Fresnel zone are particularly pronounced – more than in traveltime kernels. This means that a P wave speed anomaly that is constant across the kernel may cause a negligible amplitude perturbation because the positive and negative sensitivities compensate each other. This suggests that body wave amplitudes are more sensitive to smaller scale perturbations that fit within the first Fresnel zone.

Still clearly visible are the very strong sensitivities in the vicinity of the source and the receiver. These are general features of sensitivity kernels for point-wise measurements on wave fields that were emitted by a point source.

Another persistent feature is the asymmetry relative to the ray that is particularly visible in the cross section perpendicular to the ray at 20°E, i.e. close to the source. The asymmetry again results from the radiation pattern of the direct P wave. Its existence is thus primarily controlled by the forward field and not by the specifics of the measurements that determine the adjoint field.

The physical meaningfulness of the amplitude kernels is not as obvious as in the case of the phase-shift kernels. This is because amplitudes can depend rather non-linearly on the near receiver structure instead of depending almost linearly on the structure around the complete ray path. The best known example is strong amplifications of seismic ground motion in sedimentary basins (e.g. Frankel & Vidale, 1992). From a purely mathematical point of view, the Fréchet kernel in Fig. 12.9 represents a correct first derivative. However, the extent to which the first derivative indeed represents amplitude changes resulting from finite perturbations seems to depend on the particular type of seismic wave, the frequency content and the near receiver geology.

12.2 Surface Waves

Following the analysis of body wave Fréchet kernels, we focus on the surface wave part of longer period seismograms. The finite-frequency sensitivity of surface waves has been studied extensively in recent years. Friederich (1999) calculated 3D Fréchet kernels in radially symmetric Earth models for both body and surface waves within a normal mode framework and applied his method to the imaging of S velocity structure in the East Asian upper mantle (Friederich, 2003). Also based on normal mode theory, Zhou et al. (2004) derived Fréchet kernels for multi-taper measurements on fundamental-mode surface waves. The kernels were then used to compute a global surface wave tomographic model (Zhou et al., 2005). An extension to higher mode surface waves can be found in Zhou (2009). Working with the potential representation of surface waves, Yoshizawa & Kennett (2005) were able to derive surface wave kernels for laterally variable media, thus highlighting the importance of using the proper kernels in order to account for the heterogeneity in the real Earth. A further improvement was made by Sieminski et al. (2007b) who computed surface wave sensitivity with respect to anisotropic parameters using global spectral-element simulations.

The setup of our numerical modelling is the same as in the previous section (see Fig. 12.1), the only difference being that the source is shallow (50 km depth) so that strong fundamental-mode surface waves are excited. Figure 12.10 displays the three-component displacement velocity recorded at an epicentral distance of 25.23°. The dominant period is 50 s. Rayleigh waveforms are clearly visible on both the E–W and vertical components. The Love wave is restricted to the N–S component because the propagation is strictly in E–W direction.

12.2.1 Isotropic Earth Models

To compute the sensitivity of the surface wave trains with respect to perturbations in isotropic Earth structure, we isolate the Love and Rayleigh waves using a cosine taper, the boundaries of which are displayed in the form of vertical dashed lines in Fig. 12.10. The tapered waveforms mostly consist of fundamental-mode surface

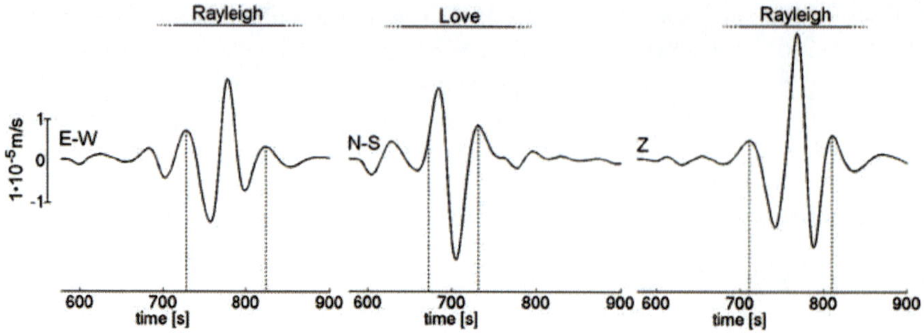

Fig. 12.10 Synthetic surface wave trains with a dominant period of 50 s. The source–receiver geometry is the same as in Fig. 12.1, but the source depth is only 50 km, so that strong fundamental-mode surface waves are excited. The *dashed vertical lines* indicate the boundaries of the tapers used to isolate waveforms for the computation of Fréchet kernels

waves, but we do not make any special effort to eliminate higher modes using, for instance, multi-taper techniques (e.g. Zhou et al., 2004).

Figure 12.11 shows horizontal and vertical cross sections through the S wave speed traveltime kernel, K_{v_S}, corresponding to the isolated Love waveform in Fig. 12.10. Sensitivity extends in a bent cigar-shaped region (or banana-shaped, depending on personal preference) from the source to the receiver, and it is restricted to the uppermost 200 km of the Earth model. The kernel exhibits the typical alternating positive and negative sensitivity bands that are separated by zero-sensitivity surfaces where first-order scattering has no effect on the cross-correlation measurement. The characteristic doughnut hole found in the body wave traveltime kernels (e.g. Figs. 12.4 and 12.5) is not present in the surface wave kernels. This is due to the 2D propagation nature of surface waves (e.g. Zhou et al., 2004). The side band structure of the kernels is most affected by the details of the measurement process, such as the width and slope of the taper used to isolate the waveform. However, within the first Fresnel zone the kernels are rather independent of the measurement details.

The S wave speed sensitivity of the vertical-component Rayleigh wave is shown in Fig. 12.12. The anatomy of the kernel parallel to the propagation direction is sim-

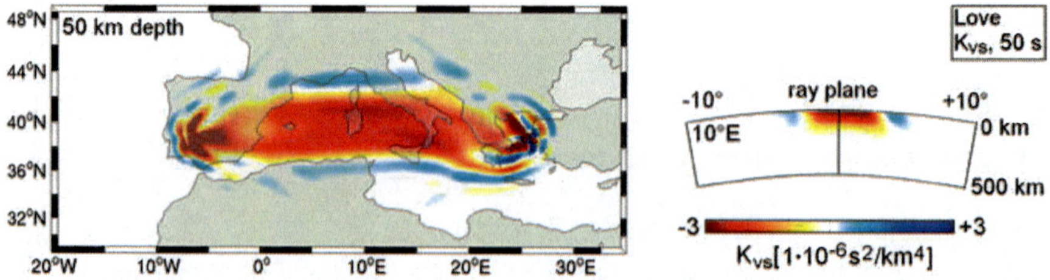

Fig. 12.11 Traveltime sensitivity kernel with respect to the S wave speed, v_S, for the 50 s Love wave from Fig. 12.10. *Left*: *Horizontal slice* at 50 km depth. *Right*: *Vertical slice* at 10°E longitude

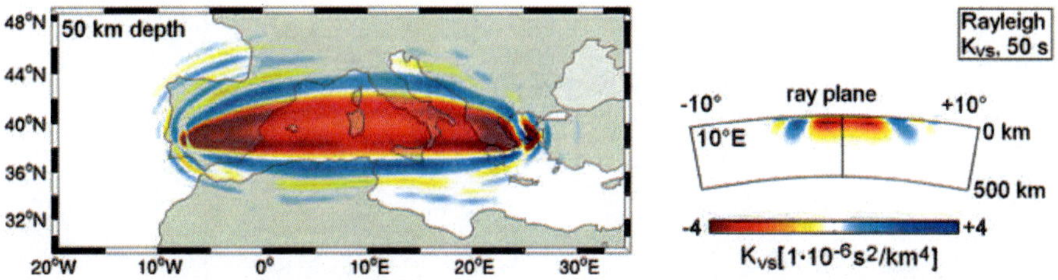

Fig. 12.12 Traveltime sensitivity kernel with respect to the S wave speed, v_S, for the 50 s vertical-component Rayleigh wave from Fig. 12.10. *Left*: *Horizontal slice* at 50 km depth. *Right*: *Vertical slice* at 10°E longitude

ilar to K_{v_S} for the Love wave (Fig. 12.11). However, the sensitivity drops to exactly zero near the surface (see also Fig. 12.14). This is in accordance with analytically derived 1D sensitivity distributions for plane waves (e.g. Takeuchi & Saito, 1972).

As can be seen in Fig. 12.13, the vertical-component Rayleigh wave is weakly sensitive to the P wave speed, v_P, within the crustal part of the model, that is, above 50 km depth. Love waves, in contrast, exhibit no P wave sensitivity whatsoever. It is therefore reasonable to restrict a surface wave tomography to the S wave structure of the Earth.

The sensitivity distributions of surface waves exhibit a characteristic frequency dependence: The depth extent of the kernels increases with increasing dominant period. This effect is visualised in Fig. 12.14, which shows vertical profiles through the v_S kernels for Rayleigh waves at periods of 50, 100 and 200 s. A 50 s Rayleigh wave is most sensitive to structure around 40 km depth, but is practically unaffected by v_S perturbations below 200 km. The 200 s kernel, in contrast, extends to more than 500 km depth, with maximum sensitivity near 200 km. It is important to note that the sharp drop in sensitivity with increasing depth is due to the dominance of the fundamental-mode surface waves within the analysis time window. The depth dependence of Fréchet kernels for higher mode surface waves is generally more complex (e.g. Takeuchi & Saito, 1972; Zhou, 2009).

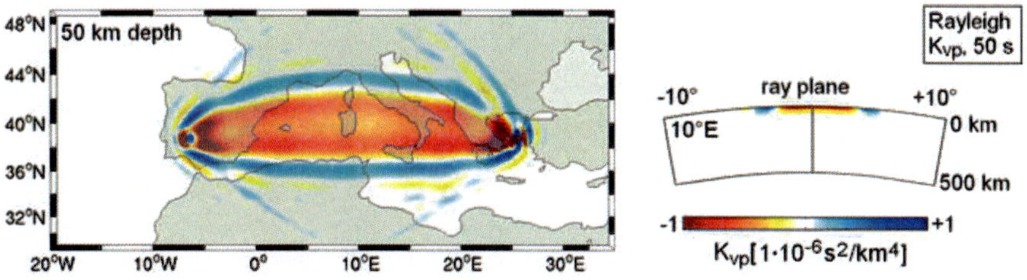

Fig. 12.13 Traveltime sensitivity kernel with respect to the P wave speed, v_P, for the 50 s vertical-component Rayleigh wave from Fig. 12.10. *Left*: *Horizontal slice* at 50 km depth. *Right*: *Vertical slice* at 10°E longitude

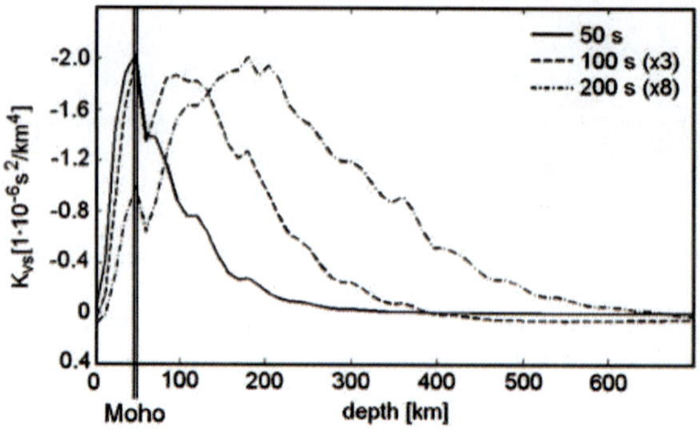

Fig. 12.14 Vertical profiles through the v_S kernels for Rayleigh waves at periods of 50 s (*solid*), 100 s (*dashed*) and 200 s (*dash dotted*). The profiles are located in the centre of the first Fresnel zone at 10°E longitude. To allow for easier comparison, the profiles for 100 and 200 s are amplified by factors of 3 and 8, respectively. The *vertical double line* indicates the location of the Moho at around 40 km depth

12.2.2 Radial Anisotropy

With the help of Eqs. (9.22) we can calculate the sensitivity of the surface wave trains with respect to anisotropic perturbations, for instance, in v_{SH} and v_{SV}.

As expected, the vertical-component Rayleigh wave is primarily sensitive to v_{SV}, which is the propagation speed of a vertically polarised plane shear wave (Fig. 12.15). Visually, $K_{v_{SV}}$ and K_{v_S} from Fig. 12.12 are hardly distinguishable. From Eq. (9.24) we know that the difference $K_{v_S} - K_{v_{SV}}$ is equal to $K_{v_{SH}}$; and indeed, $K_{v_{SH}}$ as displayed in the lower part of Fig. 12.15 is barely visible. The small

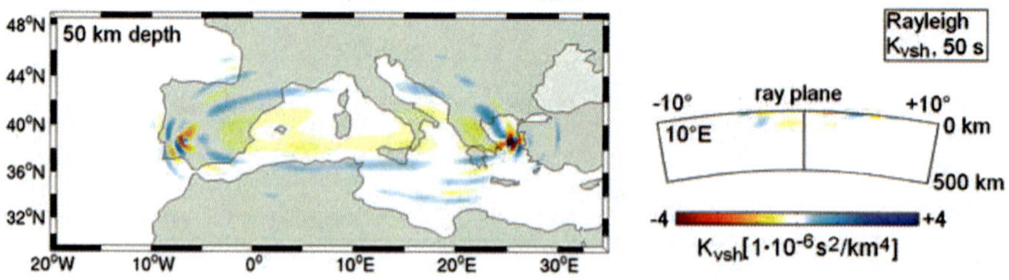

Fig. 12.15 Traveltime sensitivity kernels for the *vertical-component* Rayleigh wave from Fig. 12.10. *Top*: Sensitivity with respect to the SV wave speed, v_{SV}. *Bottom*: Sensitivity with respect to the SV wave speed, v_{SH}

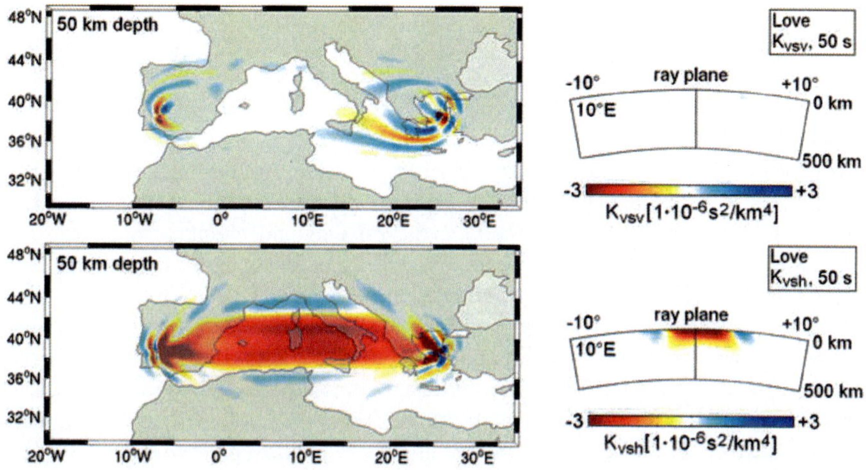

Fig. 12.16 Traveltime sensitivity kernels for the Love wave from Fig. 12.10. *Top*: Sensitivity with respect to the SV wave speed, υ_{SV}. *Bottom*: Sensitivity with respect to the SH wave speed, υ_{SH}

non-zero contributions to $K_{\upsilon_{\mathrm{SH}}}$ are mostly the result of Love–Rayleigh coupling (Sieminski et al., 2007a). A Love wave leaving the source is scattered off a $\delta\upsilon_{\mathrm{SH}}$ perturbation and partly converted into a vertically polarised shear wave that arrives within the Rayleigh wave window.

In the case of Love waves we observe a similar phenomenon (Fig. 12.16): The sensitivity with respect to υ_{SH} is largest because Love waves are horizontally polarised. From a visual comparison of $K_{\upsilon_{\mathrm{SH}}}$ and $K_{\upsilon_{\mathrm{S}}}$ from Fig. 12.11 we find that $K_{\upsilon_{\mathrm{SV}}} = K_{\upsilon_{\mathrm{S}}} - K_{\upsilon_{\mathrm{SH}}}$ should be small, which is confirmed by the slices shown in the upper part of Fig. 12.16. The υ_{SV} kernel suggests that Rayleigh–Love coupling occurs in the vicinity of the source and the receiver, but it is highly inefficient along the ray path.

12.3 Hessian Kernels: Towards Quantitative Trade-Off and Resolution Analysis

The Fréchet kernels discussed in the previous paragraphs mostly serve two purposes: First, they quantify where observations are sensitive to Earth structure, and second, they are indispensable for the computation of descent directions in gradient methods of non-linear minimisation (see Sect. 7.3).

Hessian kernels – in addition to being equally useful in minimisation algorithms – are most important as the carriers of covariance information. In the vicinity of an optimal Earth model, $\tilde{\mathbf{m}}$, we can approximate the misfit functional, χ, with the help of the Hessian, H_χ,

$$\chi(\tilde{\mathbf{m}} + \delta\mathbf{m}) \approx \chi(\tilde{\mathbf{m}}) + \frac{1}{2} H_\chi(\tilde{\mathbf{m}})(\delta\mathbf{m}, \delta\mathbf{m}) , \tag{12.3}$$

where we used $\nabla_m \chi(\tilde{\mathbf{m}}) = \mathbf{0}$. The diagonal elements of H_χ specify the resolution of individual model parameters, in the sense that the change of one single parameter leads to a strong increase of the misfit χ when the corresponding diagonal element is large. The off-diagonal elements of H_χ provide information on the trade-offs between parameters. More specifically, large off-diagonal elements indicate that changes in one parameter may be compensated by changes in another parameter, so that the total misfit, χ, remains unaffected.

As an example, we consider a long-period full waveform tomography for the European upper mantle that is summarised in Fig. 12.17. The data used in the inversion are three-component seismograms with a dominant period of 100 s that provide a good coverage of central and northern Europe (Fig. 12.17, left). The inversion was based on the measurement of time- and frequency-dependent phase differences (see Sect. 11.5; Fichtner et al., 2008). As initial model we used the 3D mantle structure from S20RTS (Ritsema et al., 1999) combined with the crustal model by Meier et al. (2007a, b). After three conjugate-gradient iterations we obtained the tomographic images shown in the centre and left panels of Fig. 12.17. One of the most prominent features is the low-velocity region beneath Iceland, that is commonly attributed to the high temperatures of a hotspot located on the Mid-Atlantic ridge.

To quantify the extent to which structure beneath Iceland is independently constrained, we consider an S velocity perturbation, inside a test volume, V_1, that plays the role of a basis function $b_1(\mathbf{x})$. The test volume is in our case a 400 km by 400 km wide box that extends from 80 to 130 km depth (see Fig. 12.18). Then, following the methodology introduced in Sect. 8.4, we compute the resolution and trade-off (RETRO) kernel, $K_{\delta m_1}$, with

$$\delta\mathbf{m}_1(\mathbf{x}) = [\delta\rho(\mathbf{x}), \delta v_{\mathrm{S}}(\mathbf{x}), \delta v_{\mathrm{P}}(\mathbf{x})] = [0, b_1(\mathbf{x}), 0] . \tag{12.4}$$

The result is shown in Fig. 12.19.

The RETRO kernel is a superposition of mostly positive arms that extend from the test volume towards several source and receiver positions. The non-zero con-

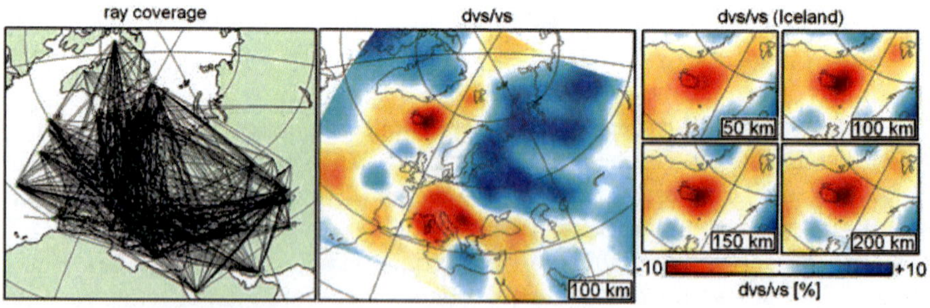

Fig. 12.17 Full waveform tomography for the European upper mantle. *Left*: Ray coverage. *Centre*: Relative S wave speed perturbations at 100 km depth. *Right*: Zoom on the Iceland hotspot at various depth levels. The background model is PREM (Dziewonski & Anderson, 1981)

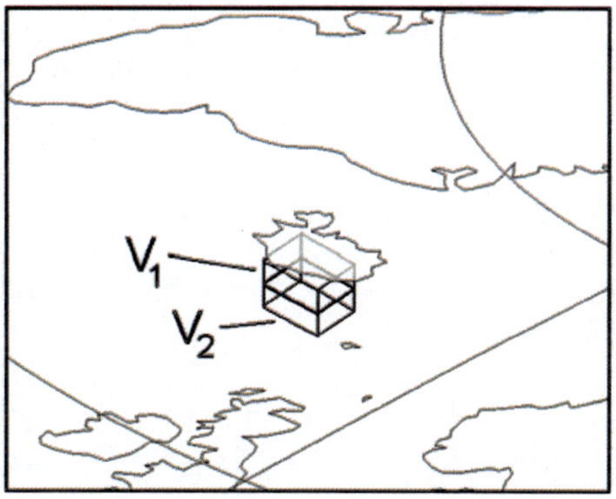

Fig. 12.18 Location of the test volumes V_1 and V_2 beneath Iceland. Both volumes are 400 km by 400 km wide. The depth range from 80 to 130 km is occupied by V_1. The test volume V_2 is located directly below, extending from 130 to 180 km depth

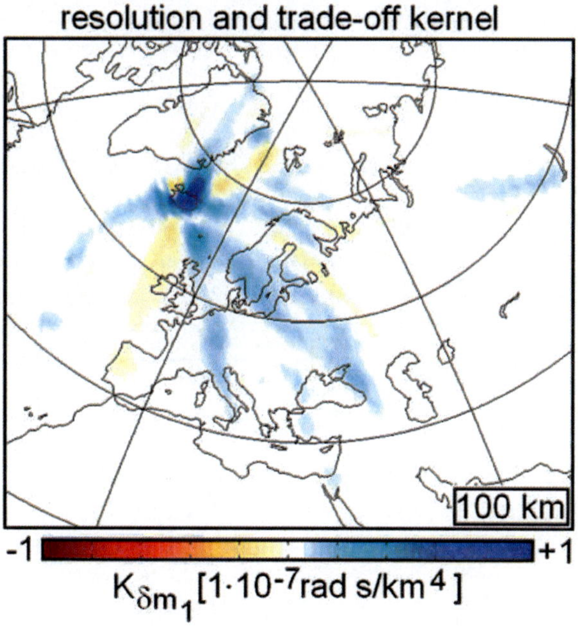

Fig. 12.19 Resolution and trade-off (RETRO) kernel, $K_{\delta m_1}$, at 100 km depth. The perturbation $\delta\mathbf{m}_1$ is a 400 km by 400 km positive S wave speed perturbation that extends from 80 to 130 km depth directly beneath Iceland (see Fig. 12.18)

tributions of $K_{\delta m_1}$ outside V_1 are those regions where v_s structure trades off with v_s structure inside V_1. It follows that the low S velocities seen between 80 and 130 km depth beneath Iceland cannot be constrained independently. In other words, the S wave speed structure in other parts of the European upper mantle affects our image of the Iceland hotspot. To see more specifically how S wave structure within V_1 trades off with deeper structure, we consider a second test volume, V_2, located directly beneath V_1, at 130–180 km depth (see Fig. 12.18). To V_2 we associate the basis function $b_2(\mathbf{x})$. Repeating the above procedure yields the RETRO kernel $K_{\delta m_2}$, with

$$\delta \mathbf{m}_2(\mathbf{x}) = [\delta\rho(\mathbf{x}), \delta v_{\mathrm{s}}(\mathbf{x}), \delta v_{\mathrm{P}}(\mathbf{x})] = [0, b_2(\mathbf{x}), 0]\,. \tag{12.5}$$

From $K_{\delta m_1}$ and $K_{\delta m_2}$ we compute the 2×2 Hessian submatrix, $\mathbf{H}_{\chi,(1,2)}$, that corresponds to the basis functions b_1 and b_2:

$$\mathbf{H}_{\chi,(1,2)} = \begin{pmatrix} H_{1,1} & H_{1,2} \\ H_{1,2} & H_{2,2} \end{pmatrix} = \int_G \begin{pmatrix} K_{\delta m_1}\, b_1 & K_{\delta m_1}\, b_2 \\ K_{\delta m_2}\, b_1 & K_{\delta m_2}\, b_2 \end{pmatrix} \mathrm{d}^3\mathbf{x}\,. \tag{12.6}$$

Then returning to Eq. (12.3), we can approximate the misfit functional in the vicinity of the global optimum (Fig. 12.17) by

$$\chi(\tilde{\mathbf{m}} + \delta\mathbf{m}) \approx \chi(\tilde{\mathbf{m}}) + \frac{1}{2} \begin{pmatrix} \delta v_{\mathrm{s},1} \\ \delta v_{\mathrm{s},2} \end{pmatrix} \cdot \begin{pmatrix} H_{1,1} & H_{1,2} \\ H_{1,2} & H_{2,2} \end{pmatrix} \cdot \begin{pmatrix} \delta v_{\mathrm{s},1} \\ \delta v_{\mathrm{s},2} \end{pmatrix}, \tag{12.7}$$

with

$$\delta\mathbf{m}(\mathbf{x}) = [\delta\rho(\mathbf{x}), \delta v_{\mathrm{s}}(\mathbf{x}), \delta v_{\mathrm{P}}(\mathbf{x})] = [0, \delta v_{\mathrm{s},1} b_1(\mathbf{x}) + \delta v_{\mathrm{s},2} b_2(\mathbf{x}), 0]\,. \tag{12.8}$$

The contour lines of the quadratic function (12.7), given in terms of the S velocity perturbations $\delta v_{\mathrm{s},1}$ and $\delta v_{\mathrm{s},2}$, are displayed in Fig. 12.20. The contour lines in

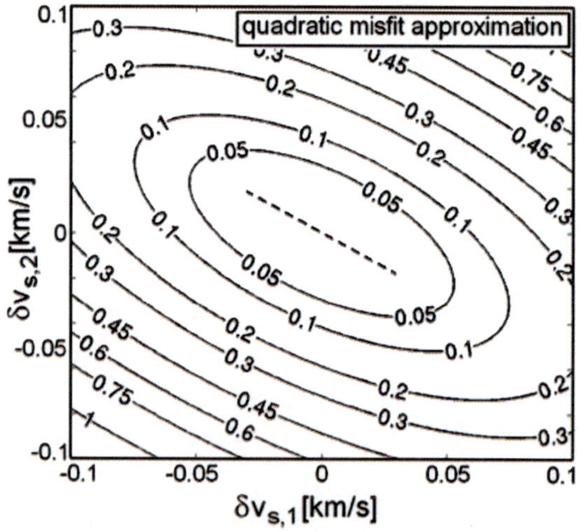

Fig. 12.20 *Contour lines* of the quadratic approximation to the misfit functional, χ, (Eq. 12.7) in terms of the S wave speed perturbations $\delta v_{\mathrm{s},1}$ and $\delta v_{\mathrm{s},2}$. Changing $\delta v_{\mathrm{s},1}$ and $\delta v_{\mathrm{s},2}$ along the *dashed line* has hardly any effect on the misfit, meaning that the shallow and deep S velocity structures trade off substantially

the direction of $\delta v_{s,2}$ are about twice as closely spaced as the contour lines in the direction of $\delta v_{s,1}$. This indicates that S velocity perturbations at greater depth (130–180 km) are better resolved than at shallower depth (80–130 km). Significant trade-offs exists between the shallower and the deeper S velocity structure. In fact, changing $\delta v_{s,1}$ and $\delta v_{s,2}$ along the dashed line in Fig. 12.20 has hardly any effect on the misfit. The RETRO kernel approach presented above allows us to study resolution and trade-offs in terms of concrete formulas that relate deviations from the optimal model to changes in the misfit. This goes far beyond chequerboard test that are commonly used for visual resolution analysis. The principal limitation is encapsulated in Eq. (12.3), which is valid and physically meaningful only when the tomographic model is indeed close to the global optimum of the misfit functional χ.

12.4 Accuracy-Adaptive Time Integration

Having examined the anatomy of a variety of Fréchet kernels, K_m, we return to a topic that we already introduced in Sect. 8.5.2: the efficient computation of the time integral

$$K_m = \int_T \mathbf{u}^\dagger \cdot \nabla_m \mathbf{L}(\mathbf{u})\, \mathrm{d}t\,. \tag{12.9}$$

Since (12.9) involves the regular field $\mathbf{u}$ and the adjoint field $\mathbf{u}^\dagger$, both have to be known simultaneously during the adjoint calculation. This implies that $\mathbf{u}$ needs to be stored at sufficiently many time steps, at least when dissipation or absorbing boundaries are present. The resulting amount of data can easily exceed conventional storage capacities. Our goal therefore is to find the minimum number of time steps where $\mathbf{u}$ needs to be stored in order to ensure the accurate representation of the sensitivity within the first m Fresnel zones.

For this, let d_m be the width of the mth Fresnel zone. We can expect the sensitivity kernel to be accurate on length scales similar to d_m when the regular wave field is stored at least once while propagating over the distance d_m. Hence, an optimistic estimate of the maximum possible storage interval is $t_s = \frac{d_m}{v_s}$, where v_s is a representative S wave speed. This means that the regular wave field needs to be stored at least every $n = \frac{t_s}{\Delta t} = \frac{d_m}{\Delta t\, v_s}$ time steps, where Δt is the time increment. The maximum value of Δt is dictated by the CFL stability condition, and ideally we have $\Delta t \approx c\frac{\min h}{\max v_{\mathrm{P}}}$, with a grid spacing parameter h and a constant c that is typically on the order of 0.3 (see Sect. 2.5). An approximation of Δt in terms of v_s and the dominant period T_d is

$$\Delta t \approx c\,\frac{\min h}{\max v_{\mathrm{P}}} \approx \frac{c\,\lambda_\mathrm{d}}{10\,\max v_{\mathrm{P}}} \approx \frac{c\,v_s\,T_\mathrm{d}}{10\,\max v_{\mathrm{P}}}\,, \tag{12.10}$$

where λ_d is the dominant wavelength. Equation (12.10) is based on the assumption that the minimum grid spacing, min h, is approximately equal to one tenth of the dominant wavelength, λ. Combining (12.10) and the average width of the mth Fresnel zone,

$$d_m \approx \frac{1}{4}\sqrt{v_s T_d \ell}(\sqrt{m} - \sqrt{m-1}), \tag{12.11}$$

yields the following estimate for n:

$$n \approx \frac{5}{2}\frac{\max\, v_P}{c}\sqrt{\frac{\ell}{T_d\, v_S^3}}\,(\sqrt{m} - \sqrt{m-1})\,, \tag{12.12}$$

where ℓ denotes the length of the ray path. Using suitable parameter values $\max\, v_P = 7$ km/s, $c = 0.3$, $\ell = 1,000$ km, $T_d = 50$ s, $v_S = 4$ km/s and $m = 2$, gives $n \approx 14$. This means that the regular wave field should be stored at least every 14 time steps in order to ensure that the sensitivity in the first and second Fresnel zones is computed accurately. Of course, Eq. (12.12) is a rule of thumb, derived under simplistic assumptions. In nevertheless proves useful in most applications.

Part III
Applications

There are many variants of full waveform inversion that are characterised by their combination of forward problem solver, misfit functional and minimisation algorithm. The usefulness of a specific combination depends on the spatial scale (local, regional, global), the type of Earth structure to be detected (reflectors, scatterers, long-wavelength heterogeneity), the distribution of sources and receivers, data quality and the available computational resources. The extreme problem dependence of full waveform inversion makes exemplary case studies particularly valuable.

Upper mantle tomography on regional to global scales (Chap. 13) critically relies on accurate numerical solutions for surface waves, as they can be obtained, for instance, with the help of the spectral-element method. A challenging problem is the implementation of the crust, because thin crustal layers require small elements, which lead to high computational costs. An elegant alternative is the construction of long-wavelength equivalent crustal models that allow us to employ larger elements, thus reducing computational costs to a more feasible level. Teleseismic waveforms at periods above 30 s are most strongly influenced by the transmission properties of the Earth. This requires a misfit measure that explicitly extracts traveltime information, such as time–frequency phase misfits. To ensure the convergence of the iterative minimisation towards the global minimum, a suitable initial model needs to be implemented. Ideally, the initial model should represent the long-wavelength structure of the study region.

The acquisition geometry of active-source experiments often allows us to study a 2D problem, where the equations of motion can be solved efficiently in the frequency domain. Exploiting only the phases of the first-arriving P waves furthermore justifies the acoustic approximation. The resulting computational efficiency makes 2D acoustic full waveform inversion a powerful tool in local-scale geologic, engineering and exploration applications (Chap. 14). Gradually increasing the frequency of the observed waveforms during the iterative inversion helps to guide the gradient-based minimisation scheme towards the global optimum. Inversion results often agree remarkably well with independent constraints, based for instance on sonic logs.

Full waveform inversion remains computationally expensive compared to classical tomographic techniques, despite advances in numerical mathematics. Data reduction schemes that lower the computational costs are therefore likely to play a key role in future applications. An efficient way of reducing the number of forward simulations is to add or stack at each receiver the seismograms corresponding to the different sources. The new data set, consisting of one stacked time series per station, can then be modelled with one single forward simulation, where all sources act at the same time. This procedure, referred to as source stacking, has been shown to work for long-period seismograms at a global scale (Chap. 15), and modifications of it are now being applied to local-scale data.

Chapter 13
Full Waveform Tomography on Continental Scales

This chapter describes the development of a technique for full seismic waveform tomography and its application to the imaging of upper-mantle structure in the Australasian region.

The forward problem is solved using an implementation of the spectral-element method that operates in the natural spherical coordinate system. We pay special attention to the construction of a long-wavelength equivalent model of the crust that can be sampled by a comparatively coarse numerical mesh without losing solution accuracy within a pre-defined period range. To ensure the convergence of the iterative minimisation, we implement a 3D initial model that has both an elastic and an anelastic component. After 19 iterations we obtain a radially anisotropic model of the Australasian upper mantle that explains entire 30 s waveforms in great detail. The estimated resolution lengths above 200 km depth are around 3° laterally and 40 km vertically. We discuss technical issues including the dependence on the initial model and the effects of non-linearity on the anisotropic part of the model.

The principal advantages of the full waveform inversion are (1) the accurate modelling of elastic wave propagation through realistically heterogeneous 3D Earth models, (2) the incorporation of all types of seismic waves, including body waves, surface waves and unidentified phases, (3) the iterative improvement of the tomographic images and (4) the elimination of the need for crustal corrections.

13.1 Motivation

In recent years, developments in both theoretical seismology and numerical mathematics have led to substantial progress in seismic tomography. It is today widely accepted that wave propagation effects such as multi-pathing, scattering or wavefront healing are important in the 3D Earth (Williamson, 1991; Williamson & Worthington, 1993; Spetzler et al., 2001). Theories and methods that allow us to account for such phenomena in seismic tomography, and to go beyond classical ray theory, have been developed by several authors (e.g. Bamberger et al., 1982; Tarantola, 1988; Yomogida, 1992; Marquering et al., 1998; Dahlen et al., 2000, Zhao et al., 2000). The evidence that the results of seismic tomography do indeed improve

A. Fichtner, *Full Seismic Waveform Modelling and Inversion*, Advances in Geophysical and Environmental Mechanics and Mathematics,
DOI 10.1007/978-3-642-15807-0_13, © Springer-Verlag Berlin Heidelberg 2011

when finite-frequency effects in 3D media are included is steadily increasing (Yoshizawa & Kennett, 2004; Zhou et al., 2005; Boschi, 2006; Chen et al., 2007). Moreover, the amount of data that we can exploit has grown immensely, thanks to efficient numerical techniques that enable us to simulate seismic wave propagation through realistic Earth models (e.g. Seriani, 1998; Komatitsch & Tromp, 2002; Dumbser et al., 2007; de la Puente, 2007). Some of them have been described in the first part of this book. We could, in principle, use complete seismic waveforms for the purpose of seismic tomography. Yet, full waveform tomography on continental to global scales – as envisioned by Capdeville et al. (2005) – remains an exception. This is in contrast to applications in engineering and exploration seismology where full waveform inversion has been used, at least in 2D, since the early 1980s (Bamberger et al., 1982; Crase et al., 1990; Igel et al., 1996; Pratt & Shipp, 1999; Bleibinhaus et al., 2007). While the equations of motion are scale independent, at least within the macroscopic world, there are nevertheless fundamental differences between local and global tomography that explain this discrepancy: (1) Engineering and exploration problems can often be reduced to dependence on just one or two dimensions. The computational costs are therefore comparatively moderate. (2) On smaller scales, seismograms are strongly affected by the scattering properties of the Earth. On larger scales, however, the transmission properties primarily determine the character of a seismogram in general, and the phases of seismic waves in particular. This requires the development of full waveform misfits that explicitly quantify phase differences between observed and synthetic seismograms.

The high-resolution images obtained in local-scale full waveform inversion encourage applications on continental and global scales, where more detailed tomographic models are expected to play a major role in the resolution of outstanding societal and geologic problems. These include, but are not limited to, the following:

1. *The seismic monitoring of the Comprehensive Nuclear Test Ban Treaty* where highly resolved models of the Earth's crust and upper mantle are needed to reproduce complete seismic waveforms as accurately as possible.
2. *Reliable tsunami warnings* based on the real-time determination of earthquake source characteristics, the quality of which depends on the ability of the Earth model to predict seismic waveforms.
3. *Tectonic reconstructions* especially in areas where geodetic observations are sparse and where the neotectonic development is marked by the rapid subduction of micro-plates (e.g. the Banda sea region).
4. *Detection and characterisation of small-scale features* such as deep mantle plumes, lateral heterogeneities within the D$''$ layer and deeply subducted oceanic slabs.
5. *Correlation of seismic velocities with independent observations and models* in order to obtain a more comprehensive picture of the Earth's rheology. Examples of independent observations are heat flow, gravity, isotope signatures, seismicity, volcanism and the mineralogical properties of xenoliths. Models derived

independently from seismic observations are, for instance, paleo-geotherms and profiles of electrical conductivity.

6. *Characterisation of the asthenosphere* in terms of its depth and thickness. The quantification of seismic anisotropy within the asthenosphere and its relation to flow patterns in the vicinity of mantle plumes is crucial in order to understand the transport of deep mantle material into the asthenosphere.
7. *Quantification of seismic anisotropy* in general and its relation to texture formation, mantle deformation and rheologic properties.
8. *Thickness of the lithosphere* as a function of age and in relation to the formation and properties of the lithospheric mantle.
9. *Characterisation of Precambrian to Phanerozoic transitions* (e.g. the Tasman Line and the Tornquist Line) in terms of sharpness and depth extent.
10. *Identification and location of geologic units* as, for instance, Precambrian cratons and micro-continents.

13.2 Solution of the Forward Problem

The numerical solution of the forward problem has two components: First, the spatio-temporal discretisation of the underlying partial differential equations, and second, the numerical implementation of a continuously defined structural model. In Sect. 13.2.1 we introduce a variant of the spectral-element method that operates in the natural spherical coordinates and that proves efficient for continental-scale wave propagation. The emphasis in Sect. 13.2.2 is then on the discretisation of finely layered crustal models that can be difficult to represent on a relatively coarse numerical grid.

13.2.1 Spectral Elements in Natural Spherical Coordinates

Teleseismic waveforms at periods above 30 s are dominated by surface waves that bare invaluable information about the structure of the upper mantle. A numerical method that can be used in continental-scale full waveform tomography should thus be able to produce accurate surface waveforms at the lowest possible computational cost. This requirement argues in favour of the spectral-element method that we described in Chap. 4.

The spectral-element method is based on the weak formulation of the equations of motion. The free surface boundary condition is therefore implicitly fulfilled. This ensures that surface waves can be modelled accurately. The required number of grid points per minimum wavelength is approximately 7.

One strength of the spectral-element method is that the elements can be designed to match topography and internal structure. The use of arbitrarily deformed elements results, however, in a large number of summands that contribute to the Galerkin projections of the stress divergence, $\nabla\cdot\boldsymbol{\sigma}$ (Eq. 4.61), and the product of the elastic tensor with the displacement gradient, $\mathbf{C}:\nabla\mathbf{u}$ (Eq. 4.68). The resulting computational costs are the prize that one needs to pay for evaluating the integrals over the elements, G_e, in Cartesian coordinates. In some local- to continental-scale applications one may reduce the number of summands involved in the Galerkin projections by working in the natural spherical coordinates (Fichtner et al., 2009). This means that the elements are defined by

$$G_e = [\theta_{e,\min}, \theta_{e,\min} + \Delta\theta] \times [\phi_{e,\min}, \phi_{e,\min} + \Delta\phi] \times [r_{e,\min}, r_{e,\min} + \Delta r], \quad (13.1)$$

for $e = 1, \ldots, n_e$, and with constant increments $\Delta\theta$, $\Delta\phi$ and Δr. The transformations from the reference cube, $\Lambda = [-1, 1]^3$, to the physical domain element, G_e, are then defined by

$$\theta = \theta_{e,\min} + \frac{\Delta\theta_e}{2}(1+\xi_1), \quad \phi = \phi_{e,\min} + \frac{\Delta\phi_e}{2}(1+\xi_2), \quad r = r_{e,\min} + \frac{\Delta r_e}{2}(1+\xi_3). \quad (13.2)$$

The geometry of the transformation $G_e \rightarrow \Lambda$ is visualised in Fig. 13.1. Note that this transformation is possible only when the spherical section excludes both the centre of the Earth and the poles, which is the case on regional to continental scales.

A realistic application of the spectral-element method in spherical coordinates – henceforth referred to as SES3D – is illustrated in Figs. 13.2 and 13.3. There we consider a source located at 90 km depth near the Solomon Islands and a station in eastern Australia. We use the radially symmetric AK135 (Kennett et al., 1995) as Earth model. The spectral-element synthetics can therefore be compared to semi-analytical solutions (Friederich & Dalkolmo, 1995). The elements are 0.5° × 0.5° wide and 38 km deep, with the Lagrange polynomial degree set to 6. This setup allows us to compute accurate waveforms with periods as low as 7 s. In general, the agreement between the spectral-element and the semi-analytical solutions is excellent (Fig. 13.3). The difference seismogram, plotted as dashed curve amplified by a factor of 25, is smallest for the early arriving P waves. Surface waves, appearing later in the seismograms, are more difficult to model due to their comparatively short wavelengths. A remarkable detail is the well-modelled small-amplitude waveform in the interval between 340 and 500 s. It partly originates from reflections at the surface and at the discontinuities located at 410 and 660 km depth.

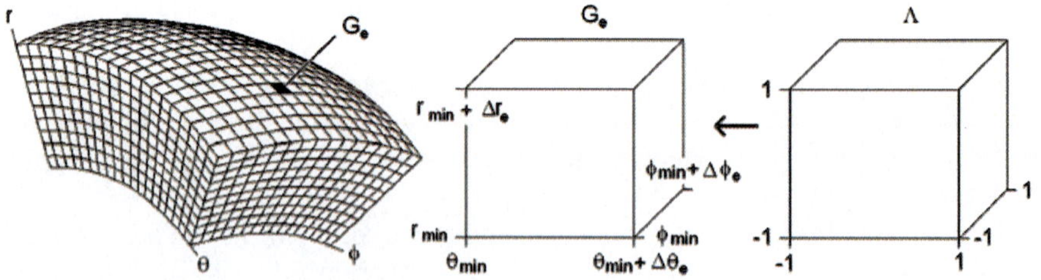

Fig. 13.1 Illustration of the transformation defined in Eq. (13.2). Each physical domain element, G_e, is a small spherical section that is related to the reference cube, Λ, via the transformation defined in Eqs. (13.2)

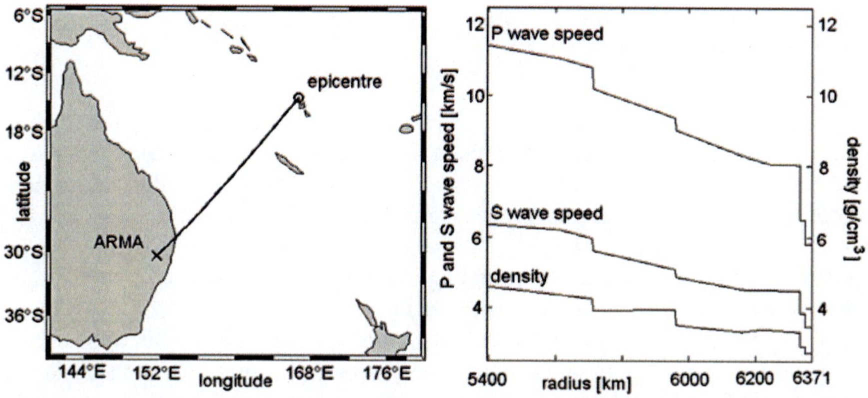

Fig. 13.2 *Left*: Source–receiver configuration. The distance between the source (o) and station ARMA (×) is 21.1°. *Right*: Radially symmetric Earth model AK135 (Kennett et al., 1995) parameterised in terms of ρ, v_P and v_S. (With permission of the *Journal of Numerical Analysis, Industrial and Applied Mathematics*)

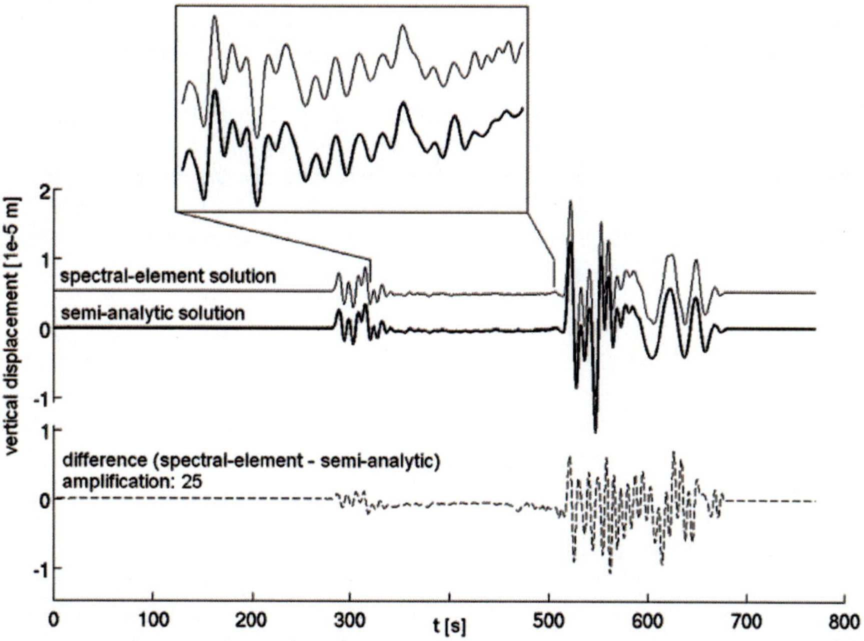

Fig. 13.3 Comparison of the semi-analytical solution (*bold line*) with the spectral-element solution (*thin line*) for station ARMA, located at an epicentral distance of 21.1° from the epicentre. The difference between the two solutions, amplified by a factor of 25 to enhance visibility, is plotted as a *dashed line* below the seismograms. Also the magnified parts of the seismograms, between 340 and 500 s, agree remarkably well. (With permission of the *Journal of Numerical Analysis, Industrial and Applied Mathematics*)

13.2.2 Implementation of Long-Wavelength Equivalent Crustal Models

All numerical methods that find approximate solutions to the wave equation rely on the replacement of a continuously defined Earth model $\mathbf{m}$ by a discrete model $\mathbf{m}_d$. Due to this discretisation, small-scale structures from $\mathbf{m}$, such as thin layers and discontinuities, can often not be represented accurately in $\mathbf{m}_d$, unless the grid spacing is reduced. This can lead to large numerical errors, simply because the *effective medium parameters* imposed by sampling the continuous model do not generally coincide with the physically effective medium at a given wavelength.

In the specific case of the spectral-element method, material discontinuities need to coincide with element boundaries for the solution to be correct. When a discontinuity is located inside an element, the rapidly varying stress and displacement fields cannot be represented accurately by the smooth Lagrange polynomials. The result of such an inaccurate model discretisation is unacceptably large numerical errors. A reduction of the grid spacing can, in principle, eliminate these errors. However, the increasing computational costs can render this option impractical.

Thin layers present even more general difficulties than isolated discontinuities because their width may be smaller than the smallest computationally feasible grid spacing. Backus (1962) and Schoenberg & Muir (1989) demonstrated that a stack of thin layers can be replaced by an equivalent slowly varying Earth model when the widths of the individual layers are much smaller than the dominant wavelength. The equivalent smooth, or effective, Earth model was shown to be transversely isotropic even when the original stack of layers is isotropic. The long-wavelength equivalence of smooth Earth models can clearly be used for the benefit of numerical efficiency because modest variations of the model parameters neither require reductions of the grid spacing nor additional algorithmic complexities. A concept similar to the one advocated by Backus (1962) and Schoenberg & Muir (1989) is the homogenisation of the equations of motion (e.g. Stuart & Pavliotis, 2007; Capdeville & Marigo, 2007, 2008; Capdeville et al., 2010a, b). This technique explicitly yields a smoothed model. The solution for the slowly varying model is close to the correct solution when the length scale of the variations in $\mathbf{m}$ is much smaller than the dominant wavelength.

In the following paragraphs we present an alternative approach for the construction of smoothed and long-wavelength equivalent Earth models that can be discretised through direct sampling (Fichtner & Igel, 2008). This helps to overcome difficulties with the discrete representation of discontinuities and thin layers as long as a pre-defined maximum frequency is not exceeded. The construction of the smoothed Earth models is based on the matching of surface wave dispersion curves corresponding to the original model $\mathbf{m}$ and a smoothed model $\mathbf{m}_s$. This means that we search in a pool of smooth models for the one that best reproduces the dispersion curves of Love and Rayleigh waves in $\mathbf{m}$. The frequency band considered is the one for which a given grid can, in principle, yield accurate numerical solutions.

13.2.2.1 The Importance of Accurately Implemented Crustal Models

We start by illustrating the effects of an improperly implemented crustal model on the accuracy of numerically computed waveforms. The example is intended to serve as a motivation for the subsequent developments. In order to establish a reference scenario, we start with a demonstration of the accuracy achievable with SES3D. For this, we compare the spectral-element solutions to semi-analytical solutions that are available for spherically symmetric Earth models. Our Earth model consists of a homogeneous single-layered crust underlain by the isotropic version of PREM (Dziewonski & Anderson, 1981). The width of the elements is chosen such that the crust is represented by exactly one layer of elements, as shown in Fig. 13.4. Black dots represent the irregularly spaced Gauss–Lobatto–Legendre points which serve as grid points in the spectral-element method.

For the simulations we use a source located at 80 km depth, and the cutoff period of the seismograms is 15 s. We compute the semi-analytical reference seismograms using the programme package GEMINI, developed and described in detail by Friederich & Dalkolmo (1995). East-component displacement seismograms for a receiver located at an epicentral distance of 21.3° are shown in Fig. 13.5. The spectral-element and semi-analytical solutions are in excellent agreement, also for the north and vertical components that are not displayed. This demonstrates that the large discrepancies that we will find in the next example are not numerical errors per se, but the result of an inaccurate numerical model representation.

We now change the Earth model so that the crust comprises two layers, as shown in Fig. 13.6. Two discontinuities are now located inside the upper layer of elements. It would, in this particular example, be possible to adapt the numerical grid to the new model, though at the expense of strongly increased computational costs.

Therefore, we deliberately do not adopt this option. Instead, we use this example to illustrate the effects of such an improper model implementation. Since the rapidly varying wave field in the vicinity of the crustal discontinuities cannot be represented

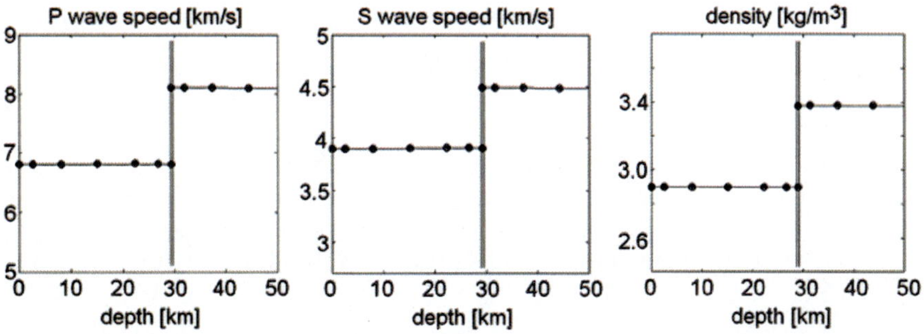

Fig. 13.4 Details of the vertical discretisation of the homogeneous crust overlying PREM. *Dots* symbolise the SES3D *grid points* (Gauss–Lobatto–Legendre points) obtained by sampling the continuously defined model plotted as a *solid line*. One layer of elements, each comprising 7^3 points, is used to represent the crust. The *vertical double line* marks the boundary between the two upper layers of elements

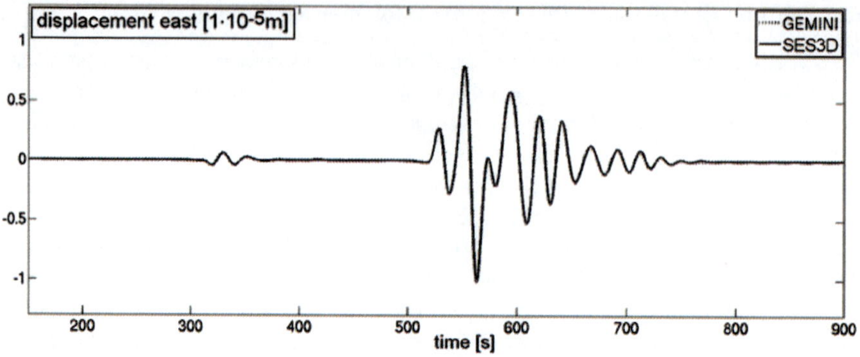

Fig. 13.5 Comparison of east-component synthetic displacement seismograms computed with GEMINI (*dotted curve*) and SES3D (*solid curve*) for the homogeneous crust model from Fig. 13.4 and a station at an epicentral distance of 21.3°. The *dotted line* is barely visible, thus indicating that the two solutions are in excellent agreement

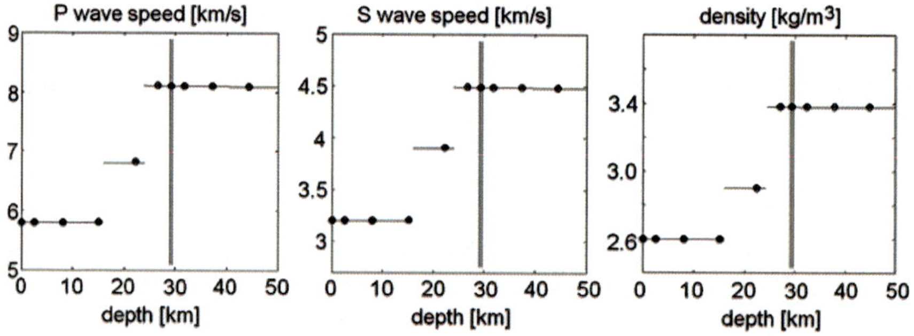

Fig. 13.6 Details of the vertical discretisation of the crust with two layers. *Dots* symbolise the spectral-element grid points (Gauss–Lobatto–Legendre points). Two discontinuities are located in the interior of the upper element layer where displacement and stress fields are represented by smooth polynomials. The exact solution, however, varies rapidly near the discontinuities due to reflection and refraction phenomena. The *double vertical line* marks the boundary between the two *upper layers* of elements

by the smooth polynomials inside the elements we can expect larger discrepancies between the GEMINI and the SES3D solutions. That they indeed occur can be seen in Fig. 13.7. While the body waves remain largely unaffected, the errors in the surface wave train become unacceptably high. Using the spectral-element synthetic from Fig. 13.7 in a full waveform inversion would lead to meaningless tomographic images.

13.2.2.2 Existence of Long-Wavelength Equivalent Crustal Models

We want to tackle the problem illustrated above by replacing the upper part of the original Earth model by an equivalent smooth model. For this we exploit the fact

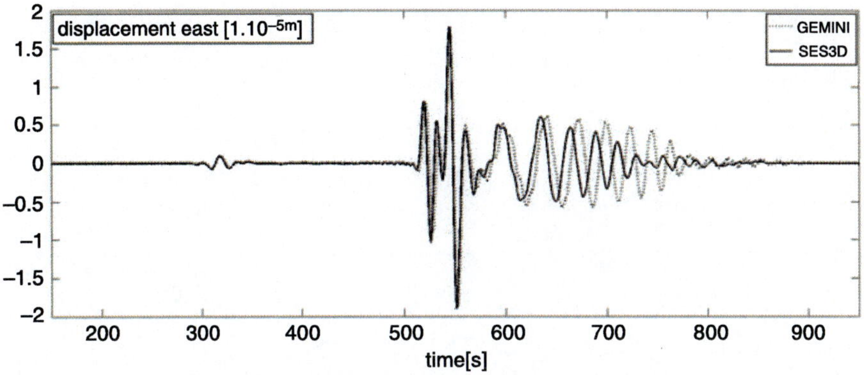

Fig. 13.7 Comparison of east-component synthetic displacement seismograms computed with GEMINI (*dotted line*) and SES3D (*solid line*) for the two-layer crust model from Fig. 13.6 and a station at an epicentral distance of 21.3°. While the inadequate implementation of the crust has little effect on the body waves, the surface waves in the spectral-element solution are severely distorted

that surface waves in a limited frequency range do not uniquely determine an Earth model. That is, there is a non-empty set of Earth models all of which produce surface wave solutions that differ from each other by less than a subjectively chosen small value. From this set of models we can therefore choose one that is numerically advantageous, i.e. smooth.

To find such a smooth model one should focus on those aspects of the surface wave train that are most important in the real data analysis and that facilitate the model construction. We therefore construct smooth Earth models by matching the surface wave dispersion curves of the original model to the dispersion curves of models where the upper part is defined in terms of low-order polynomials. In the general case, the smoothed models will be transversely isotropic (see Backus, 1962). The minimisation of the dispersion curve misfit is done non-linearly by simulated annealing (SA) (Kirkpatrick et al., 1983). To improve the uniqueness of the smooth models we impose that they be close to the original Earth model. Before delving into the methodological details, we first justify the approach by its results shown in Figs. 13.8, 13.9 and 13.10.

A slowly varying crustal model that reproduces almost exactly the surface wave dispersion curves from the original model is displayed in Fig. 13.8. That the model is mildly anisotropic inside the upper layer of elements can be seen from the differences between the wave speeds of SH and SV waves and the wave speeds of PH and PV waves. Moreover, the parameter η (Eq. 9.19) is different from 1. For the fundamental modes of Love and Rayleigh waves the mean difference in the phase velocities for the two models is less than 4×10^{-4} km/s in the period range from 17 to 67 s. Details of the dispersion curves and the associated misfits are shown in Fig. 13.9. For both Love and Rayleigh waves the absolute phase velocity differences rarely exceed 1×10^{-3} km/s and are mostly below 0.5×10^{-3} km/s.

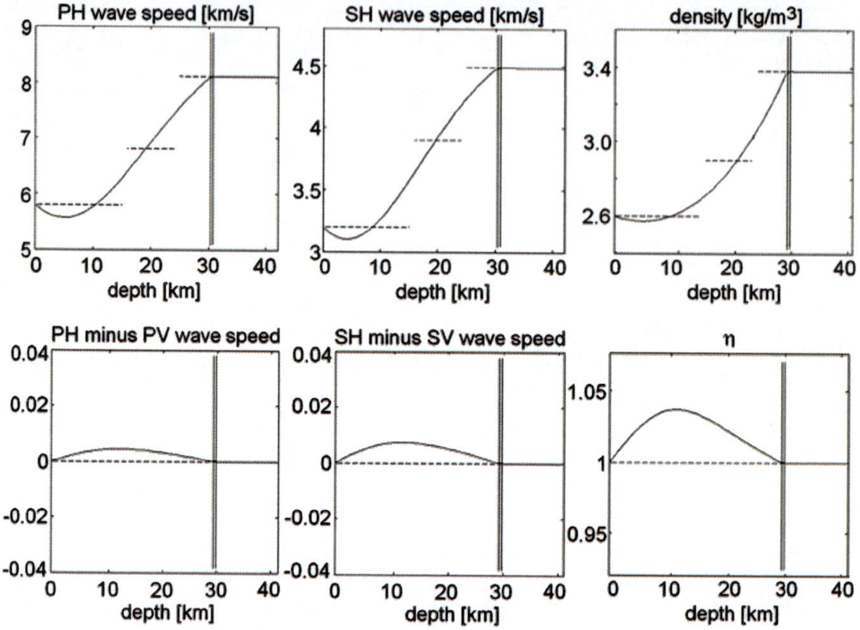

Fig. 13.8 Upper part of the original Earth model (*dashed line*) with two crustal layers (Fig. 13.6) and the smoothed Earth model (*solid line*). The smooth model is slightly anisotropic and parameterised in terms of 4th-order polynomials. The corresponding dispersion curve differences are displayed in Fig. 13.9

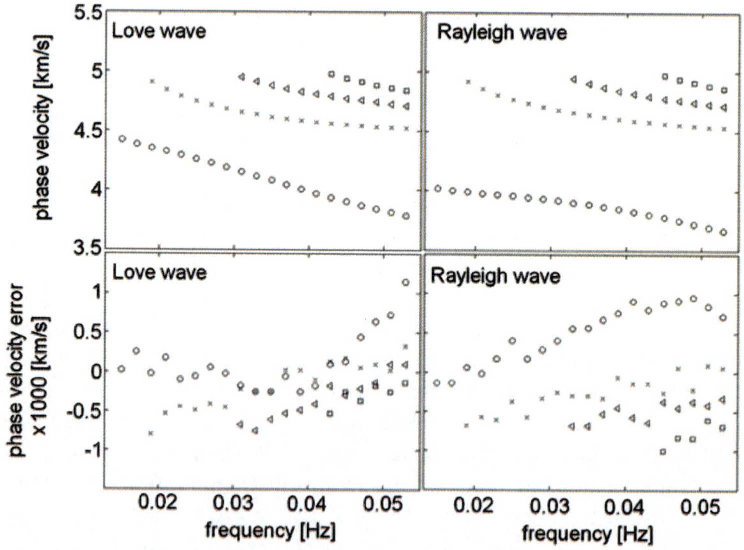

Fig. 13.9 *Top*: Dispersion curves for Love and Rayleigh waves for the original Earth model. *Circles* denote the respective fundamental modes and *crosses* (×), *triangles* (◁) and *squares* (□) represent the higher modes. *Bottom*: Corresponding dispersion curve errors (smoothed minus original, multiplied by 1,000) that are due to the smoothing of the crust

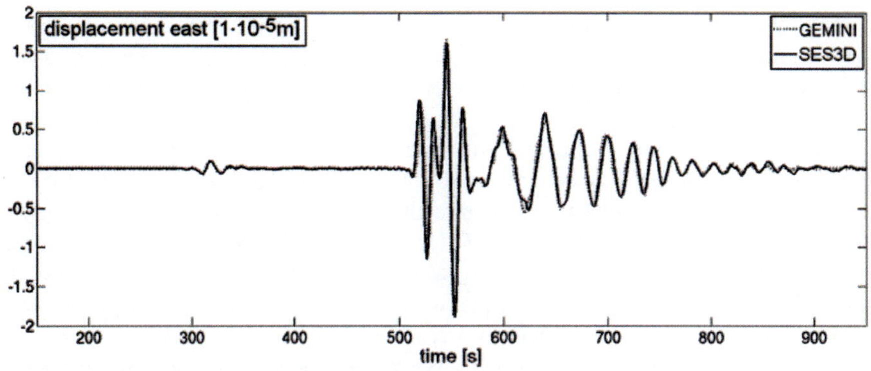

Fig. 13.10 Comparison of synthetic seismograms computed with GEMINI (*dotted line*) and SES3D (*solid line*) for the same station considered in Fig. 13.7. The GEMINI seismograms were computed for the Earth model with the two-layered crust (*dashed lines* in Fig. 13.8) and the SES3D seismograms were computed for the corresponding smoothed model (*solid lines* in Fig. 13.8). The surface wave part is now well modelled

That the new model does indeed remove most of the discrepancies between the semi-analytical and the spectral-element solutions can be seen in Fig. 13.10. Both the amplitudes and the phases are now close to the ones observed in the case of the correctly implemented single-layered crustal model. This demonstrates that the small dispersion curve misfit shown in Fig. 13.9 is indeed a reliable proxy of significantly smaller numerical errors. Small discrepancies between the two solutions remain. These are, however, practically irrelevant because the differences are much smaller than commonly observed differences between data and synthetics.

Since the construction of the slowly varying crustal model is based on a simulated annealing algorithm the result is different in each run. This leads to a whole family of models that produce essentially the same surface wave dispersion characteristics within the pre-defined period range. This non-uniqueness can be reduced significantly by requiring that the smooth model be close to the original one. This additional constraint is also in the interest of accurately modelled amplitudes.

13.2.2.3 Methodological Details

The modification of discrete Earth models for the benefit of numerical accuracy can be interpreted as the optimisation of a discretised differential operator. Let us denote by $L_{ij}\mathbf{u}_j = \mathbf{f}_i$ a discretised version of the wave equation $L\mathbf{u} = \mathbf{f}$, where L (L_{ij}) is the (discrete) wave equation operator, $\mathbf{u}$ ($\mathbf{u}_j$) the (discrete) displacement field and $\mathbf{f}$ ($\mathbf{f}_i$) the (discrete) external force density. We are interested in finding L_{ij} such that $\mathbf{u}_j$ is as close as possible to the exact solution $\mathbf{u}$.

Generally, the discrete operator L_{ij} depends on both the numerical method and the discrete version $\mathbf{m}_\mathrm{d}$ of the continuously defined Earth model $\mathbf{m}$. Since we wish to leave the numerical method unchanged, we are left with the option of modifying $\mathbf{m}_\mathrm{d}$ to improve the numerical solution.

Ideally, one would directly apply the numerical method – in our case the spectral-element method – to a set of smoothed discrete test models. One could then choose the discrete model that best satisfies the strong criterion that the wave fields $\mathbf{u}_j(\mathbf{m}_\mathrm{d})$ and $\mathbf{u}(\mathbf{m})$ be close to each other. Yet, in practice, exact solutions $\mathbf{u}(\mathbf{m})$ are often unavailable and solving the discrete problem sufficiently many times is computationally too expensive. We therefore replace this strong and direct criterion by a weaker and indirect criterion: We require that $\mathbf{m}$ and a smooth and continuously defined Earth model $\mathbf{m}_\mathrm{s}$ produce essentially identical surface wave dispersion curves in the period range of interest. Then $\mathbf{m}_\mathrm{s}$ is sampled at the numerical grid points to give the discrete model $\mathbf{m}_\mathrm{d}$. Since $\mathbf{m}_\mathrm{s}$ is by construction slowly varying, it is represented accurately by $\mathbf{m}_\mathrm{d}$. Consequently, the surface wave dispersion of the numerical solution $\mathbf{u}_j(\mathbf{m}_\mathrm{d})$ is close to the surface wave dispersion of the exact solution $\mathbf{u}(\mathbf{m})$. This approach is justified by our observation that merely sampling $\mathbf{m}$ to obtain $\mathbf{m}_\mathrm{d}$ leads to unacceptably large numerical errors mainly in the surface wave trains. That dispersion curve matching indeed reduces the numerical errors significantly has already been demonstrated by comparison with semi-analytical solutions.

In the following paragraphs we address the dispersion curve matching procedure and the generation of random models for the simulated annealing algorithm.

Non-linear Dispersion Curve Matching by Simulated Annealing

We start with the assumption of a spherically symmetric Earth model where surface wave dispersion can be computed analytically (e.g. Takeuchi & Saito, 1972). Let us denote by $v_\mathrm{L}^i(\omega, \mathbf{m})$ and $v_\mathrm{R}^i(\omega, \mathbf{m})$ the frequency-dependent phase velocities of the ith mode of the Love and Rayleigh waves in the Earth model $\mathbf{m} = (\rho, \mu, \lambda, a, b, c)$. Our aim is to find a continuously defined smooth model $\mathbf{m}_\mathrm{s} = (\rho_\mathrm{s}, \mu_\mathrm{s}, \lambda_\mathrm{s}, a_\mathrm{s}, b_\mathrm{s}, c_\mathrm{s})$ such that

$$
\begin{aligned}
E = &\sum_{i=0}^{n} \alpha_i \int_{\omega=\omega_1}^{\omega_2} |v_\mathrm{L}^i(\omega, \mathbf{m}) - v_\mathrm{L}^i(\omega, \mathbf{m}_\mathrm{s})|\, d\omega \\
&+ \sum_{i=0}^{n} \alpha_i \int_{\omega=\omega_1}^{\omega_2} |v_\mathrm{R}^i(\omega, \mathbf{m}) - v_\mathrm{R}^i(\omega, \mathbf{m}_\mathrm{s})|\, d\omega + \beta\, ||\mathbf{m} - \mathbf{m}_\mathrm{s}||_M
\end{aligned} \tag{13.3}
$$

is minimised. The misfit measure E includes the dispersion curve misfits for Rayleigh and Love waves in a pre-defined frequency interval $[\omega_1\ \omega_2]$ and the relative L_1 distance between the two models $\mathbf{m}$ and $\mathbf{m}_\mathrm{s}$. We decided to use the L_1 norm because it produced the best results in the numerical experiments. Typically, n is 3 or 4. Including higher modes hardly improves the results because they are mostly sensitive to deeper structure. The scalars α_i and β are weighting factors.

Including the term $||\mathbf{m} - \mathbf{m}_\mathrm{s}||_M$, and therefore forcing the smooth model to be close to the original model, serves a variety of purposes. First, it reduces the non-uniqueness of the problem. Uniqueness is in principle not required but still intuitively desirable. Keeping the differences between $\mathbf{m}$ and $\mathbf{m}_\mathrm{s}$ as small as possible is also important in the context of smoothing Earth models with lateral heterogeneities. In the presence of sources inside the smoothed region, one would in principle have to modify the moment tensor. This complication can be avoided when $||\mathbf{m} - \mathbf{m}_\text{s}||_M$ is small.

Minimising E with respect to $\mathbf{m}_\text{s}$ is a non-linear problem that can be solved by a variety of standard techniques. Since the calculation of surface wave dispersion curves in a 1D medium is computationally inexpensive we decided to employ a simulated annealing algorithm, introduced by Kirkpatrick et al. (1983). In addition to being easy to implement, simulated annealing also offers the possibility to impose constraints on the solution simply by means of the random model generation.

Random Model Generation

The simulated annealing algorithm requires random test models that are compared with the currently best model. To generate test models that serve their purpose we use two different parameterisations: In the lower part of the model where smoothing is unnecessary we retain the original model parameterisation. In the upper part of the model, however, we employ low-order piecewise Lagrange polynomials defined on the elements used for the spectral-element simulations. The collocation points are the Gauss–Lobatto–Legendre points. Therefore, the models vary more rapidly near the edges of the elements than near the centre of the elements. This corresponds well to the parameterisation of the displacement and stress fields. Different sets of collocation points may work equally well, depending on the numerical method used. It is then natural to generate random models through the perturbation of the polynomial coefficients. To ensure that the test models are physically reasonable we impose the following conditions: (1) The smoothed model and the original model are identical at the surface. This restriction is motivated by the fact that the amplitudes of seismic waves are very sensitive to the structure directly beneath the receiver. (2) There are no jumps at the element boundaries. Hence, there are no additional discontinuities that may generate undesirable reflections. (3) The smoothed mass density distribution is strictly positive, i.e. $\rho_s > 0$.

13.2.2.4 Extension to the 3D Heterogeneous Case

The very purpose of numerical wave propagation is the computation of wave fields in heterogeneous Earth models for which analytical solutions are mostly unavailable. So far, however, our analysis has been restricted to purely spherically symmetric Earth models, because dispersion curves cannot be computed exactly in the presence of lateral heterogeneities. A direct application of the dispersion curve matching to an arbitrary 3D model is, therefore, not possible.

Yet, in the case of long-wavelength lateral heterogeneities – such as those present in the crustal model Crust2.0 (Bassin et al., 2000) – dispersion curve matching can be applied regionally (Fichtner & Igel, 2008). For this, the crust is partitioned into domains with similar structure. In each of the domains, we compute a new long-wavelength equivalent model. As an additional constraint, we impose that the individual smooth models for the different regions be close to the long-wavelength equivalent model for the average crust. This ensures that there are no abrupt lateral parameter changes in the complete model. As shown by Fichtner & Igel (2008), the regionalisation approach yields satisfactory results when the wavelength of lateral heterogeneities are longer than the shortest wavelengths of the surface waves.

13.3 Quantification of Waveform Differences

Finding a suitable misfit functional for full waveform inversion means to find a reasonable balance between the restrictions imposed by the physics of the problem and the exploitation of maximum information.

To extract the wealth of information contained in complete seismograms, time-domain full waveform inversion often attempts to minimise the objective functional $\sum_{r=1}^{N} \int [u(\mathbf{x}^r, t) - u^0(\mathbf{x}^r, t)]^2 \, dt$, i.e., the cumulative L_2 distance between the complete data u^0 and the complete synthetics u recorded at N receivers denoted by $\mathbf{x}^r$ (e.g. Bamberger et al., 1982; Tarantola, 1984; Gauthier et al., 1986; Chap. 11).

The advantages and disadvantages of the pure L_2 difference become most apparent through its close relation to diffraction tomography. In diffraction tomography (Devaney, 1984; Wu & Toksöz, 1987), the scattered wave field, i.e. the difference $\Delta u(t) = u(t) - u^0(t)$, is linearly related to the spectrum of the heterogeneity that caused the incident wave field to be scattered. This relationship is reminiscent of the well-known projection slice theorem from X-ray tomography. Diffraction tomography yields accurate images even of small-scale structural heterogeneities when the following two conditions are satisfied:

1. The assumed *background structure* is close to the true structure in the sense that the first-order Born approximation holds. This requires the remaining differences to be small compared to the dominant wavelength.
2. The heterogeneity is *illuminated from all directions* (Mora, 1989) either by direct waves as in medical tomography or by waves reflected off major discontinuities.

Diffraction tomography is qualitatively equivalent to the first iteration in a time-domain full waveform inversion where the misfit is quantified via the L_2 waveform difference. This interpretation is confirmed by numerical results (Gauthier et al., 1986; Wu & Toksöz, 1987; Mora, 1988), and it suggests that the two tomographic methods are applicable under similar conditions. However, meeting those conditions is problematic when the Earth is studied on continental or global scales.

A sufficiently accurate reference model, that ensures the validity of the Born approximation, is usually not available. This is due to the nature of the tomography problem with limited data. Even 1D models of the whole Earth (Dziewonski & Anderson, 1981; Morelli & Dziewonski, 1993; Kennett et al., 1995) can differ by several percent, especially in the upper mantle and near discontinuities. The 1D density structure of the Earth has rather limited constraints (Kennett, 1998). There are three immediate consequences of this dilemma:

1. The remaining differences between any presently available Earth model and the true Earth can usually not be treated as scatterers. This implies that the linearisation of the L_2 waveform difference is mostly not appropriate.
2. The observed waveform residuals at periods above several seconds are mostly due to transmission and interference effects, which result in phase shifts between observed and synthetic seismograms that can only be quantified using a misfit measure that does not mix phase and amplitude information.
3. Time-domain full waveform inversion on continental or global scales based on the L_2 waveform difference is highly non-linear. The strong non-linearity promotes non-uniqueness and requires heavy regularisation (see Chap. 7).

The data coverage necessary for diffraction tomography is not achievable in a 3D Earth. Vast regions of the Earth's surface are practically inaccessible, and sufficiently strong sources are confined to a few seismogenic zones. Strong reflectors that could, in principle, improve this situation (Mora, 1989) are not present in the Earth's mantle. As a result diffraction tomography cannot work on continental or global scales, unless very long-period data are used (Capdeville et al. 2005; Chap. 15).

An additional complication is that some of the most important information about the Earth's structure is contained in the *phases* of waveforms with comparatively small amplitudes; a classical example is provided by P body waves. The L_2 waveform difference is dominated by the largest amplitudes that typically correspond to surface waves. Thus, when surface waves are not muted artificially, tomographic resolution is limited to the uppermost regions in the Earth where surface wave sensitivity is large. On the other hand, eliminating surface waves also constitutes a major loss of information on shallow structure.

Phase differences are known to be quasi-linearly related to structural variations and so are well-suited for an iterative, gradient-based misfit minimisation. In contrast, the dependence of amplitudes on variations in the medium properties is often highly non-linear. An iterative inversion algorithm may therefore converge slowly or not at all; see Gauthier et al. (1986) for an example. Amplitudes depend strongly on the local geology near receivers that may not be well controlled. Information about the deeper Earth can thus be masked by shallow structures such as hidden sedimentary basins.

In the frequency range used for continental- and global-scale waveform inversion, seismograms are mostly affected by the transmission properties of the Earth and only to a lesser extent by its diffraction properties. The transmission properties manifest themselves in the time- and frequency-dependent phases of seismic waveforms.

In consequence, we conclude that phases and amplitudes need to be separated and weighted depending on their usefulness for the solution of a particular tomographic problem. One approach that allows us to extract full waveform information while meeting the requirement of phase and amplitude separation consists in the construction of independent phase and envelope misfits, as proposed in Sect. 11.5 (Fichtner et al., 2008). The phase misfit in particular introduces a transmission tomography component that is crucial on continental and global scales where the distribution of sources and receivers is sparse – in contrast to exploration scenarios that are characterised by spatial over-sampling. Moreover, the time–frequency misfits do not rely on the isolation of particular phases. This property is important at short epicentral distances where body and surface waves often do not have a clearly separate identity. Time–frequency misfits naturally combine body and surface wave analysis.

Figure 13.11 illustrates the measurement of the phase misfit for vertical-component waveforms from an event in the Loyalty Islands region (25 March 2007; latitude: −20.60°, longitude: 169.12°, depth: 41 km) recorded at stations CAN

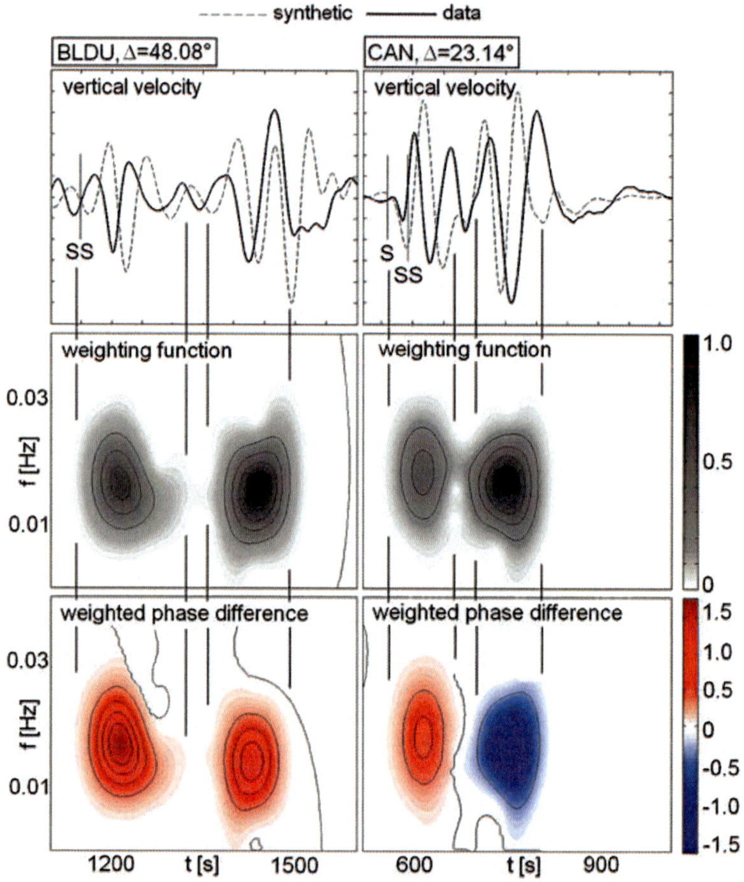

Fig. 13.11 *Top*: Vertical-component velocity seismograms from an event in the Loyalty Islands region recorded at the stations BLDU (*left*; lat.: −30.61°, lon.: 116.71°) and CAN (*right*; lat.: −35.32°, lon.: 148.99°). Data are plotted as *solid* and synthetics as *dotted lines*. Both are low-pass filtered with the cutoff frequency $f_c = 0.02$ Hz. *Centre*: The phase weighting function, W_p, defines the time–frequency range that contributes to the measurement. It excludes cycle skips, waveforms with amplitudes below the noise level and frequencies that are too high to be modelled accurately. *Bottom*: Weighted phase differences, $W_p(\omega, t)\Delta\phi(\omega, t)$, corresponding to the seismograms above. *Contour lines* are plotted at multiples of 20% of the maximum value (figure modified from Fichtner et al. (2008))

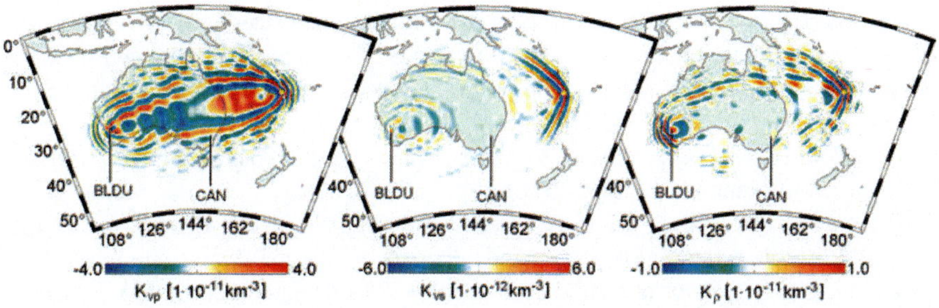

Fig. 13.12 *Horizontal slices* at 100 km depth through the Fréchet kernels K_{v_S} (*left*), K_{v_P} (*centre*) and K_ρ (*right*). A *yellow circle* indicates the position of the epicentre in the Loyalty Islands region. The kernels with respect to the P velocity, K_{v_P}, and density, K_ρ, attain small absolute values, compared to the kernel with respect to the S velocity, K_{v_S}. This implies that 3D P velocity and density variations cannot be constrained in the inversion. Note that the colour scales are different (figure modified from Fichtner et al. (2008))

(latitude: −35.32°, longitude: 148.99°) and BLDU (latitude: −30.61°, longitude: 116.71°). The positions of both stations are shown in Fig. 13.12. The advance of the observed waveforms with respect to the synthetic waveforms at station BLDU (Fig.13.11, top left) maps to a positive phase difference (Fig. 13.11, bottom left). The time–frequency range of interest is determined by the phase weighting function, W_P, (Fig. 13.11, centre left) as introduced in Eq. (11.37). The phase weighting function ensures that cycle skips are excluded and that measurements are restricted to frequencies where the synthetic seismograms are known to be sufficiently accurate. At station CAN, both a phase advance (positive phase difference) and a phase delay (negative phase difference) can be observed. The bottom row of Fig. 13.11 highlights the fact that the phase misfit is a multi-frequency measurement. It includes, in our case, frequencies between $\sim 5 \times 10^{-3}$ and $\sim 3 \times 10^{-2}$ Hz.

Horizontal slices through the sensitivity kernels corresponding to the phase misfit measured at station BLDU (Fig. 13.11) are shown in Fig. 13.12. Since the measurement is dominated by shear waves, the sensitivities with respect to the P wave speed, v_P, and density, ρ, are small compared to the sensitivity with respect to the S wave speed, v_S. Lateral variations in both v_P and ρ are therefore unlikely to be resolvable. Significant sensitivity far from the geometrical ray between source and receiver can generally be observed. This, and the complexity of the sensitivity kernels, is due to the length of the analysed time window and the frequency band limitation of the waveforms.

13.4 Application to the Australasian Upper Mantle

The crust and upper mantle of the Australasian region have been the object of seismological studies since the 1940s when de Jersey (1946) estimated the average crustal thickness from Rayleigh wave dispersion and PP/P amplitude ratios. His results were in remarkable agreement with those for continental crust of similar age in North America (Gutenberg, 1943). Observations of P and S arrivals from nuclear explosions (Cleary et al., 1972) revealed early on the deep structural division of the Australian continent: the Precambrian central and western parts are characterised by fast seismic wave speeds, whereas the wave speeds in the Phanerozoic east are lower than average (Bolt et al., 1958; Cleary, 1967; Cleary et al., 1972). Those results were confirmed and extended by the analysis of surface wave dispersion (Bolt, 1957; Bolt & Niazi, 1964; Goncz & Cleary, 1976).

The amount of seismic data increased dramatically with the installation of the SKIPPY portable array that was operated between 1993 and 1996 by The Australian National University (van der Hilst et al., 1994). SKIPPY enabled the construction of tomographic images with unprecedented resolution and it motivated methodological developments that shaped modern seismic tomography. The discovery and characterisation of numerous deep structural elements in the Australasian region significantly advanced our understanding of continental formation and evolution: Deep cratonic roots in the centre and west, a pronounced low-velocity zone around 140 km depth in the east and a sharp contrast between Precambrian and Phanerozoic lithosphere (Fig. 13.13) were clearly imaged using surface wave tomography

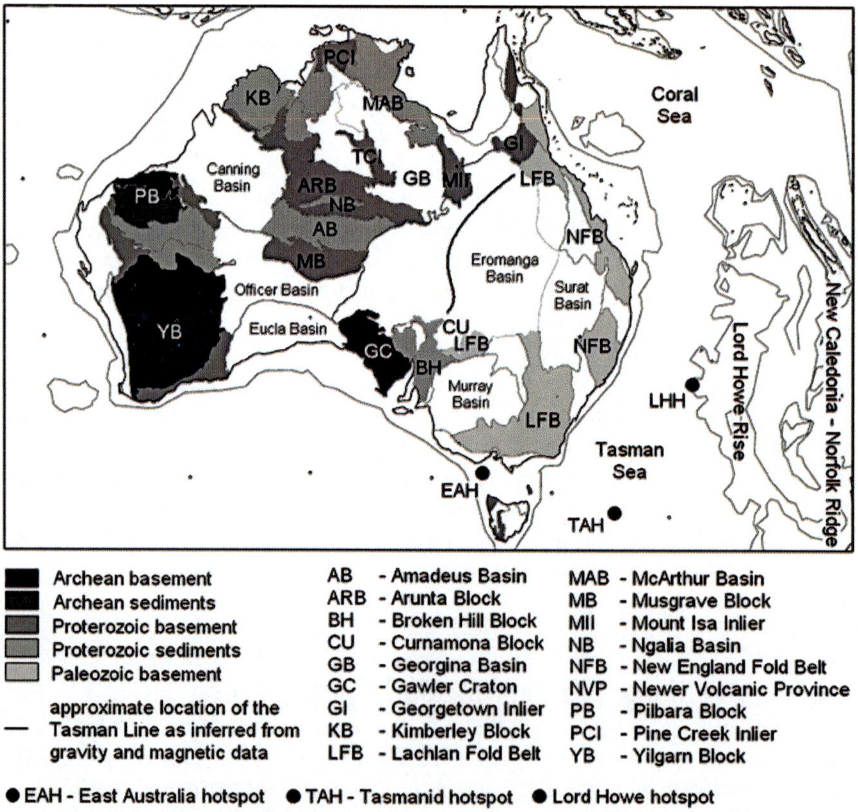

Fig. 13.13 Map of major geologic features in the study area, adapted from Myers et al. (1996). Present hotspot locations, indicated by *filled circles*, are from McDougall & Duncan (1988)

(Zielhuis & van der Hilst, 1996; Simons et al., 1999; Yoshizawa & Kennett, 2004; Fishwick et al., 2005, 2008, Fishwick & Reading, 2008). Significant azimuthal anisotropy has been shown to exist in the uppermost mantle under Australia by Debayle & Kennett (2000) and Simons et al. (2002). Constraints on the locations of seismic discontinuities and the attenuation structure were obtained through the analysis of body wave arrivals (e.g., Gudmundsson et al., 1994; Kaiho & Kennett, 2000).

The quality of the tomographic images has improved continuously over the years thanks to increasing data coverage and data quality, technological developments and advances in theoretical and numerical seismology. While de Jersey (1946) analysed seismograms from three stations, broadband data from several hundred recording sites in the Australasian region are available today. The advent of numerical wave propagation allows us to make optimal use of these data and to exploit full waveform information. This is intended to yield more detailed tomographic images that contribute to our understanding of the Earth's interior constitution and allow us to predict seismic ground motion.

In the following paragraphs we elaborate on the different steps to be taken in a successful full waveform tomography on continental scales. We pay special attention to the selection of suitable data, the construction of an initial model, and the analysis of resolution and waveform fit.

13.4.1 Data Selection and Processing

The data selection in full waveform inversion deserves special attention because the misfit measures tend to be rather susceptible to noise and because the computational resources are limited. In the framework of this study, we have adopted the following data selection criteria:

1. The estimated signal-to-noise ratio in the frequency band from $1/250$ to $1/30$ Hz is required to be higher than 20. This ensures that the waveform misfit is dominated by the discrepancies between the mathematical model and the Earth. The choice of a high signal-to-noise ratio is motivated by the properties of the phase misfit measurements, where small-amplitude waveforms contribute as much as large-amplitude waveforms.
2. The adjoint method enables us to compute event kernels, i.e. the cumulative sensitivity of all recordings corresponding to one event, with just one adjoint simulation. This means that the adjoint method operates most efficiently when a large number of recordings per event are available. We therefore accept only those events with more than 10 high-quality recordings in the Australasian region.
3. The event magnitude is required to be smaller than $Ms = 6.9$; this choice allows us to neglect finite-source effects in seismogram modelling.
4. To further reduce the influence of near-source effects we exclude data recorded at epicentral distances of less than 5°.

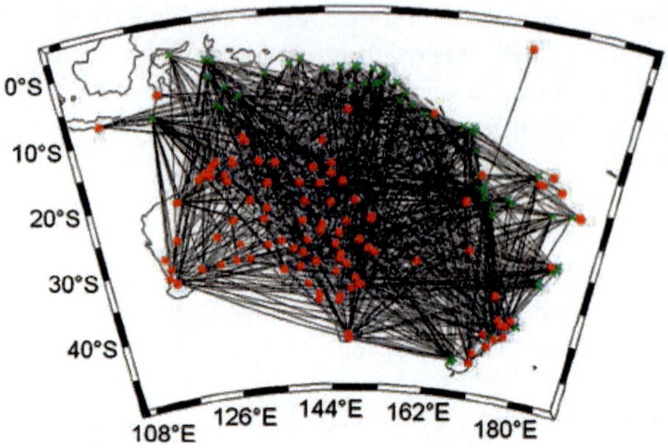

Fig. 13.14 Ray coverage. Epicentres are plotted as *green crosses* and stations as *red stars*. The coverage is particularly dense in central and eastern Australia and in the Tasman and Coral Seas, east of the continent. The western part of Australia is less densely covered

5. The event distribution is chosen such that it leads to a ray coverage that is as even as possible. We try to avoid any clustering of events that may produce artefacts in the tomographic images (see Fig. 13.14).
6. After the first iteration, we exclude events where neither a relocation nor an additional moment tensor inversion leads to satisfactory waveform fits.
7. After each iteration we visually examine the agreement between observed and synthetic waveforms, and we incorporate new recordings where the waveform fit has improved sufficiently to allow for a meaningful misfit quantification. Recordings where the misfit is above a pre-defined threshold are excluded. This selection procedure is, at least to some degree, subjective but based on experience.

The final data set comprises 2, 137 three-component recordings from 57 earthquakes that occurred in the Australasian region between 1993 and 2009. About 80% of the data were recorded at permanent stations operated by Geoscience Australia, IRIS and GEOSCOPE. The remaining 20% originate from the SKIPPY, KIMBA, QUOLL and TASMAL experiments, undertaken by The Australian National University. The periods range from 30 to 250 s.

From the observed seismograms we manually select waveforms that show a clear correspondence to synthetic waveforms, i.e. a sufficient waveform similarity. Fundamental-mode surface waves account for nearly 60% of the selected waveforms. The remaining 40% are S waves and higher mode surface waves. A small fraction of the waveforms is not clearly identifiable. The ray coverage (Fig. 13.14) is good throughout the eastern part of the continent and decreases towards the west.

Our data set differs from those used in previous surface wave studies (e.g. Zielhuis & van der Hilst, 1996; Simons et al., 1999, 2002; Debayle & Kennett, 2000; Yoshizawa & Kennett, 2004; Fishwick et al., 2005) in that it contains mostly recordings from permanent stations, many of which were not operational prior to

2006. The improved permanent station coverage allows us to invert a comparatively large number of high-quality waveforms with dominant periods >60 s.

For the first iteration, all data are band-pass filtered between 150 and 250 s. In the course of the inversion we successively decrease the lower cutoff period to 30 s. This procedure ensures that the quasi-linearity of the waveform misfit with respect to Earth model perturbations is maintained.

13.4.2 Initial Model

The implementation of an initial model that is close to the expected global optimum is crucial in full waveform inversion, for a variety of reasons:

1. Full waveform inversion is a large non-linear minimisation problem that can only be solved efficiently with iterative gradient-based minimisation schemes such as the conjugate-gradient or Newton algorithm (see Chap. 7). All gradient-based algorithms require an initial model that is close to the global minimum in order to avoid convergence towards a local minimum that is potentially far from the true Earth.
2. The convergence rate of gradient-based minimisation schemes tends to improve for initial models that are close to the true model.
3. The amount of exploitable data is larger for initial models that already capture the principal geologic features of the study region. This is because more synthetic waveforms will be sufficiently similar to the observed waveforms to allow for a meaningful misfit quantification.

In the following paragraphs we describe the construction of the mantle and crustal parts of the initial model.

13.4.2.1 Mantle Part

The initial mantle model represents the long-wavelength features of the Australasian upper mantle. The 3D elastic part of the model, shown in the top row of Fig. 13.15, is isotropic. It is based on a smoothed version of the v_S model by Fishwick et al. (2005). Clearly visible are the seismically fast Precambrian lithosphere in central and western Australia, and the low seismic wave speeds beneath Phanerozoic eastern Australia, the Coral and the Tasman Seas. We set the initial v_P variations to 0.5 times the initial v_S variations, consistent with results from refracted body wave studies (Kaiho & Kennett, 2000). As elastic 1D reference model we use the isotropic version of PREM (Dziewonski & Anderson, 1981) with the 220 km discontinuity replaced by a gradient, to avoid biases in the 3D tomographic images.

To minimise the effect of undiscovered 3D anelastic structure on the tomographic images, we implement a 3D model of shear wave attenuation that was derived from multi-frequency measurements of body wave amplitudes (Abdulah, 2007). A horizontal slice through the model is shown in the bottom row of Fig. 13.15. It features a

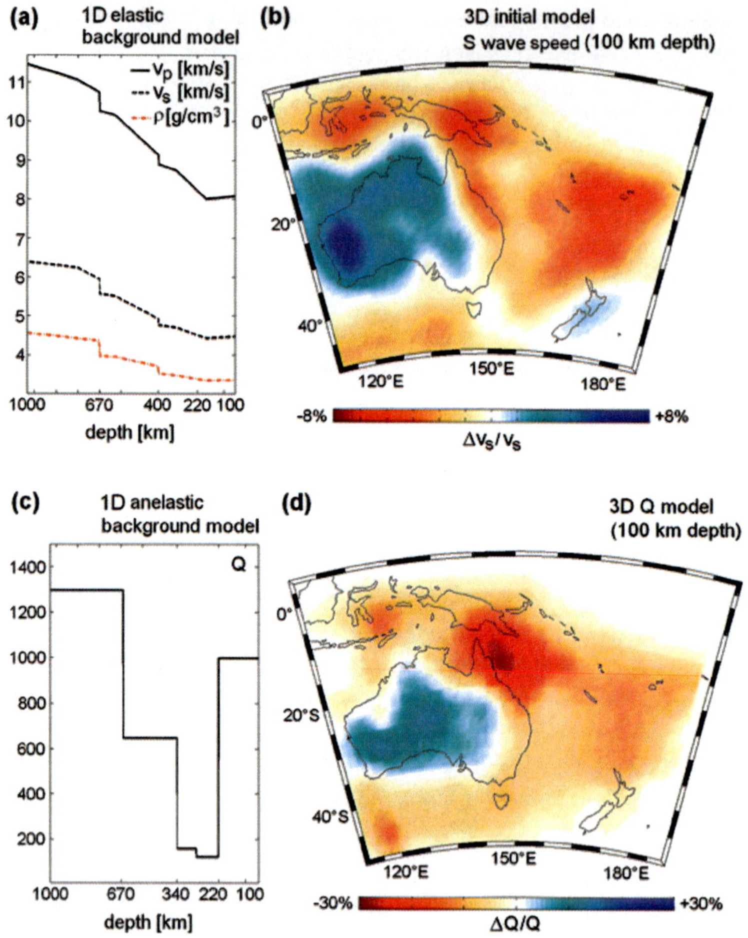

Fig. 13.15 *Top panels*: (**a**) 1D elastic background model of the P wave speed, v_P, the S wave speed, v_S, and density, ρ. The model is derived from the isotropic version of PREM (Dziewonski & Anderson, 1981). The discontinuity at 220 km depth is removed to avoid biases in the 3D tomographic images. (**b**) *Horizontal slice* at 100 km depth through the initial 3D v_S model, derived from the surface wave tomographic images of Fishwick et al. (2005). *Bottom panels*: (**c**) 1D background model of shear attenuation. (**d**) *Horizontal slice* at 100 km depth through the 3D shear attenuation model, as derived by Abdulah (2007) on the basis of multi-frequency amplitude measurements on body waves. The anelastic model does not change during the waveform inversion. 3D variations of P wave attenuation are not included (figure modified from Fichtner et al. (2010))

weakly attenuative continental lithosphere, strong attenuation beneath the Coral Sea and a pronounced low-attenuation zone between 220 and 340 km depth. We disregard 3D variations of P wave attenuation, because their influence on long-period seismograms is negligible.

13.4.2.2 Crustal Part

Our data set mostly consists of surface waveforms at periods of 30 s and above. These are sensitive to the structure of the crust while not being able to resolve its characteristic features: the strengths and locations of discontinuities. A realistic crustal structure is therefore required as part of the initial model. That the importance of crustal models can hardly be overestimated was demonstrated by Bozdağ & Trampert (2008). Inaccurate crustal models or crustal corrections can lead to apparent anisotropy that can be as strong as mineralogic anisotropy in the upper mantle.

The implementation of a realistic crust is, however, complicated by our insufficient knowledge concerning its thickness and elastic properties. In the Australasian region the crust has mostly been studied on the basis of reflection/refraction profiles (e.g. Lambeck et al., 1988; Klingelhoefer et al., 2007) and receiver functions (Shibutani et al., 1996; Clitheroe et al., 2000; Chevrot & van der Hilst, 2000). Three-dimensional crustal models can therefore only be obtained by interpolation that may not capture the strong lateral variations found along some isolated seismic lines (Lambeck et al., 1988).

The lack of information on the 3D characteristics of the crust makes the selection of one particular crustal model to some degree subjective. We use the model crust2.0 (Bassin et al., 2000; http://mahi.ucsd.edu/Gabi/rem.html) that is displayed in the right panel of Fig. 13.16.

To implement a long-wavelength equivalent of crust2.0, we proceed as follows: (1) For a set of $5° \times 5°$ grid cells across the Australasian region we compute a smooth long-wavelength equivalent version of the original crustal profile using the methodology introduced in Sect. 13.2.2. We include the dispersion curves of the fundamental and first three higher modes in the period range from 15 to 100 s. A representative profile through a continental structure is displayed in the left panel of Fig. 13.16. (2) To assess the quality of the smooth profiles we compare the corresponding synthetic seismograms to the synthetic seismograms for the original crustal profiles. The errors introduced by the smoothing are negligible for periods longer than 25 s. (3) A smooth 3D crustal model is then constructed by laterally interpolating the individual smooth profiles.

13.4.3 Model Parameterisation

Our elastic model is parameterised in terms of density, ρ, the speeds of vertically and horizontally polarised S waves, v_{sv} and v_{sh}, the speeds of horizontally and vertically travelling P waves, v_{ph} and v_{pv}, and the parameter η. The data are mostly sensitive to variations in v_{sv} and v_{sh}. Sensitivity with respect to v_{ph} and v_{pv} is small, and sensitivity with respect to ρ and η is negligible. This implies that v_{ph}, v_{pv}, ρ and η can hardly be resolved. We therefore set the variations of v_{ph} and $v_{\mathbf{pv}}$ to 0.5 times the variations of v_{sh} and v_{sv}, as is common practice in global tomographic studies (e.g. Nettles & Dziewonski, 2008). We furthermore disregard lateral variations in ρ

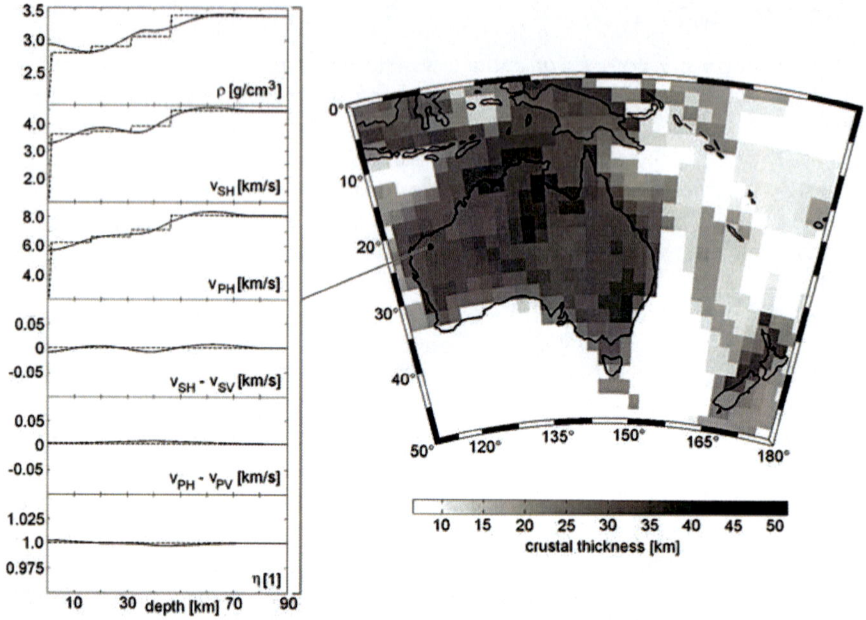

Fig. 13.16 *Right*: Map of the crustal thickness in the Australasian region, according to the model crust2.0 (Bassin et al., 2000; http://mahi.ucsd.edu/Gabi/rem.html). Within the continent, the crustal thickness varies between 25 km and more than 50 km. *Left*: Profiles of ρ, v_{SH}, v_{PH}, $v_{SH} - v_{SV}$, $v_{PH} - v_{PV}$ and the dimensionless parameter η (Eq. 9.19) from the surface to 90 km depth for a location in NW Australia. The original profile from crust2.0 is plotted with *dashed lines* and the long-wavelength equivalent model with *solid lines* (figure modified from Fichtner et al. (2009))

and set $\eta = 1$, consistent with the isotropic initial model (Fig. 13.15). The restrictions imposed on the variations of the parameters ρ, η, v_{ph} and v_{pv} are, to some degree, subjective. Their small sensitivities, however, ensure that they have nearly no influence on the lateral variations of v_{sh} and v_{sv}.

As basis functions we use blocks that extend $1° \times 1°$ laterally and 10 km vertically. This parameterisation is chosen such that structures we expect to be resolvable can be represented with sufficient accuracy. The expected resolution lengths are 2–5° laterally and 20–50 km vertically. This estimate is based on previous tomographies where similar amounts of data were used (e.g. Debayle & Kennett, 2000a; Fishwick et al., 2005).

Our parameterisation does not include azimuthal anisotropy, because the limited data coverage (Fig. 13.14) does not allow us to simultaneously determine source locations and azimuthal wave speed variations.

13.4.4 Tomographic Images and Waveform Fits

We have, at this point, prepared all the necessary ingredients for a continental-scale full waveform inversion: (1) an efficient solver of the elastic wave equation consisting of the spectral-element method combined with a long-wavelength equivalent of the crust, (2) high-quality waveform data with a good coverage of the target region, (3) the phase misfit as a physically meaningful measure of waveform differences that exploits as much information as possible while conforming to the restrictions imposed by the physics of the problem, (4) an initial model that contains the large-scale structural elements and (5) the adjoint method for the computation of Fréchet kernels combined with a conjugate-gradient algorithm for the non-linear misfit minimisation (see Sect. 7.3.4).

Combining all ingredients, we inverted for the radially anisotropic upper-mantle structure in the Australasian region. The inversion procedure terminated when the decrease of the cumulative phase misfit from one iteration to the next dropped below 5%. This was the case after 19 iterations. The threshold of 5% was chosen to prevent overfitting the data. The resulting 3D model, AMSAN.19, is presented in Fig. 13.17. For a detailed interpretation of the tomographic model the reader is referred to Fichtner et al. (2008, 2009).

A sufficiently large number of iterations are essential to obtain images that are stable in the sense that they do not change much when additional iterations are performed. The 19th iteration modifies the SH and SV models by less than 0.1% of the reference isotropic S wave speed. However, the changes are mostly above 0.5% during the first 12 iterations. Differences between the SH and SV models after one iteration and after 19 iterations locally exceed 6%. Iterative changes in the anisotropy are nearly as large.

An exemplary collection of the final waveform fits is displayed in Fig. 13.18 for a variety of source–receiver geometries. Model AMSAN.19 explains the waveforms of the major seismic phases, including higher and fundamental-mode surface waves and long-period body waves. The fit is good for periods above and including 30 s, where the influence of crustal scattering is small. We point out that complete waveforms – and not only aspects of them, such as dispersion curves – can be modelled accurately.

AMSAN.19 provides a fit of relative amplitudes that is significantly better than for the initial model, even though no information on absolute amplitudes was used in the inversion. This observation suggests (1) that lateral contrasts are imaged accurately and (2) that the time–frequency phase and amplitude misfits are rather strongly related, in agreement with recent results by Tian et al. (2009).

In the course of the iteration, the synthetic waveforms become increasingly similar to the observed waveforms. This allows us to successively include waveforms that were initially not usable due to excessive dissimilarities between data and synthetics. The total number of exploited waveforms increases from about 2, 200 in the first iteration to nearly 3, 000 in iteration 19. The final model can thus explain data that were initially not included in the inversion. This is strong evidence for the effectiveness of the inversion scheme and the physical consistency of the tomographic model.

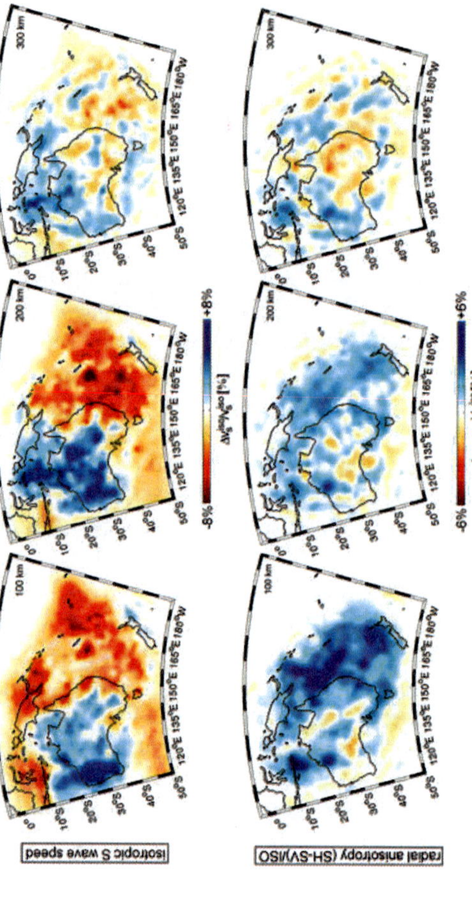

Fig. 13.17 *Top*: *Horizontal slices* through the isotropic S wave speed, $v_{s,iso} = 2v_{SH}/3 + v_{SV}/3$, of the tomographic model AMSAN.19 at depths of 100, 200 and 300 km. *Bottom*: *Horizontal slices* through the radial anisotropy, $(v_{SH} - v_{SV})/v_{s,iso}$, at depths of 100, 200 and 300 km (figure modified from Fichtner et al. (2010))

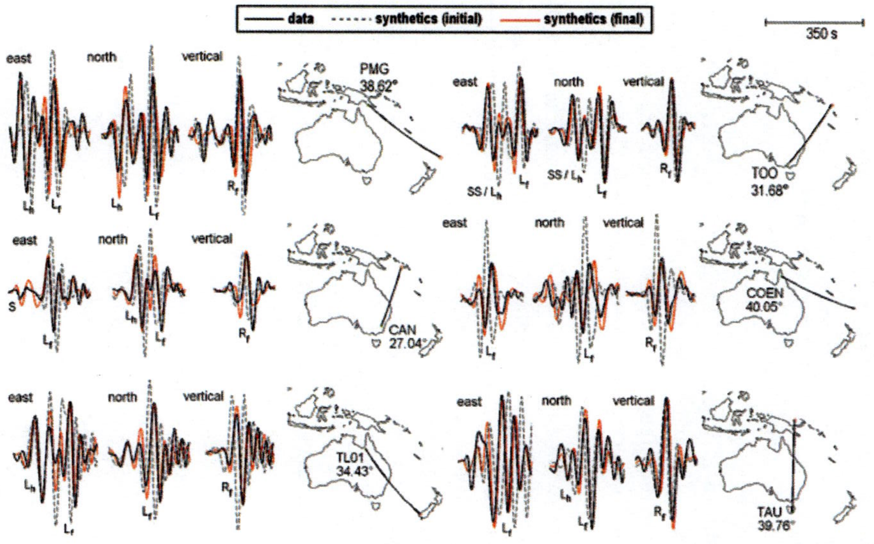

Fig. 13.18 Exemplary waveform comparisons for a variety of source–receiver geometries. *Black solid lines* are data, *black dashed lines* are synthetics for the initial model (Fig. 13.15) and *red solid lines* are synthetics for the final model (Fig. 13.17). The dominant period is 30 s. A time scale is plotted in the *upper-right corner*. Identifiable phases are marked for reference ($L_{f/h}$ = fundamental/higher mode Love wave, $R_{f/h}$ = fundamental/higher mode Rayleigh wave, SS= surface-reflected S wave). While significant discrepancies exist between data and the initial synthetics, the final synthetics accurately explain both the phases and the amplitudes of the observations (figure modified from Fichtner et al. (2010))

13.4.5 Resolution Analysis

One of the major advantages of the use of the full waveform modelling scheme for 3D varying media is that we are able to carry out resolution analysis incorporating genuine 3D velocity structure. No approximations are required in the construction of the artificial data. The reconstructions of the input models therefore provide realistic estimates of the resolution capabilities of our full waveform tomographic method. We performed a series of combined patch recovery and chequerboard tests to estimate the spatial resolution in the tomographic images. For this, a set of 3D anomalies in the SH and SV wave speeds was introduced into the 1D background model and synthetic seismograms were computed using the same set of events and stations as in the real-data inversion. To obtain conservative resolution estimates, we did not use a 3D initial model.

In the specific test shown in Fig. 13.19, the input model consists of a high-velocity patch located in central and western Australia at a depth of 100 ± 25 km. This is intended to mimic the high velocities of the Precambrian lithosphere. A chequerboard with a dominant lateral wavelength of about 3° is superimposed on the high-velocity patch.

Both the long- and short-wavelength features of the input model are well recovered by the full waveform tomographic method. For most of central and eastern Australia and the adjacent Tasman and Coral Seas we estimate that the resolution lengths are around 3° laterally and 40 km vertically, above 200 km depth. Additional tests indicated that below 200 km depth the estimated resolution lengths are around 7° laterally and 60 km vertically. We note that structures outside the area covered by rays can also be recovered to some degree. This is the case, for example, beneath the South Australian Basin and north of the Solomon Islands. We nevertheless recommend not to interpret structures in such regions, because the assessment of their reliability may require a more detailed resolution analysis.

In a longer series of resolution tests (Fichtner et al., 2009), we found that the amplitudes of positive chequerboard anomalies generally seem to be less well recovered than their neighbouring negative anomalies – a phenomenon for which we can currently not provide a definite explanation.

There are, furthermore, clear non-linear effects in the sense that the quality of the reconstructions improves when the amplitudes of the perturbations are increased while keeping their geometry constant. This phenomenon, also observed in global-scale synthetic inversions (see Chap. 15), is in contrast to linearised tomography. An intuitive explanation is that small-amplitude perturbations essentially translate to phase shifts only. Larger amplitude perturbations, however, lead to both phase shifts and substantial waveform distortions. In the latter case, the waveforms seem thus to contain more information even though the geometry of the anomalies has not changed. This additional waveform information seems to improve the resolution.

After 19 iterations, the synthetic waveforms and the artificial data are practically identical. This implies that the imperfections seen in the reconstructions are mostly the result of insufficient data coverage and the frequency band limitation. Potential

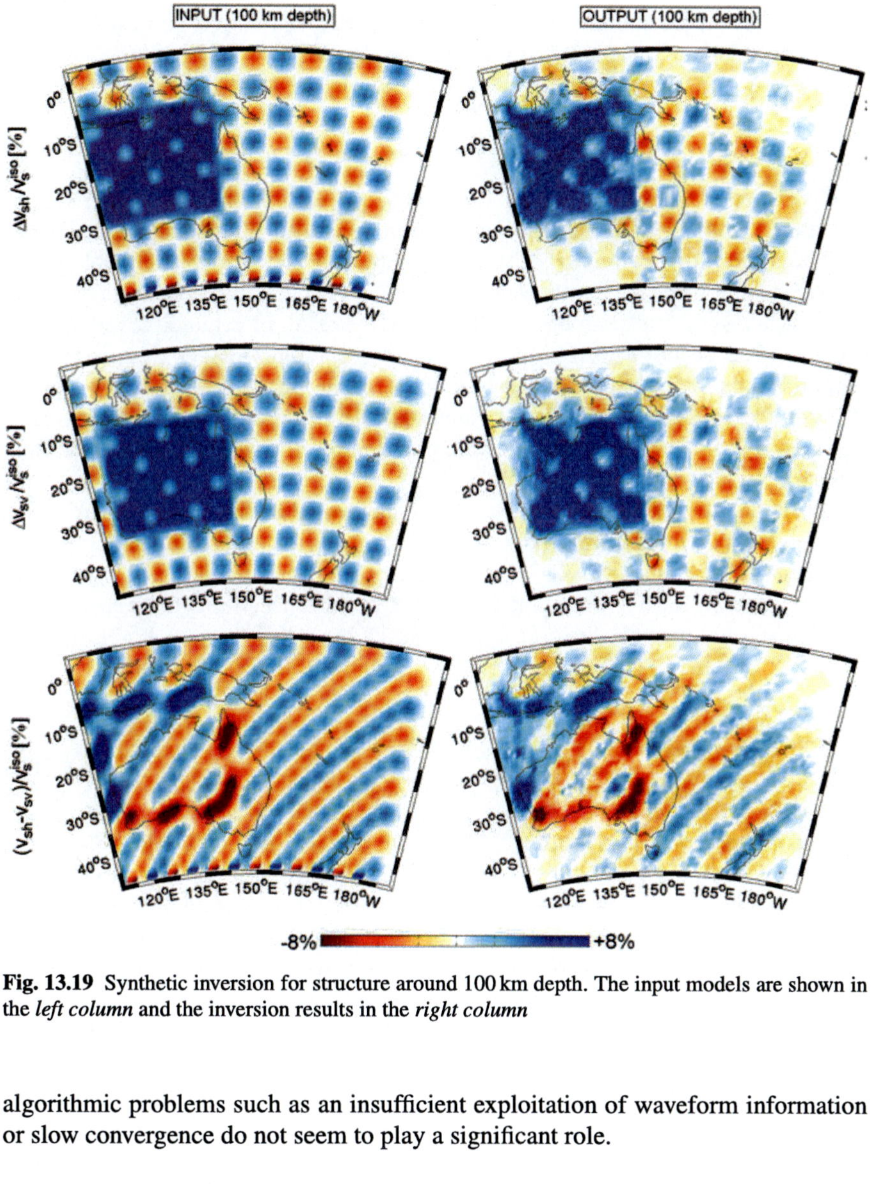

Fig. 13.19 Synthetic inversion for structure around 100 km depth. The input models are shown in the *left column* and the inversion results in the *right column*

algorithmic problems such as an insufficient exploitation of waveform information or slow convergence do not seem to play a significant role.

13.5 Discussion

In the previous paragraphs we presented a full waveform tomographic technique that can operate on continental scales. We described the numerical solution of the forward problem, the data selection procedure, the quantification of waveform misfits, their iterative minimisation, the resulting tomographic images and the achieved waveform fit and resolution. To conclude this chapter, we return to some of the most important technical aspects of full waveform inversion:

13.5.1 Forward Problem Solution

The correct solution of the equations of motion in realistically heterogeneous Earth models is the most significant advantage of the full waveform tomography. The numerical solution of the elastic wave equation ensures that the differences between observed and synthetic seismograms are indeed the result of yet undiscovered Earth structure. The occurrence of artefacts in the tomographic images that are due to simplifications of the wave propagation process can thus be avoided. The advantages of the full waveform method are particularly important for strongly heterogeneous regions of the Earth, such as the thermal boundary layers or ocean–continent transitions.

The number of events that we have used in the inversion was limited by the available computational resources and not by the available high-quality data. However, the comparatively small number of seismograms is in part compensated by extracting as much waveform information as possible, for instance, through the application of the time–frequency misfits (Sects. 11.5 and 13.3).

13.5.2 The Crust

Bozdağ & Trampert (2008) demonstrated that crustal corrections influence the inferred upper-mantle anisotropy. Most previous tomographic studies on continental or global scale computed crustal corrections from either crust2.0 (Bassin et al., 2000) or 3SMAC (Nataf & Ricard, 1995). One of the outstanding advantages of full waveform tomography is that no crustal corrections are required. This is because numerical solutions of the elastic wave equation are also accurate in the presence of a laterally varying crust. In the process of the iterative misfit minimisation, the crustal and the mantle parts are treated equally, meaning that both are updated as required by the data. We can therefore expect less artificial anisotropy in the tomographic images.

13.5.3 Time–Frequency Misfits

The time–frequency misfits as defined in Sect. 11.5 (see also Fichtner et al., 2008) have several advantages in the context of full waveform inversion: (1) the separation of phase and envelope information, (2) the applicability to any type of seismic wave, (3) a quasi-linear relation to Earth structure and, most importantly, (4) the use of complete waveform information.

There are, however, two disadvantages that the time–frequency misfits share with *any* other measure of full waveform differences, such as the classical L_2 norm or the generalised seismological data functionals (Gee & Jordan, 1992). First, the comparatively high susceptibility to noise, and second, the difficulty of assessing noise effects on the misfit measures and the tomographic images. Since the amount of data in full waveform tomography is small, we can reduce the influence of noise by choosing data of exceptional quality.

The phase misfit is meaningful only when the observed and synthetic waveforms are sufficiently close to avoid phase jumps. This criterion is usually satisfied when the phase differences are less than $\pi/2$. To ensure that no phase jumps occur, useful time windows need to be chosen in each seismogram. We currently pick and weight the time windows manually – a process that is unavoidably subjective, but efficient in that it allows us to incorporate a trained seismologist's experience that can greatly accelerate the convergence of the minimisation algorithm.

13.5.4 Dependence on the Initial Model

Tomographic images generally depend on the choice of the initial model, and the full waveform tomography presented in the previous paragraphs is no exception. One can, however, argue that this dependence is weaker in a non-linear iterative inversion than in a linearised inversion: When the initial model, $\mathbf{m}_0$, is outside the basin of attraction of the global optimum, then both an iterative non-linear and a linearised inversion will produce incorrect results that depend strongly on $\mathbf{m}_0$. Otherwise, the non-linear iterative inversion closely approaches the global optimum irrespective of the actual location of $\mathbf{m}_0$ within the basin of attraction. In this sense, an iterative inversion depends less on the initial model than a linearised inversion that can never approach the true solution arbitrarily closely due to the non-linearity of the problem.

13.5.5 Anisotropy

Iterative improvement strongly affects radial anisotropy. Its amplitude tends to decrease with a growing number of iterations. Apparent anisotropy is thus increasingly well explained by isotropic structure as the inversion proceeds. This effect, also noted by Nettles & Dziewonski (2008), highlights the non-linearity of the tomographic problem.

Further methodological contributions to long-wavelength differences of the tomographically inferred radial anisotropy include the choices of the initial model and the parameterisation. The neglect of azimuthal anisotropy may result in discrepancies, where azimuthal coverage is poor. The influence of these factors requires further research, and the different approaches taken in this and previous studies can all be justified.

13.5.6 Resolution

Thanks to the accurate modelling of seismic wave propagation in a 3D heterogeneous Earth, the tomographic images presented in Fig. 13.17 can be considered more realistic than similar images obtained using more restrictive methods. The effect of full wave field modelling on the resolution of the tomographic images is less obvious. Concerning the comparison between ray tomography and finite-frequency tomography, arguments both in favour (e.g. Yoshizawa & Kennett, 2004; Montelli, 2004; Chen et al., 2007) and against (e.g. Sieminski et al., 2004; van der Hilst & de Hoop, 2005; Trampert & Spetzler, 2006) a higher resolution of the latter have been presented.

There are several reasons for the absence of a general consensus on this issue: Most importantly, the resolution increase of finite-frequency and full waveform tomography, if present, is not always visually obvious. This is a subjective but nevertheless powerful impression. An objective comparison of resolution capabilities is difficult because realistic tomographies are computationally expensive and because there is no universally valid definition of resolution for deterministic non-linear inverse problems. Our limited computational resources merely allow us to consider a small number of synthetic inversions – typically chequerboard tests – that may not be representative (Lévêque et al., 1993). A further complication is due to the fact that ray theoretical sensitivities are distributed along infinitesimally thin rays. The regularisation in ray tomography smears the sensitivity into a region around the ray path. The width of this region is mostly chosen as a function of data coverage, and it may therefore be thinner than the actual influence zone of a wave with a finite frequency content. Thus, the apparent resolution in ray tomography may be higher than the resolution that is physically possible. This phenomenon becomes most apparent in the form of the central slice theorem (Cormack, 1963) which ensures perfect resolution in the case of sufficient ray coverage.

An encouraging observation is that structure in some regions with no ray coverage whatsoever can be resolved at least to some degree. This is due to the large lateral extent of Fréchet kernels for measurements on finite-frequency waveforms. Intuitively, however, we would not recommend to interpret such structures.

In the light of those difficulties, we can currently not make a quantitative and objective statement concerning the comparative resolution capabilities of full waveform tomography. We conjecture, however, that the advantages will become more apparent as the frequency band broadens and as the amount of exploitable information in individual seismograms increases.

We suggest to make a clear distinction between *realistic* and *well-resolved*. Major factors that determine the resolution of a tomographic model are the data coverage, the data quality and the frequency bandwidth. Given a specific data set, the model can be made more realistic by more accurately accounting for the true physics of wave propagation.

A promising approach to resolution quantification is based on the RETRO kernel concept introduced in Sects. 9.3.1 and 12.3. This method is, however, still in its experimental stage.

At this stage of its development, more realistic and physically consistent Earth models are the principal advantage of full waveform tomography. This improvement is crucial on our way towards a more quantitative interpretation of tomographic images in terms of geodynamic processes (Bunge & Davies, 2001; Schuberth et al., 2009a,b).

Chapter 14
Application of Full Waveform Tomography to Active-Source Surface-Seismic Data

This chapter presents a case study of acoustic waveform tomography applied to active-source long-offset data acquired across the San Andreas Fault near Parkfield, California. A multi-scale frequency-domain adjoint method was employed to invert the early, forward-scattered, P wave arrivals. The inversion uses the lower part of the available frequency spectrum to increase the resolution of a kinematic model about four times. The final waveform tomography model reveals the shape of a large intrusion with unprecedented detail, and it indicates the position of two major faults, although it falls short of resolving their structure. The quality of the solution is controlled by the data fit and a sonic log from the San Andreas Fault Observatory at Depth research borehole. Data pre-conditioning and weighting, the inversion strategy and the residual reduction are presented in detail. The discussion focuses on the implications of the signal bandwidth for the required accuracy of the initial model and on the problem of decreasing robustness at higher frequencies.

14.1 Introduction

The application of waveform tomography is controlled to a large extent by the computational expense of the forward solution. Naturally, the first problems to be tackled with were 2D. Active-source exploration profiles, cross-hole surveys and VSP surveys provide appropriate data, if sources and receivers are densely distributed and confined to a plane.

In order to efficiently solve the inverse problem for active-source data, an acoustic frequency-domain multi-scale method was developed that specifically aims at inverting the early arrivals, i.e. the P wave and its coda (Pratt & Worthington, 1990; Woodward, 1992). These forward-scattered waves are the most linear part of a seismogram, because there is little interference from other phases at early times and because multiple scattering is less likely. In many cases, the restriction to the early

Written by Florian Bleibinhaus, Department of Geography and Geology, Salzburg University, Salzburg, Austria

A. Fichtner, *Full Seismic Waveform Modelling and Inversion*, Advances in Geophysical and Environmental Mechanics and Mathematics,
DOI 10.1007/978-3-642-15807-0_14,

arrivals also allows us to work in the acoustic approximation (see Sect. 2.3). For controlled source data this has advantages besides the computational savings: No initial S wave velocity model is required, and the inversion can be focused on the P wave velocities, which are much better constrained by most active sources, because pressure pulses and vertical forces produce little shear wave energy.

The goal of multi-scale reconstruction is to mitigate the non-linearity of waveform inversion by inverting the lowest data frequencies first to constrain the long wavelengths of the model, before sequentially proceeding to higher data frequencies and shorter wavelengths (Bunks et al., 1995; Sect. 7.3). This approach is greatly facilitated by modelling the seismic data in the frequency domain. For small models and few frequency components, solving the Helmholtz equation, i.e. the frequency-domain version of the acoustic wave equation, is highly efficient (Pratt, 1999). In addition, computing the gradient of the objective function directly from frequency-domain wave fields requires minimum storage capacities and input/output operations.

For lab data acquired in a water tank on a physical scale model, Pratt et al. (1999) demonstrated the feasibility and the value of this method. Other groups have developed similar methods and applied them to 2D active-source cross-hole data (e.g. Song et al., 1995; Zhou et al., 1995; Pratt & Shipp, 1999; Wang & Rao, 2006) and surface-seismic data (e.g. Ravaut et al., 2004; Operto et al., 2006; Sheng et al., 2006; Gao et al., 2007; Malinowski & Operto, 2008; Jaiswal et al., 2009). This chapter presents the application of Pratt's method to a land-seismic long-offset survey across the San Andreas Fault in California.

14.2 Data

In 2003 a seismic survey was conducted across the San Andreas Fault (SAF) near Parkfield, California, to provide models of the reflectivity and of the seismic velocity at seismogenic depth for the San Andreas Fault Observatory at Depth (SAFOD). A 50-km-long seismic line was deployed from the Salinian terrane across the SAF over the Franciscan into the Great Valley (Fig. 14.1)

The Salinian is dominated by a granite that is overlain by 1–3 km sediments. The SAF juxtaposes this granite against the Franciscan, an accretionary wedge of folded and faulted sediments that is famous for strong attenuation of seismic waves.

Sixty three inline dynamite shots of 30–90 kg were recorded by the full spread of 50-m-spaced receivers. To avoid a large gap in the line and also to save computation time, the first 10 km was not considered for waveform inversion, resulting in a 37-km-long line of 822 receivers with an average shot spacing of 650 m.

Figure 14.2 displays two exemplary shot records. They are dominated by the direct wave (Pd) and a refraction from the Salinian granite (Pg) to the southwest. Due to the rough topography, especially due to P–Rayleigh conversions at the surface, the data are strongly scattered. The bandwidth of the data is 3–30 Hz (Fig. 14.3), but only frequencies below 15 Hz were considered, for computational

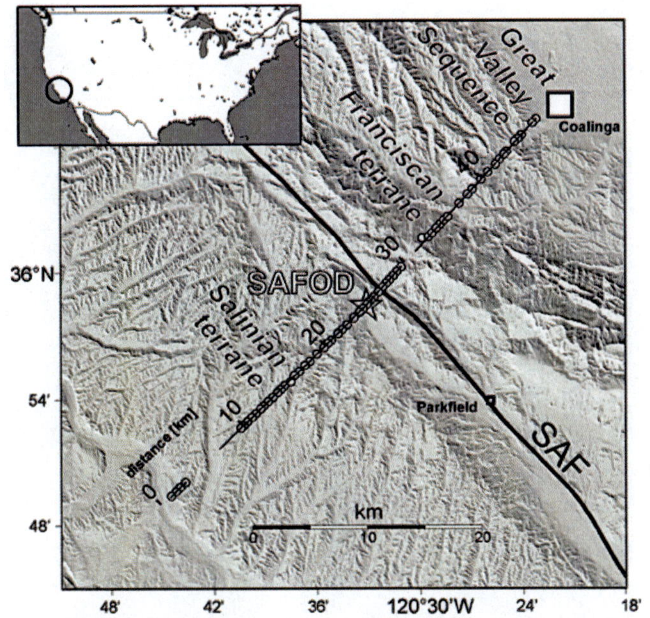

Fig. 14.1 Map of the 2003 seismic survey across the SAFOD site (*white star*) near Parkfield , CA. *Circles* are shots; coincident *black line* indicates receivers. For more details on the geology and the local fault structure see Bleibinhaus et al. (2007)

reasons, and because the inversion stops gravitating towards the global minimum. At the dominant frequency of 8 Hz the signal-to-noise ratio (S/N) drops from 8 at near offsets to 4 at intermediate offsets to 1.5 at long offsets. At the low frequencies and long offsets, S/N is 1 at most.

For constructive interference of the adjoint or back-projected waveforms, the distance between measurements should be half a wavelength at most. At a velocity of 4 km/s, this corresponds to a maximum spacing of 0.1–0.7 km for the highest and lowest frequency, respectively. For the receivers, this condition is met at all frequencies, except for a 2-km-wide gap in the line at 30 km distance. In the vicinity of this gap, imaging artefacts must be expected. For the shots, which are spaced 0.3–1.5 km, not considering the gap, this condition is never fully met. Already at the lowest frequency, some shots are more than two times the required minimum apart. Due to reciprocity, however, the dense receivers can compensate for spatially aliased shot points to a large extent. In a synthetic study, Brenders & Pratt (2007) showed that artefacts from sparse sources deteriorate the inversion significantly, only when the spacing is larger than three times the required minimum at the lowest frequency. This is not the case here, although for real, noisy data, the threshold might be overly optimistic. Therefore, artefacts from the sparse sources cannot be excluded, especially not at higher frequencies.

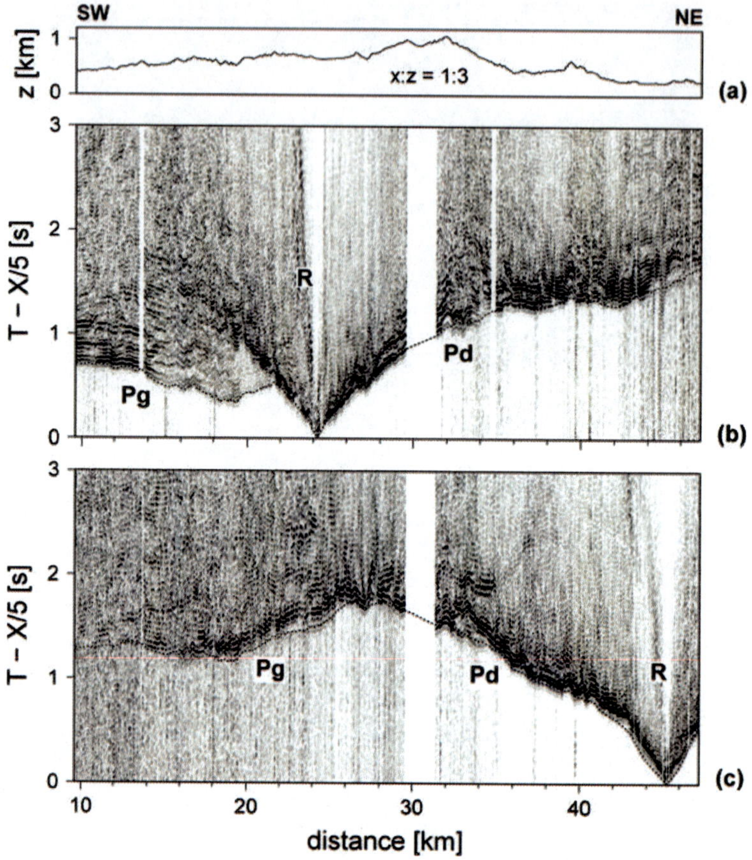

Fig. 14.2 (**a**) Elevation, three times vertically exaggerated. (**b**) Trace-normalised, time-reduced display of the vertical component of shot record 34 and (**c**) of shot record 3. *Dashed line* is travel-times computed in the model of Hole et al. (2006). Pd: direct P wave travelling through sediments, Pg: refracted P wave travelling through granite, R: Rayleigh waves

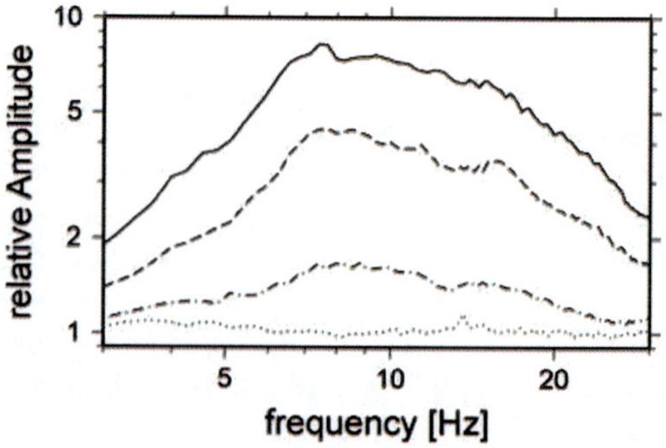

Fig. 14.3 Average signal spectrum at near offsets (5–15 km, *solid line*), intermediate offsets (15–25 km, *dashed*) and long offsets (>25 km, *dash dotted*). *Dotted line* is noise

14.3 Data Pre-conditioning and Weighting

In order to prepare the data for acoustic waveform tomography, noisy traces were deleted, and the surface waves and any direct S waves were muted. To focus the inversion on the early arrivals, a 1.4-s-long time window was selected, the start time of which was computed by subtracting 0.25 s from the kinematic first-arrival traveltimes. The length of this window was chosen to contain at least three cycles at the lowest frequency considered for inversion (3 Hz). Traces recorded at less than 2 km offset were removed because the separation of P and S phases is insufficient to provide a meaningful acoustic signal in the considered frequency range. To further reduce the non-linearity, an exponential time-scaling with a half-decay time of 0.25 s was applied to the data. Later, back-scattered, or multiply scattered, arrivals are suppressed by that means, and the forward-scattered tomography data are being enhanced. The pre-conditioned data were then Fourier-transformed, and a few spectral components were picked for the inversion. A preprocessed shot gather and the corresponding phase of the frequency-domain data at 4 Hz is displayed in Fig. 14.4.

Due to the dense receiver spacing with respect to the wavelength, the phase variations from one receiver to the next are mostly smooth, if they are not generated by random noise. The degree of smoothness was used to compute weighting factors (Fig. 14.4c) (Bleibinhaus et al., 2009). Unfortunately, the sparse and irregular shot spacing prevents the application of 2D smoothness constraints or 2D filters to further mitigate the impact of noise.

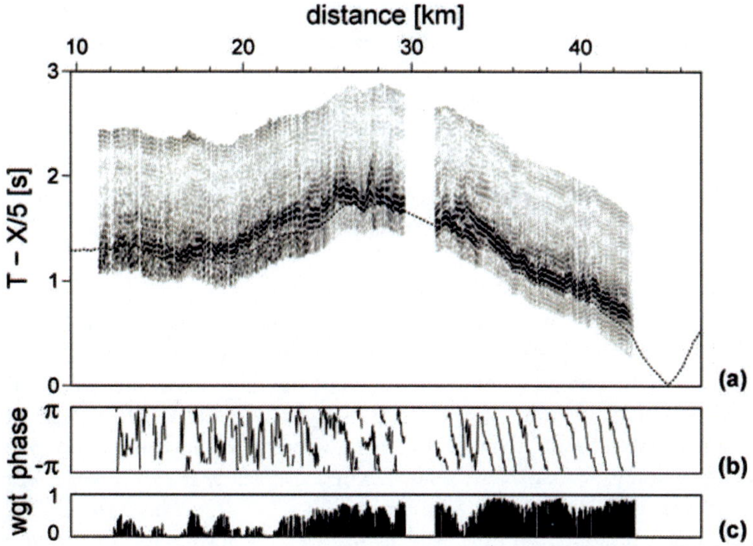

Fig. 14.4 (**a**) Preprocessed shot gather 3 of Fig. 14.2c, (**b**) corresponding phase at 4 Hz, (**c**) corresponding weighting factors at 4 Hz

14.4 Misfit Functional

Before forming the residual, trace normalisation and spectral whitening were applied to the observations and to the synthetics. This means that the amplitude of the frequency-domain data was not used for the inversion. This phase-only inversion is successful because it is the phase that is sensitive to the velocity, while the amplitude is mostly sensitive to the spatial derivative of the velocity. In addition, amplitudes are also strongly affected by attenuation, near-surface structure and receiver coupling. Those factors vary strongly for the SAFOD data (Bleibinhaus et al., 2007) and are poorly controlled. Moreover, two approximations are made in the forward modelling that affect the amplitudes much, but hardly the phases. First, the modelling is 2D, and it cannot reproduce 3D geometric spreading. Second, the acoustic modelling yields pressure, which is independent of a wavefront's angle of incidence, while the data were recorded on the vertical component. It was therefore decided that saving computational resources, and avoiding time-consuming true-amplitude processing, outweighs the potential benefit of inverting amplitudes, which, according to a study of Shin & Min (2006), is small anyway. Indeed, Brossier et al. (2009) demonstrated with a synthetic study that for noisy data, back-projecting the phase difference, only, yields even better results than using true amplitudes.

14.5 Initial Model

The initial model plays a crucial role in most non-linear inverse problems, especially when higher derivatives of the misfit functional are not considered. For gradient methods to converge, it is required that the starting point is in the same valley as the global minimum (see Sect. 7.3). This means for waveform tomography that the initial model must be accurate enough to match most observations within less than half a period ($T_0/2$) deviation. If the predicted phase at a certain frequency is off by more than $\pm\pi$, the inverse algorithm will attempt to match the wrong cycle of the seismogram (Pratt, 1999), and the correlated forward and adjoint wave fields will not focus at the location of the scatterer, as demonstrated, for instance, by Gauthier et al. (1986) and Pratt et al. (1996).

For acoustic active-source full waveform inversion, ray tomography is the preferred choice to derive an initial model. It does not require additional data, the rays cover much of the model space and the quality is generally good enough, at least in the well-constrained, central portions of the model. For this study, starting velocities were derived from a traveltime tomography model of Hole et al. (2006). Some ray path artefacts in this model were reduced by a smoothing operator, and a separate inversion for the structure below 2 km depth below sea level was performed to improve the fit at long offsets (Bleibinhaus et al., 2007). The resulting velocity model is displayed along with the waveform inversion results in Fig. 14.6a. It has an RMS traveltime residual of 40 ms, or $T_0/8$ at 3 Hz, so that the bulk of the observations is indeed matched within less than $T_0/2$ at the lowest frequency considered for inversion. At long offsets, however, there is no significant low-frequency signal (Fig. 14.3), and the dominant frequency of 8 Hz should be considered to judge whether the waveforms computed in the initial model match the observations with $T_0/2$, or 62.5 ms, accuracy. If one accounts for the degraded picking accuracy of ~100 ms at the long offsets (Hole et al., 2006), then it becomes clear that a significant portion of the observations is not matched within half a cycle. Consequently, the deeper structure may not be resolved by waveform inversion.

In order to analyse the resolving power at the low-data frequencies, monochromatic 4 Hz kernels were computed in the initial model for a source–receiver pair at 20 km distance (Fig. 14.5). The first Fresnel zone of the kernel, which corresponds to a $T_0/2$-long time range after the first arrival, is the region that is best constrained by early arrival waveform tomography. Outer lobes correspond to back-scattered arrivals at later times in the seismogram. The width of the first Fresnel zone (3 km) is a rough indicator for the longest wavelengths that can be constrained by the waveforms. Even longer wavelengths must be contained in the initial model for

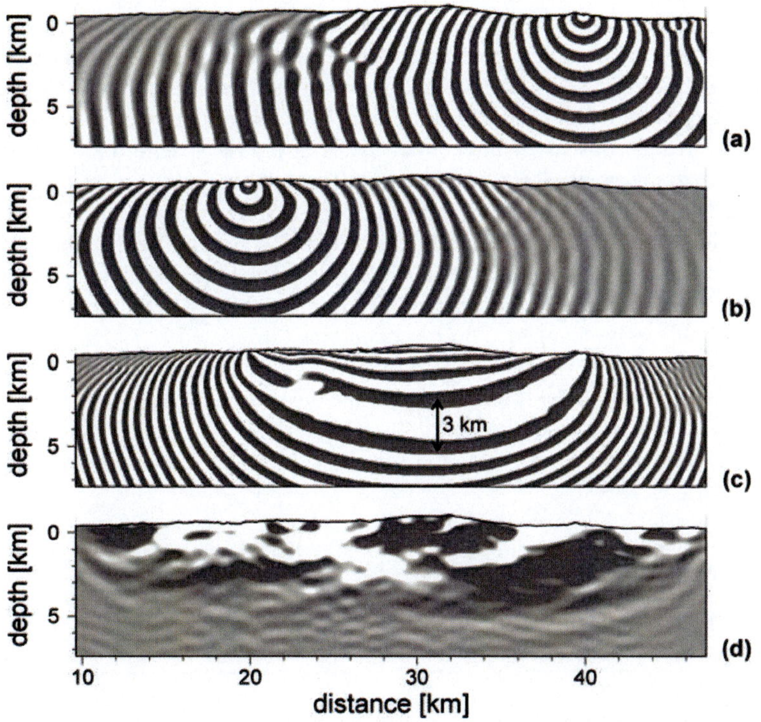

Fig. 14.5 (**a**) Real part of the 4 Hz wave field for a shot point at 40 km. (**b**) Real part of the adjoint wave field for a receiver at 20 km offset. (**c**) The correlation of the regular and adjoint fields yields the 4 Hz gradient for this source–receiver pair. The *arrow* denotes the width of the first Fresnel zone. (**d**) The total gradient is the weighted sum of the gradients of all source–receiver pairs in the 3–4 Hz range

the waveform inversion to succeed. (This condition is essentially the wave number correspondent of the cycle-skipping problem discussed above.) The total gradient in Fig. 14.5d illustrates the resolving power and the subsurface coverage from the low frequency data. Its resolution is 1–3 km in the vertical direction, and 2–20 km in the horizontal direction, which is lower than the resolution of the initial model. This wavelength overlap of traveltime tomography and waveform inversion at the low frequencies compensates for potential inaccuracies of the initial model, thereby mitigating the dependency of the solution on the starting point of the iterative inversion. The limited penetration depth of the total gradient of ~3 km depth below sea level is mostly a result of the lower S/N at long offsets, but is also due to the velocity structure with a shallow high-velocity body (the Salinian granite) that impedes deep penetration.

14.6 Inversion and Results

In order to derive the model perturbations, the gradient was low-pass wave number filtered to reduce short-wavelength artefacts that appear mostly near the sources and receivers, and a step length was computed by a line-search method and a conjugate-gradient algorithm (Sect. 7.3.4). After six iterations, the inversion was halted, and the resulting model was used as initial model for the next higher set of frequencies. Each group consists of three frequencies at 0.4 Hz spacing below 8 Hz and at 0.8 Hz spacing above 8 Hz (Table 14.1). The total residual reduction for each group amounts to ~30%, decreasing from ~10% at the first iteration to ~1% at the last iteration. As the tomography progresses from low to high frequencies, increasingly finer detail is added to the model (Fig. 14.6). The long-wavelength structures, which are also confirmed by other data (Thurber et al., 2003; Malin et al., 2006), are not significantly changed. The inversion was terminated at 14 Hz because the resulting model did not improve the overall match of seismograms anymore. The preferred waveform model is obtained after the inversion of frequencies up to 12 Hz (Fig. 14.6d). It has a resolution of 400 m in horizontal direction and of 200 m in vertical direction (Table 14.1), and it reveals several features that are not con-

Table 14.1 Groups of simultaneously inverted frequencies, spacing used for the forward and inverse grids, gradient regularisation in horizontal direction (half this value was used in the vertical direction) and residual reduction achieved for the corresponding frequency range after Bleibinhaus et al. (2007)

Frequencies (Hz)	Grid spacing (m)	Wavelength lowcut (m)	Residual reduction (%)
3.2/3.6/4.0	100	1,250	21.3
4.4/4.8/5.2	50	1,000	28.6
5.6/6.0/6.4	50	800	32.3
6.8/7.2/7.6	50	700	33.8
8.0/8.8/9.6	25	500	24.6
10.4/11.2/12.0	25	400	23.7
12.8/13.6/14.4	25	250	25.9

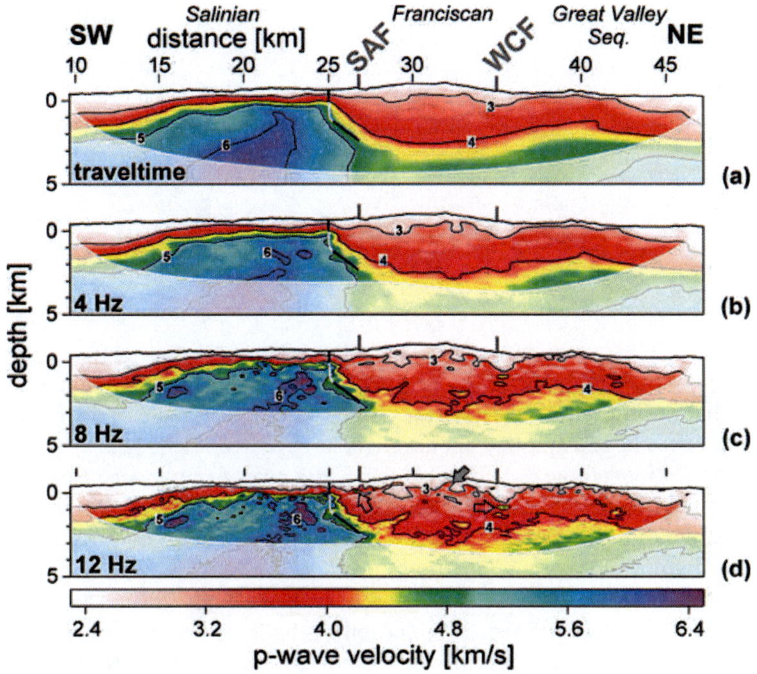

Fig. 14.6 (**a**) Starting velocity model derived from traveltime tomography and (**b**) waveform tomography model after the inversion of 4.0 Hz data, (**c**) 7.6 Hz data and (**d**) 12.0 Hz data. Note the improvement of the resolution with increasing frequency. A *gray arrow* points at a ray-shaped feature that could be a smearing artefact from the gap in the line, and *hollow arrows* point at the two strongest scatterers within the sediments that correlate with vertical reflectors. The *line* at 25 km is the SAFOD borehole (*black* – sediments, *white* – granite). The *masked area* in (**a**) is the ray coverage after Hole et al. (2006), and in (**b–d**) it is the wave path coverage as determined from the gradient of Fig. 14.5. SAF – San Andreas Fault, WCF – Waltham Canyon Fault. (**a–c**) after Bleibinhaus et al. (2007)

strained by the traveltimes. The increased resolution generally provides a much sharper image of the Salinian granite. A pronounced syncline of the 5 km/s contour at 18–20 km distance correlates with a sharp decrease of RMS receiver amplitudes (see Fig. 14.4 of Bleibinhaus et al., 2007) that are more typical for the Franciscan than for the Salinian, suggesting that the rocks that fill this syncline might be as strongly deformed, too. A few shallow, small anomalies near the position of the SAF are probably caused by reflected refractions. These phases are part of the early arrivals, and they were used for reflection imaging of the SAF zone. The anomalies at 35 km distance at 1–2 km depth below sea level could be related to the Waltham Canyon Fault (WCF), which was also imaged from reflected refractions (Bleibinhaus et al., 2007). However, individual fault strands are supposed to be 50–150 m wide (Li et al., 2004), and the reflection of 12 Hz (i.e. ∼300 m wavelength) data is a complicated superposition of two phase-shifted signals that depend on the width of the fault and on the properties of the fault and host rocks. This complexity could

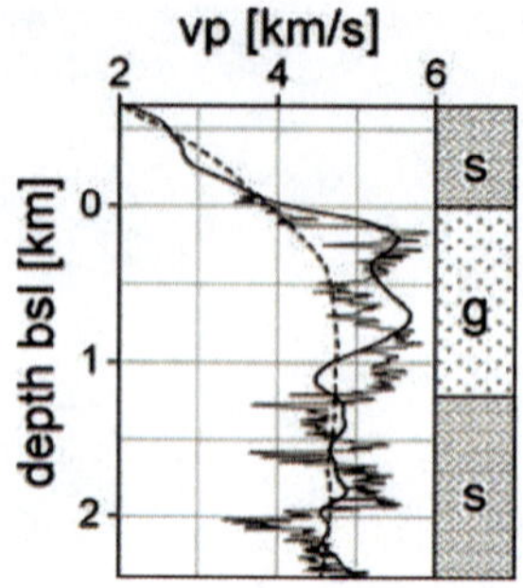

Fig. 14.7 Sonic log of the SAFOD borehole (*dotted gray line*), kinematic model (*dashed*) and 12 Hz waveform model (*solid*). g: granite, s: sediments

cause imaging problems, particularly with spatially aliased data, and it may explain why the shape of the anomalies does not conform to vertical faults and why the sign of the anomalies is positive, and not negative, as expected for faults.

At the position of the SAFOD borehole, the waveform model can be compared with a sonic log (Fig. 14.7). The velocities match well, particularly the sharp increase at the top of the granite. The waveform model also shows a low-velocity zone (LVZ) within the granite at 0.4 km depth below sea level and the transition back to sediments near 1 km depth below sea level, although 200 m shallower than in the sonic log. Those differences might reflect the lower lateral resolution of the waveform model, without necessarily indicating inaccuracy. The deeper LVZs in the sonic log of ~100 m width are too thin for the resolution of the waveform models.

14.7 Data Fit

Figure 14.8 shows a comparison of observed and synthetic data. For offsets closer than 20 km (distance range 25–45 km) the waveform match in the 12 Hz waveform tomography model has significantly improved compared to the initial model. At longer offsets, however, the initial data match is poor, and there is virtually no improvement through waveform tomography: The major change is that the first arrivals appear blurred, probably reflecting the poor S/N. Those observations at more than 20 km offset are mostly sensitive to structures below ~3 km depth below sea level, and the mismatch confirms that this part of the model is not constrained by the waveforms.

Comparing seismograms is crucial to control the data fit, but it is a qualitative measure only, and it is also tedious due to the large amount of data and the great amount of detail in the waveforms. In contrast, the residual reduction listed in Table 14.1 is a quantitative measure, but it provides no insights into the distribution of the residual, and it is also restricted to the frequency components within the group and thus not representative of the overall waveform fit. A more detailed and representative statistical overview is provided by maps of the weighted RMS phase misfit as a function of offset and frequency at different stages of the inversion (Fig. 14.9). In the initial model, the residual at low frequencies (3–4 Hz) increases

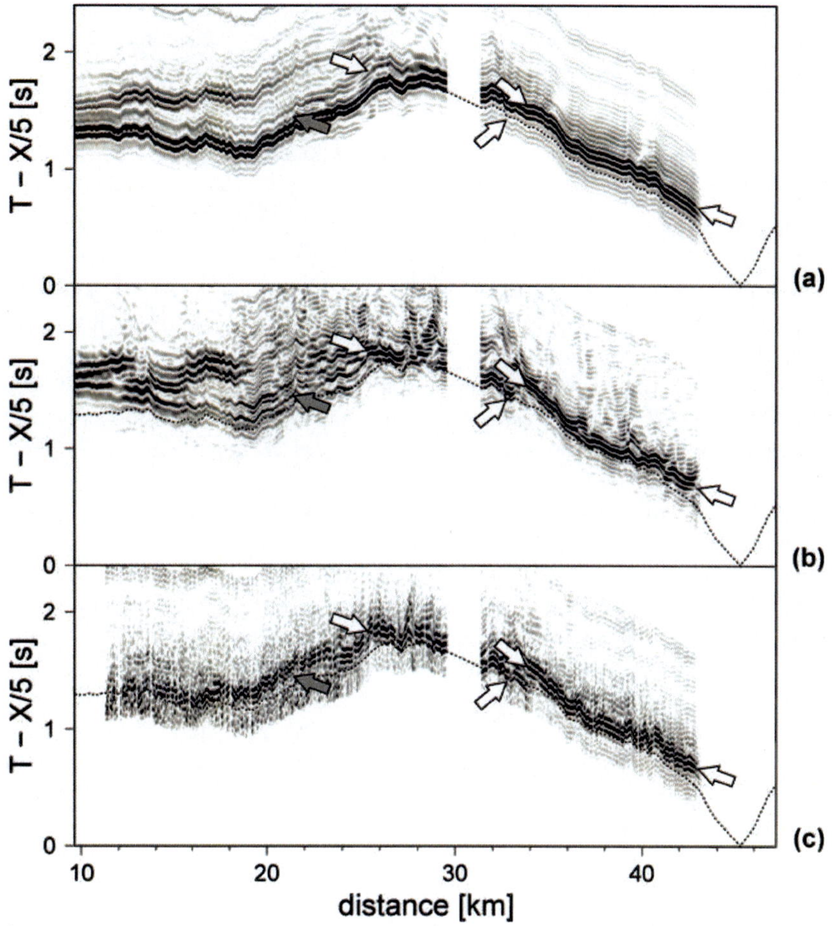

Fig. 14.8 (**a**) Synthetic seismograms for shot gather 3 computed in the initial model and (**b**) in the 12 Hz waveform model. (**c**) Preconditioned observations. All data are 3–12 Hz band-pass filtered. *White arrows* denote improved data fit, *gray arrows* denote no improvement. Note that all corresponding *arrows* are at the same (x, t) position

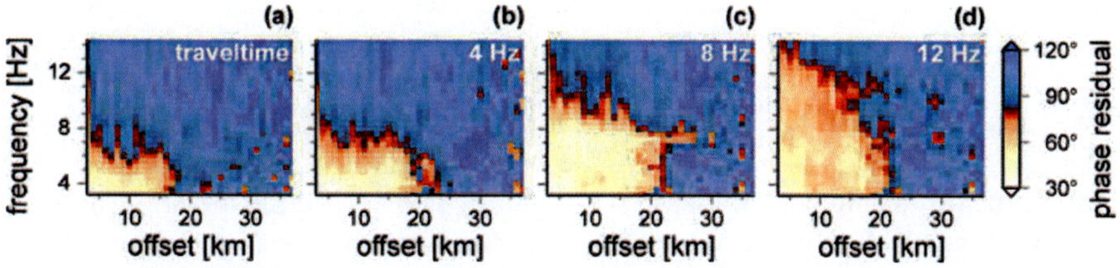

Fig. 14.9 (**a**) Weighted RMS phase residual as a function of offset and frequency for the initial model and (**b**) the waveform models after the inversion of 4 Hz, (**c**) 7.6 Hz and (**d**) 12 Hz. (**a–c**) After Bleibinhaus et al. (2009)

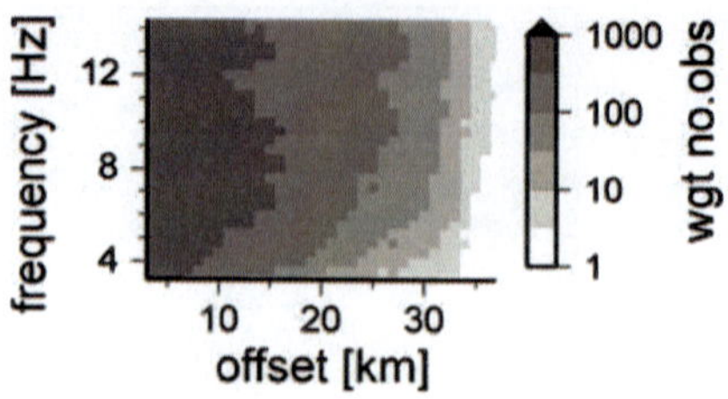

Fig. 14.10 Weighted number of observations as a function of offset and frequency after Bleibinhaus et al. (2009)

from ~45° ($T_0/8$) at near offsets to over 90° ($T_0/4$) at intermediate offsets. Beyond 20 km offset, the values start to fluctuate due to the small number of weighted observations (Fig. 14.10). As the inversion progresses, the overall misfit is steadily being reduced at offsets smaller than 20 km. An improvement at longer offsets after the inversion at 8 Hz (Fig. 14.9c) is restricted to a narrow band. This indicates cycle skipping at those offsets, otherwise the match at lower frequencies would have to be better.

At 12 Hz, the ability to fit waveforms is reduced to the near offsets (Fig. 14.9d). In addition, the match at the low frequencies starts to degrade again. This is surprising because the inversion of the 10–12 Hz frequency components hardly alters the long-wavelength structure of the model as determined from the inversion of the low frequencies. What is happening is that an increasing number of small, shallow scatterers attenuate the low frequencies propagating in the shallow part of the model. Low-frequency seismograms, in turn, are dominated by waves propagating in the deeper, smoother parts of the model, where scattering attenuation is insignificant. However, those parts of the model are poorly constrained, and the corresponding phases are erroneous, thus decreasing the low-frequency waveform fit and preventing the inversion of higher frequencies.

14.8 Discussion

In order to better understand the reasons for the increasing growth of artefacts beyond 12 Hz, it is instructive to analyse the kernels at higher frequencies. Figure 14.11 depicts a monochromatic 14 Hz wave field and kernel computed in the 14 Hz waveform model for the same source and receiver as in Fig. 14.5. At this frequency, 20 km offset corresponds to more than 70 wavelengths. Along its path, the wave is being scattered several times, and the complexity of the kernel reflects that. The single, broad Fresnel zone of a low-frequency kernel in a smooth model (Fig. 14.5c) is broken down into several Fresnel zones, which, in the high-frequency approximation, would correspond to multiple ray paths. So while the non-linearity has grown considerably, the ratio of data to unknown parameters is decreasing: The maximum source spacing exceeds the aliasing limit 12 times at 14 Hz, not accounting for the gap in the line. In addition, the size of scatterers potentially resolved by this frequency is ~100 m. It seems overly optimistic to believe that at this level of detail, out-of-plane scattering in a real geologic environment would

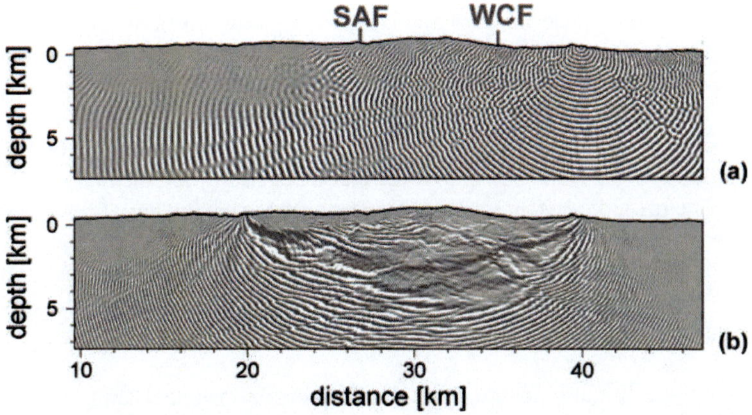

Fig. 14.11 (**a**) Monochromatic 14 Hz wave field and (**b**) 14 Hz wave path for a source and receiver at 20 km offset computed in the 14 Hz waveform model. Note the impact of scatterers near prominent faults on the wave field. SAF – San Andreas Fault, WCF – Waltham Canyon Fault

still be negligible. Even if it is a small effect, corresponding artefacts might become amplified over several iterations. This means that in order to further improve the resolution, 3D variations probably have to be accounted for by the modelling, and 3D data need to be acquired to constrain them.

There are a number of other problems related to the forward modelling, which potentially affect the convergence of the inversion. P–S and P–Rayleigh converted energy is part of the early arrivals, but it is not accounted for by the acoustic modelling. Attenuation and anisotropy were ignored, too. Scattering from topography is not accounted for, because receivers and shots were embedded in the medium, and an absorbing boundary was put above the receivers. The degradation from ignoring the free surface has been tested, and it is relatively moderate, despite the strong topography. Most of the P–Rayleigh scattered energy remains in the data space, and it is not projected onto the waveform model (Bleibinhaus & Rondenay, 2009).

The limited accuracy of the deeper parts of the initial model could also pose an impediment for the waveform tomography, especially because low S/N at long offsets and at low frequencies led to cycle skipping at long offsets that was not fixed during the inversion. In the absence of better data, several methods have been proposed to alleviate the dependency of the waveform inversion on the initial model, most prominently the so-called Laplace-domain inversion (Shin & Cha, 2008), which employs variable time damping with a series of increasing half-decay times that progressively shift the weight to later arrivals during the inversion. In the case of the SAFOD data, however, the S/N is not high enough at the long offsets to employ smaller half-decay times. Instead of amplifying the first arrival, they boost the noise before the first arrival and lead to a degradation of the overall waveform match. Offset-dependent weights to amplify the contribution of the long offsets have also been tried, but without improving upon the deep structure.

In any case, the phase misfit maps (Fig. 14.9) show no indication of leakage from the poorly resolved deep parts to the shallow structure down to ~2 km below sea level. To the contrary, the resolution of the shallow structure improved considerably from the waveform inversion of frequencies up to 12 Hz. The resulting model reveals several features that were not constrained by the traveltimes, and it matches the sonic log well. The spatial correlation of strong velocity anomalies with imaged reflectors indicates that, even at frequencies that are too low to resolve faults, waveform models can be used to infer their position.

Resolving more details from the inversion of higher frequencies was not possible. Spatial aliasing probably contributes to this limitation, but the complexity of high-frequency wave paths suggests that out-of-plane scattering could be a decisive factor.

Despite these limitations, the SAFOD case study shows that applying acoustic 2D waveform tomography to surface-seismic data has significant merits. It is essentially a subsequent step to traveltime tomography, it does not require additional data or models and it is computationally feasible on a small cluster. The gain in resolution helps considerably with the interpretation, and it allows for better predictions for drilling or other geo-technical problems.

Chapter 15
Source Stacking Data Reduction for Full Waveform Tomography at the Global Scale

In this chapter, we present a full waveform tomographic method based on a special data reduction scheme. This is intended to overcome the high numerical cost of a full waveform inversion with large data sets. Most numerical methods used to solve the forward problem in seismology allow us to trigger several sources at the same time within one simulation with no incremental numerical cost. Doing so, the resulting synthetic seismograms are the sum of seismograms from each individual source for a common receiver and a common origin time, with no possibility to separate them afterwards. The summed synthetic seismograms are not directly comparable to data, but using the linearity of the problem with respect to seismic sources, we can sum all data for a common station and a common zero time, and we perform the same operation on synthetics. Using this data reduction scheme substantially reduces the computational requirements because we can model the complete data set with one single forward simulation, instead of requiring as many simulations as there are events. The optimisation scheme used to find the minimum of the objective functional is a Gauss–Newton method, and the spectral-element method is once again used to solve the forward modelling problem. The approximate Hessian of the Gauss–Newton method is either computed with the spectral-element or with a normal mode method. Several tests confirm the feasibility of the method. It appears that this approach can work owing to the combination of two factors: the off-path sensitivity of the long-period waveforms and the presence of multiple scattering, which compensate for the loss of information in the summation process. We discuss the advantages and drawbacks of such a scheme.

15.1 Introduction

In the previous chapters, full waveform tomography methods have been applied based on the comparison of individual seismograms and their corresponding synthetics. The possibility of such a comparison rests on the solution of the forward problem for each seismic source, which can be a computationally expensive task.

Written by Yann Capdeville, Institut de Physique du Globe, Équipe de Sismologie, Paris, France

A. Fichtner, *Full Seismic Waveform Modelling and Inversion*, Advances in Geophysical and Environmental Mechanics and Mathematics,
DOI 10.1007/978-3-642-15807-0_15,

Here, we present an alternative full waveform inversion method that combines the Gauss–Newton algorithm with a data reduction scheme that drastically reduces numerical costs. This is based on data stacking for common receivers. The idea of source stacking was mentioned first by Mora (1987), though it was not further investigated at that time. For the full waveform inversion problem at the global scale, this method was first proposed by Capdeville et al. (2005) and applied, in a slightly different way, at the seismic exploration scale by Krebs et al. (2009) and Ben Hadj Ali et al. (2009a, b) for 2D cases. The idea has been very successful for synthetic tests, and the application to a global data set is work in progress (Capdeville, 2010).

The global scale is interesting because, at very long periods, the Earth is known to be weakly inhomogeneous and 1D models already provide synthetic seismograms that explain the data very well. It is therefore an 'easy' case compared to smaller scale problems. Moreover, the data coverage at long periods is good and the numerical problem is simpler for a variety of reasons: There is no absorbing boundary, no heterogeneity outside of the inverted domain can pollute the solution and the partial derivatives necessary for the partial Hessian can be efficiently computed with normal mode solutions for the 1D starting model.

The numerical method used to solve the forward problem is the coupling method of spectral elements and normal modes that we already mentioned in Sect. 4.3 (Capdeville, 2003a, b; Chaljub et al., 2003). The spectral-element solution is used for the Earth's mantle, and the normal mode solution is used for the core that is assumed to be spherically symmetric.

15.2 Data Reduction

We propose to use two properties of most numerical wave equation solvers that help us to reduce the numerical cost of a non-linear least-squares inversion: First, because the wave field is computed at any location in the Earth, the numerical cost of an inversion is independent of the number of receivers. Second, it is possible to input several sources in the scheme and to trigger them simultaneously, without increasing the numerical cost. Of course, the resulting traces on the receiver side will be the sum of the traces due to each individual source and there is no possibility to separate them once the computation is done. While we cannot recover the individual synthetic seismograms after the computation, we can perform an equivalent stack of data for common seismometers assuming a common origin time for all the events and use this reduced data set instead of traces for individual events. The stacked data are then directly comparable to the stacked synthetic seismograms. This operation is possible thanks to the linearity of the wave equation with respect to seismic sources which means that computing traces for one seismometer for each source separately and then summing them is equivalent to computing one trace of all the sources triggered simultaneously. Using this data reduction scheme, we can model the whole data set with one simulation only. This is in contrast to the classical approach where the number of simulations is equal to the number of events. Finally, we note that summing the traces with a common zero time is not necessary (i.e. the sources can be staggered in time), but it is used here in the interest of simplicity. Of course, data reduction is not without drawbacks and some information that is contained in individual seismograms will inevitably be lost in the summation process. Our hope is that this loss of information will be compensated by exploiting all the information present in a long time series for each trace.

15.3 The Source Stacked Inverse Problem

Our aim is to find an Earth model with the minimum number of parameters that explains our seismic data set, as well as data not used in the inversion but obtained under similar conditions. By Earth model, we mean the 3D variations of elastic parameters, anelasticity and density. We wish to solve the inverse problem using a classical least-squares inversion (Tarantola & Valette, 1982) and with a complete modelling theory, for instance the spectral-element method applied to the wave equation. In classical full waveform inversion, the data $\mathbf{u}$ are seismograms from N_s events, recorded by N_r three-component seismometers yielding $3 \times N_s \times N_r$ time series. Here, $\mathbf{u}$ is the stack of N_s seismic time traces recorded by N_r three-component seismometers yielding $3 \times N_r$ time series. When $\mathbf{L}$ is the explicit forward modelling function for the N_s sources triggered at the same time, it allows us to model the data for a given set of model parameters: $\mathbf{u} = \mathbf{L}(\mathbf{m})$. In our case $\mathbf{L}$ represents the coupled spectral-element method, which is able to compute an accurate set of synthetics for nearly arbitrary models. We wish to minimise the classical objective function

$$\chi(\mathbf{m}) = [\mathbf{L}(\mathbf{m}) - \mathbf{u}] \cdot \mathbf{C}_u^{-1} \cdot [\mathbf{L}(\mathbf{m}) - \mathbf{u}] + (\mathbf{m} - \mathbf{m}_0) \cdot \mathbf{C}_m^{-1} \cdot (\mathbf{m} - \mathbf{m}_0), \quad (15.1)$$

where $\mathbf{m}_0$ is the a priori value of the model parameters, and $\mathbf{C}_u$ and $\mathbf{C}_m$ are the covariance matrices of the data and the model parameters, respectively. A model $\mathbf{m}$ that minimises a similar objective functional for the unstacked data also minimises the objective functional for the stacked data. The reverse, however, is not generally true. We nevertheless wish to demonstrate that the optimal model for the stacked inverse problem is close to the optimal model for the unstacked inverse problem for physically meaningful cases. In the tests presented in this chapter, we show that the method indeed converges, under specific circumstances, towards the target model used to generate the synthetic data set. Therefore, the optimal model for the stacked objective functional is very close to the optimal model for the unstacked objective functional. For the geophysical exploration scale, Krebs et al. (2009) and Ben Hadj Ali et al. (2009b) have shown that the optimal model for the stacked case is slightly different from the optimal model for the unstacked objective functional, but still close enough.

As $\mathbf{L}$ is non-linear with respect to $\mathbf{m}$, the minimum of χ needs to be approximated by an iterative optimisation scheme such as the Gauss–Newton method (Tarantola

& Valette, 1982). Given the model at iteration i, we can obtain the model at iteration $i+1$ as follows:

$$\mathbf{m}_{i+1} = \mathbf{m}_i + \left(\mathbf{G}_i^{\dagger} \cdot \mathbf{C}_u^{-1} \cdot \mathbf{G}_i + \mathbf{C}_m^{-1}\right)^{-1} \left[\mathbf{G}_i^{\dagger} \cdot \mathbf{C}_u^{-1} \cdot (\mathbf{u} - \mathbf{L}(\mathbf{m}_i)) - \mathbf{C}_m^{-1}(\mathbf{m}_i - \mathbf{m}_0)\right], \tag{15.2}$$

where $\mathbf{G}_i$ is the partial derivative matrix

$$\mathbf{G}_i = \nabla_{\mathbf{m}} \mathbf{L}|_{\mathbf{m}=\mathbf{m}_i}. \tag{15.3}$$

Usually, the forward problem is solved using first-order approximations such as, for example, the Born approximation within the normal mode framework (e.g. Woodhouse & Dziewonski, 1984) or arrival time Fréchet kernels (e.g. Dahlen et al., 2000). This leads to a linear relation matrix ($\mathbf{G}_0$) between the set of parameters and the synthetic data. In that case, only one iteration of (15.2) is required and the partial derivative matrix is built within the forward at a relatively low numerical cost. Some tomographic approaches are slightly non-linear (e.g. Li & Romanowicz, 1996) but are still based on the Born approximation. They also have the advantage of providing naturally the partial derivative matrix with no extra numerical cost, at least at the a priori model stage ($i = 0$). When the spectral-element method is used, the partial derivative matrix or sensitivity kernels cannot be computed naturally, but we can use either a 'brute force' finite-difference formulation or the adjoint solutions. For the simple tests here, the brute force solution is possible but is not an option for more realistic cases. In this chapter we use both the brute force solution and the adjoint solution, but with normal modes (Capdeville, 2005) and not with spectral elements.

15.4 Validation Tests

In this section, we present several numerical experiments to assess the robustness of the inversion when the stack data reduction is applied. The tests are circular in the sense that the 'data' to be inverted are generated with the same forward theory as the one used to invert. We name the model used to generate the data to be inverted the *input model* or the *target model*. These tests only provide information on the ability of the process to converge towards the solution under specific circumstances (e.g. amplitude of velocity contrast, data coverage, presence of noise). They do not provide information on the behaviour of the inversion in the case of an incomplete theory, like, for example, how an isotropic inversion would map an anisotropic medium or how high-degree horizontal spherical harmonics components (or equivalent) would leak or alias into a low-degree inversion. Nevertheless, the tests provide valuable information on the feasibility of the process. At least if the synthetic inversions failed, there would be little hope that it will ever succeed in realistic cases.

In the following tests, no damping is applied ($\mathbf{C}_u = \mathbf{I}$ and $\mathbf{C}_m^{-1} = \mathbf{0}$), so that the least-squares inversion process from Eq. (15.2) simplifies to

$$\mathbf{m}_{i+1} = \mathbf{m}_i + (\mathbf{G}_i^\dagger \cdot \mathbf{G}_i)^{-1} \cdot [\mathbf{G}_i \cdot (\mathbf{u} - \mathbf{L}(\mathbf{m}_i))] \ . \tag{15.4}$$

In order to limit the numerical cost of these experiments, only $\mathbf{G}_0$ will be computed and will be used instead of $\mathbf{G}_i$ at iteration i. We will see that this approximation does not hinder the convergence, at least for these tests. Note that if the starting model is spherically symmetric, normal mode perturbation theory provides an exact and computationally efficient solution for $\mathbf{G}_0$ (Woodhouse, 1983). The normal mode method used to compute $\mathbf{G}_0$ is very close to the adjoint method and is presented in Capdeville (2005). Of course, this normal mode perturbation approach is an option only when the starting model is spherically symmetric, which may not be desirable with the present level of sophistication in tomography. Nevertheless, an interesting possibility for 3D starting models may be to combine the adjoint problem solution to compute an accurate gradient of the cost function and normal mode perturbation theory to compute the approximate Hessian $(\mathbf{G}_i^\dagger \cdot \mathbf{G}_i)$.

15.4.1 Parameterisation

Instead of spherical harmonics or block parameterisation, we use piecewise polynomials as in our spectral-element discretisation (Sadourny, 1972; Ronchi et al., 1996; Chaljub et al., 2003). The sphere is discretised in non-overlapping elements and each of these elements can be mapped onto a reference cube. On the reference cube, a polynomial basis is generated by the tensor product of a 1D polynomial basis of degree $\leq N$ in each direction. The continuity of the parameterisation between elements is assured. More details on this discretisation mesh can be found in Chaljub et al. (2003). Figure 15.1 presents the mesh of the sphere used for this parameterisation with a polynomial degree $N = 2$. The number of free parameters is 274, which roughly corresponds to a spherical harmonic degree 8 horizontally in the upper mantle and a degree 4 horizontally in the lower mantle. In practice, this parameterisation may not be a good choice, because parameters at the corner of elements have a different spatial spectral content than parameters at the centre of an element. However, for the tests presented here, as the input model is represented on the same mesh as the inversion mesh, this choice does not affect the results.

15.4.2 Experiment Setup and Input Models

The following experiments have been carried out with the mesh from Fig. 15.1 and only the S wave speed, v_s, has been inverted. We choose a realistic source-receiver configuration of 84 well-distributed events recorded at 174 three-component stations of the IRIS and GEOSCOPE networks. This source–receiver geometry is shown in Fig. 15.2.

The corner frequency is $1/160$ Hz and each trace has a duration of 12,000 s. For each test, the starting model is the spherically symmetric PREM (Dziewonski &

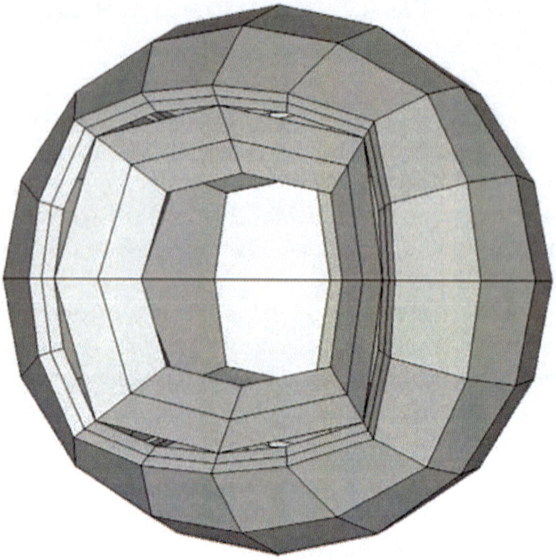

Fig. 15.1 View into the mesh of the sphere used to parameterise the velocity model. The number of free parameters is 274

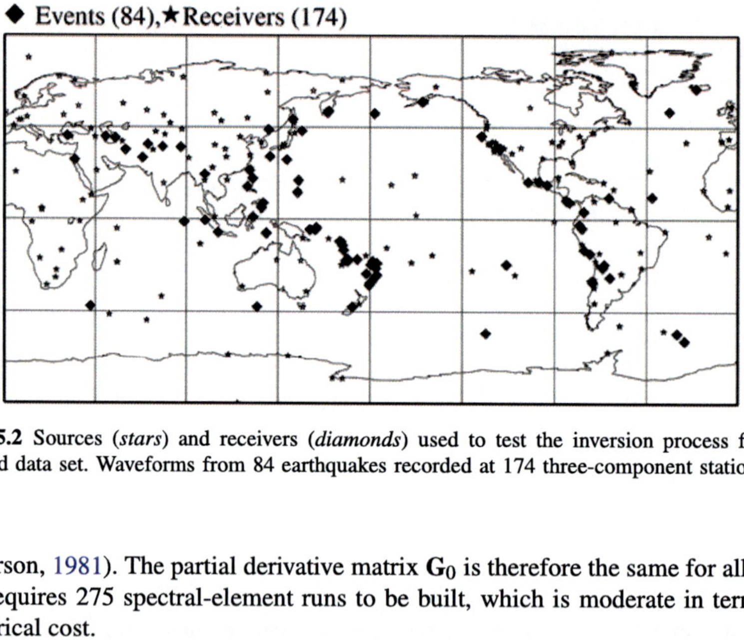

Fig. 15.2 Sources (*stars*) and receivers (*diamonds*) used to test the inversion process for the reduced data set. Waveforms from 84 earthquakes recorded at 174 three-component stations are used

Anderson, 1981). The partial derivative matrix $\mathbf{G}_0$ is therefore the same for all tests and requires 275 spectral-element runs to be built, which is moderate in terms of numerical cost.

Two input models will be used. For both of them, the reference background model is PREM. To this we add a 3D v_s velocity contrast field generated on the same mesh as the one that will be used for the inversion (Fig. 15.1).

The first model, shown in Fig. 15.3 and named BIDON, is very simple: all parameters are set to zero except for one in the upper mantle and one in the lower mantle. The maximum amplitude of the velocity perturbations, +9 %, is large compared to

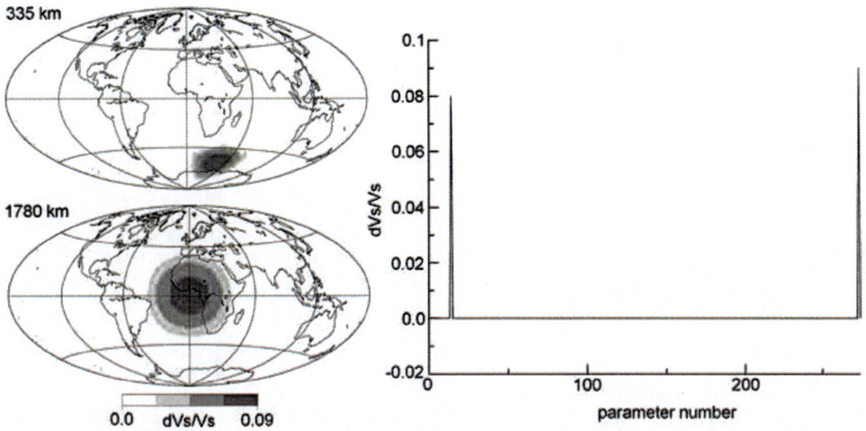

Fig. 15.3 Earth model BIDON. Only two parameters have a velocity contrast with respect to the spherically symmetric reference model (PREM). The *left panel* shows maps at two different depths and on the *right* is shown a 1D representation of the model where the S velocity contrast is plotted as a function of the parameter number (from 1 to 274)

what we expect in the Earth for such a long spatial wavelength. In Fig. 15.3 (left) we plot a depth cross section of the model and in Fig. 15.3 (right) we show the v_s perturbation as a function of the parameter number of the mesh, ranging from 1 to 274. This 1D representation does not provide a precise idea of what a map of the model would actually look like, but it provides more accurate information on the precision of the inversion than a geographical map. The parameter indices are sorted such that the lower mantle is predominantly on the left side of the plot and the upper mantle-predominantly on the right side of the plot. This provides some information about the location of potential errors.

The second model, named SAW6 and shown in Fig. 15.4, is more realistic than BIDON. It is derived from the tomographic model SAW24B16 (Mégnin & Romanowicz, 2000), truncated at degree 6 and mapped onto the 274-parameter mesh from Fig. 15.1. The maximum velocity contrast is much lower (around 3%) than in BIDON which is typical of long-wavelength mantle heterogeneity. In this case all the parameters have non-zero values, as can be seen in the right plot of Fig. 15.4.

15.4.3 Test in a Simple Two-Parameter Model

Stacked data are generated with the coupled spectral-element method in the model BIDON and are then inverted following the inversion scheme presented in the previous paragraphs. The results of the first three iterations of inversion are shown in Fig. 15.5. The first iteration already gives a velocity contrast very close to the correct value for the two parameters with non-zero velocity contrast, but for the other ones

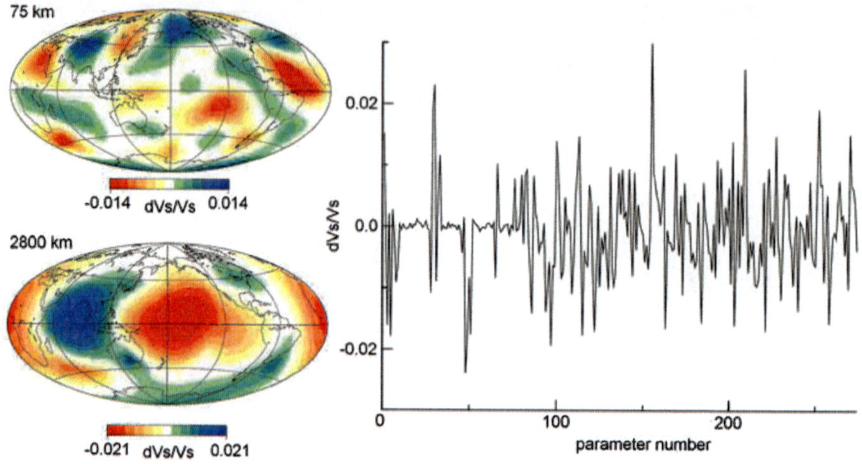

Fig. 15.4 Earth model SAW6. This model is derived from the tomographic model SAW26B16 (Mégnin & Romanowicz, 2000). Maps at two different depths (*left*) and the velocity contrast as a function of the parameter number (*right*) are presented

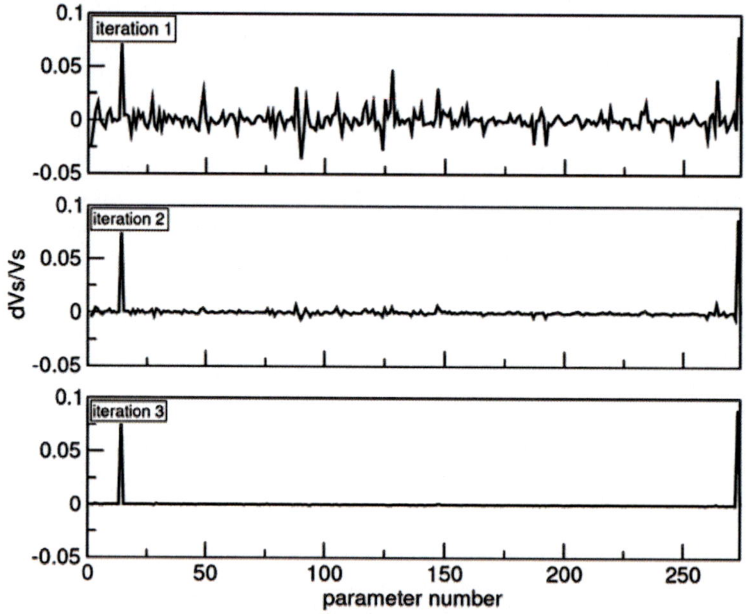

Fig. 15.5 Inversion results for the three first iterations for data generated in model BIDON (Fig. 15.3). The velocity contrast with respect to PREM is plotted as a function of the parameter number. The input model is accurately retrieved after three iterations

the result is very noisy. The second iteration gives a much better result and the third one has converged towards the correct result. This first experiment is satisfactory and shows that the process can work, at least in simple models. The fact that the first iteration is relatively far from the correct model is interesting because it means that a method based on the first-order Born approximation would give a very poor result in that case. The non-linearity is strong enough to justify a non-linear scheme, but it is weak enough to ensure convergence towards the right solution and not towards a wrong local minimum model, and this without updating the partial derivative matrix $\mathbf{G}_i$ at each iteration.

15.4.4 Tests in a Realistic Degree-6 Global Model

15.4.4.1 The Ideal Case

We now perform the same test but with data generated in the more realistic model SAW6. Results of the first two iterations of the inversion are shown in Fig. 15.6. They already reveal an acceptable convergence towards the input model for the second iteration. The fast convergence compared to the first test can be explained by the lower velocity contrast of the input model which implies smaller non-linear effects. All model parameters, from the lower mantle to the surface, are well retrieved.

15.4.4.2 Influence of Noise

The purpose of this experiment is to assess the noise sensitivity of the inversion scheme. This kind of test reflects the stability of the inversion, and in this experiment, we are not in a favourable situation. In fact, data with periods of 160 s and above have poor depth resolution, and to obtain very good results with such an

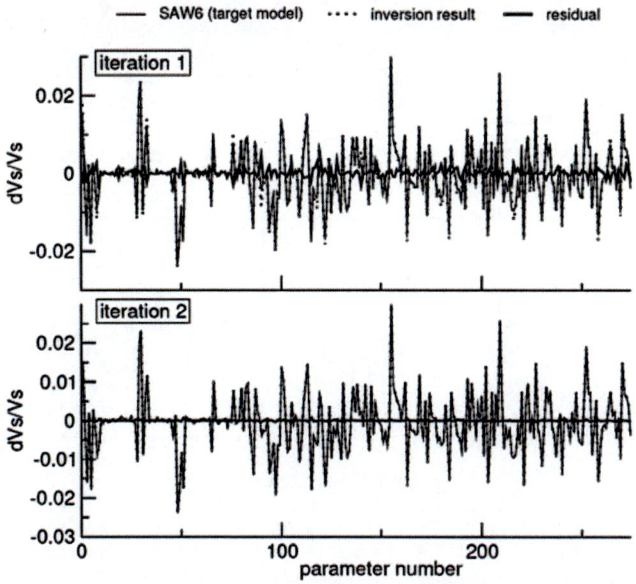

Fig. 15.6 Inversion results for the two first iterations for data generated in model SAW6 (Fig. 15.4). After two iterations, the inversion result matches very well the input model for all model parameters

experiment, one should use higher frequency data or decrease the number of vertical parameters. We perform the test with the SAW6 input model and synthetic noise added to the data. For this we generate a random signal corresponding to realistic background noise in this frequency band (the noise spectrum has a slope from -175 to -165 dB in the 100–300 s period range) and for each event–station pair, stack them and then add them to the synthetic data.

The result of the inversion is shown in Fig. 15.7. The noise affects the results of the inversion, but the scheme is still able to retrieve the target model correctly. The deepest parameters of the model are the most affected by noisy data which is not a surprise knowing the poor sensitivity to deep layers of long-period data. Figure 15.8 shows that, despite the noise, the inversion is able to retrieve a model that explains the data far beyond the noise level. The fact that we are able to fit the data so well, even though the model is not perfect, is also due to the lack of depth sensitivity of long-wavelength data.

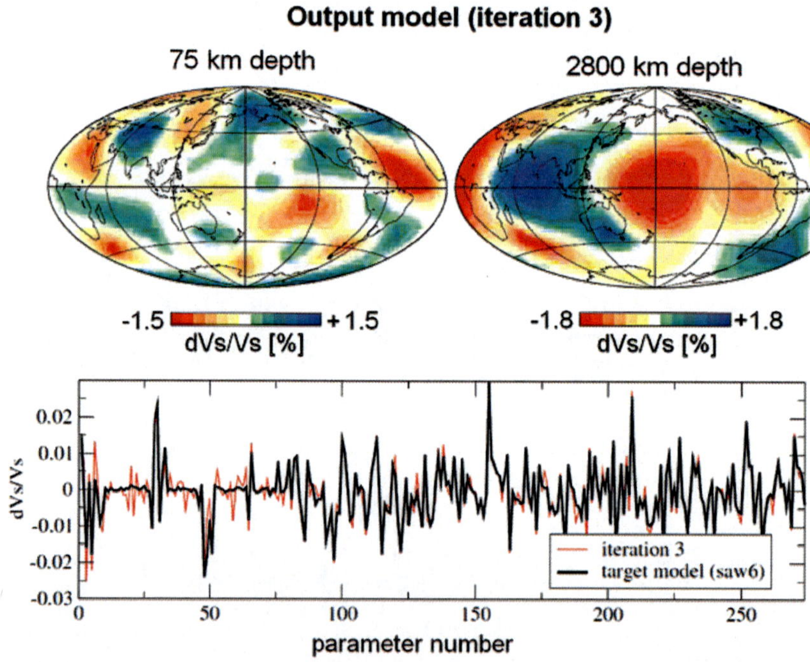

Fig. 15.7 Inversion results after three iterations for the data generated in the model SAW6 (Fig. 15.4) and with noise added. The two maps show the result of the inversion (output model) after three iterations when synthetic noise is added to the synthetic data. The *lower plot* shows the v_S velocity contrast on the input and output model as a function of the parameter number. Note that the deep parameters are more affected by the noise than the ones close to the surface. The model is nevertheless correctly retrieved by the inversion

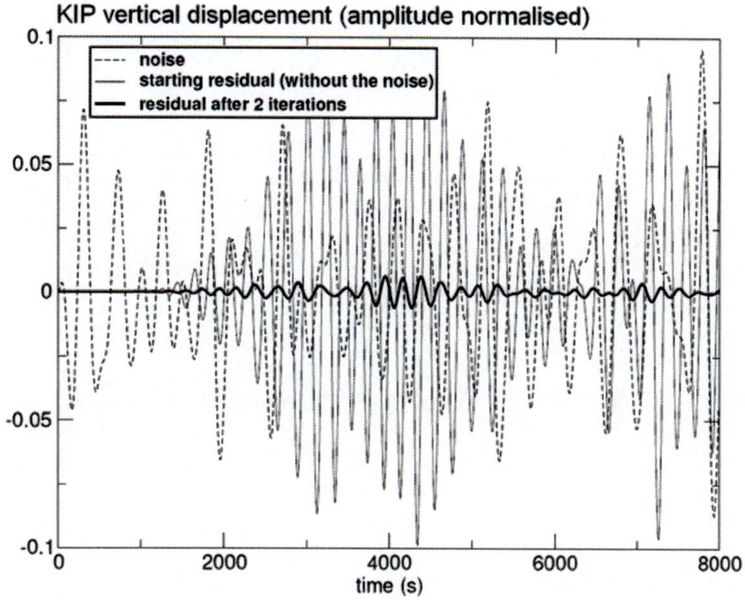

Fig. 15.8 Time series at station KIP: Noise added to the synthetic vertical-component data before inversion (*dotted line*). Starting residual without noise, i.e. synthetic data without noise minus synthetic for the reference model (*solid line*). Residual after two iterations, i.e. synthetic for the output model minus synthetic for the starting model (*bold line*). The scheme fits the synthetic data beyond the noise level, meaning that the amplitude of the last residual is smaller than the noise amplitude

15.4.4.3 One-Station Inversion

We perform an extreme test to assess the ability of the process to recover information in the case of very poor data coverage. To do so, we limit the number of receivers to 1: the GEOSCOPE station KIP in Hawaii (Fig. 15.9). This time, no noise is added

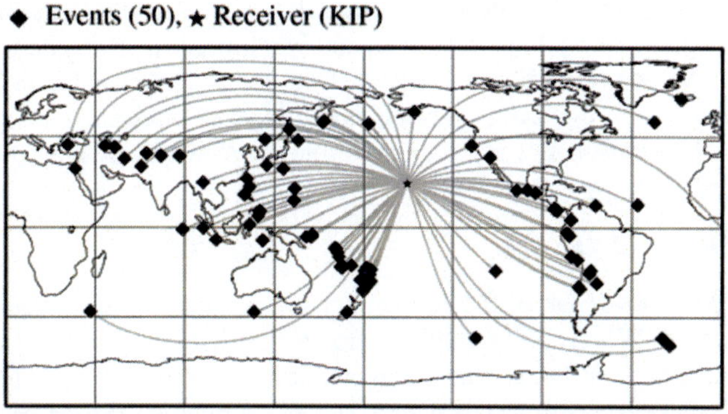

Fig. 15.9 Data coverage used in the single-station test. The same number of events (84, plotted as *diamonds*) as in the other tests is used, but with only one station (KIP)

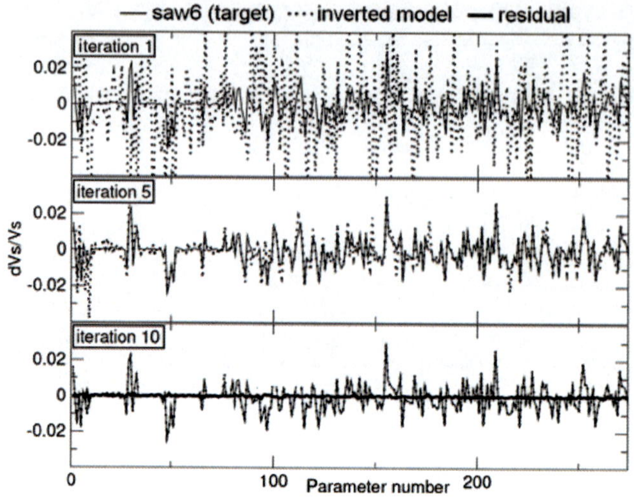

Fig. 15.10 Inversion results with only one station (KIP, see data coverage in Fig. 15.9) for synthetic data generated in model SAW6 (Fig. 15.4) for iterations 1, 5 and 10. The convergence is slow, but after 10 iterations, the output model matches the input model

to the traces. The input model is SAW6 (Fig. 15.4) and the results of the inversion are shown in Fig. 15.10 in the 1D representation for iterations 1, 5 and 10. The output model for the first iteration is very far from the input model, and at this point it seems that the inversion scheme has no chance to recover it. However, after 10 iterations, the process finally converges towards the input model. It is impressive that the process is still able to converge without updating the partial derivative matrix $\mathbf{G}_0$ at any iteration. The conclusion of this experiment is that, what allows us to retrieve the input model is not only the wide off-path sensitivity but also the non-linearity or, in other words, the multiple scattering. Indeed, a Born theory with no geometrical approximation has the same wide off-path sensitivity as a direct solution method like the spectral-element method, but would give a wrong model (similar to the one which is obtained at iteration one). Clearly, the inversion is in that case highly unstable and a very high data precision is required to allow the inversion to converge towards the right model. A one-station inversion for real data is unlikely to be successful due to the presence of noise and physical processes not included in the theory, such as anisotropy, attenuation, effects of atmospheric pressure.

15.4.4.4 Influence of Errors in the A Priori Moment Tensor

So far, a perfect knowledge of the source location, origin time and moment tensor has been assumed. When applying the method to real data, this will not be the case and significant errors on the source parameters can be expected. In order to partly address this issue, we perform a test where the a priori moment tensors are incorrect, but we keep the source locations and origin times perfectly known. This reflects the fact that, at least at very long periods, the location and origin time errors are small compared to the wavelength. In this test we generate data in the SAW6 model with each component of each a priori moment tensor randomly perturbed by up to ±30%. The moment tensors used to generate the partial derivatives and to compute the forward modelling part of the inversion are therefore not the ones that have been used to generate the synthetic data to be inverted.

The result of the inversion after three iterations is shown in Fig. 15.11. The scheme can clearly not retrieve the input model. The unknown moment tensors create large errors in the output model that cannot be overcome with the reduced number of data. A solution to this problem can be to increase the number of data and therefore, because the number of stations cannot be significantly increased, to use multiple stacked data sets. Another solution is to simultaneously invert for both the moment tensor and v_s. In this case, a difficulty due to the stacked data set is that, for sources close to each other, only the sum of these moment tensors can be retrieved, but not individual moment tensors. If the primary goal of the inversion is to retrieve v_s, an accurate sum of the moment tensors of sources very close to each other is enough, because it will give a correct prediction of the stacked displacement. In fact, this is all we need for a v_s tomography with stacked data. Now, if we are also interested in individual moment tensors, a solution can be to separate sources in the time domain by introducing time delays between close sources. Doing so, different sources located at the same place will have a different effect on stacked data. In this example, we will only focus on retrieving the v_s field correctly. In the case of sources very close to each other, we therefore wish to invert only for the sum of the moment tensors. In order to do so, we generate partial derivatives of individual components of moment tensors. The approximate Hessian matrix ($\mathbf{G}_i^\dagger \cdot \mathbf{G}_i$) for moment tensors is then built, an eigenvalue analysis of this matrix is performed and only the 75% largest eigenvalues are kept. This is equivalent to a damping that removes instabilities, but it only affects the moment tensor inversion part. Of course the choice of 75% will prevent us from explaining the signal perfectly. Therefore, we expect a

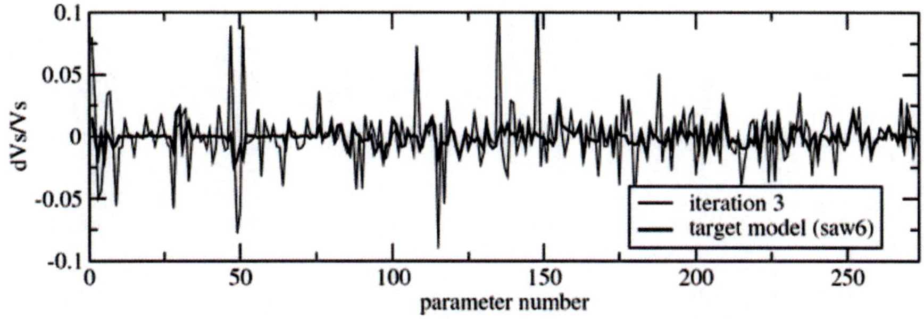

Fig. 15.11 Inversion results after three iterations for data generated in model SAW6 (Fig. 15.4) with error on the a priori moment tensors. To generate the data, a random perturbation of up to ±30% has been applied to each component of each moment tensor. The result of the inversion (without inverting for sources) after three iterations (*thin line*) does not match the input model

small error in the inverted S velocity field. Finally, we invert for v_S and the moment tensors jointly.

Figure 15.12 shows the result of the inversion. Thanks to the inversion for the moment tensors, we are able to retrieve the input model very well. The remaining errors are due to the 75% threshold on the eigenvalues for the source inversion. Indeed, this choice is not optimal. Part of the signal is not explained, thus slightly degrading the v_S inversion.

15.5 Towards Real Cases: Dealing with Missing Data

The success of all preliminary tests encourages applications to real data. When working with real data, however, a problem with the data reduction scheme immediately appears: the missing data. Indeed when trying to gather data for a sufficient number of events, say around 50, recorded at a large number of stations (around 80),

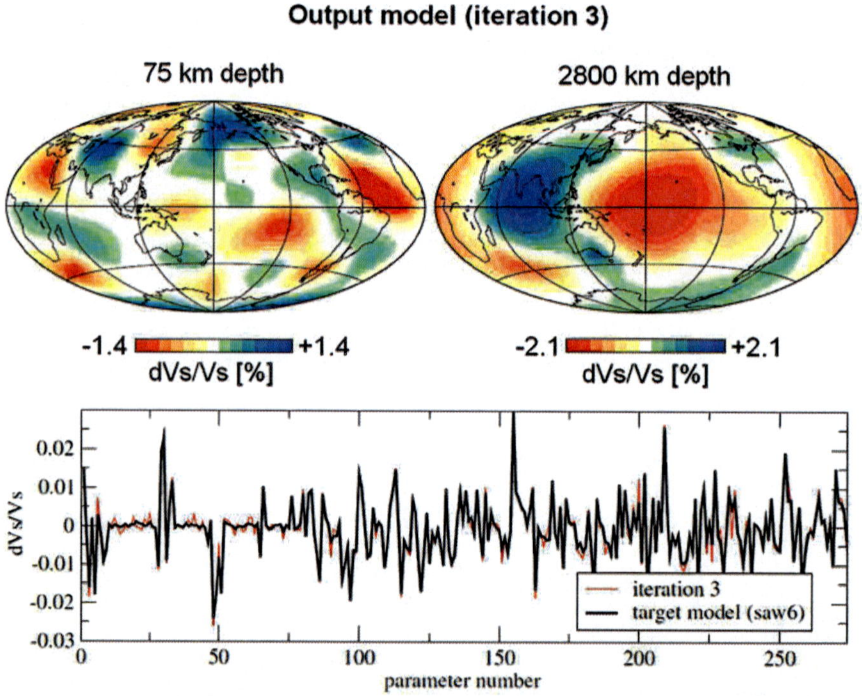

Fig. 15.12 The same test as the one presented in Fig. 15.11 with error on the a priori moment tensors, but this time v_S and the moment tensor have been inverted jointly. The maps show the result of the inversion (output model) after three iterations. The *lower plot* shows the v_S velocity contrast on the input and output model as a function of the parameter number. When moment tensors are inverted jointly with v_S, the inversion is able to retrieve correctly the input model despite large errors on a priori moment tensors

there are always between 10 and 20% of missing data whatever the configuration is. This is true even for large events with magnitudes from 6.5 to 7. The reasons why the data are sometimes not available at a given station vary from case to case, but there are very few stations that have 100% availability for 50 events. For our inversion scheme, missing data are a problem as it is easy to generate the sum of all the data in one run but impossible to remove some of them without computing each missing source individually. Since almost all sources are missing at least at one station, removing missing data would require to perform a simulation for each source, and it seems we are back at our starting point.

However, the sum of all the data $\mathbf{u}_\mathrm{t}$ can be separated into the sum of missing data $\mathbf{u}_\mathrm{m}$ and the sum of available data $\mathbf{u}_\mathrm{a}$:

$$\mathbf{u}_\mathrm{t} = \mathbf{u}_\mathrm{a} + \mathbf{u}_\mathrm{m}. \tag{15.5}$$

The total direct problem can also be separated into missing and available synthetic parts:

$$\mathbf{L}_\mathrm{t}(\mathbf{m}) = \mathbf{L}_\mathrm{a}(\mathbf{m}) + \mathbf{L}_\mathrm{m}(\mathbf{m}). \tag{15.6}$$

The main difficulty is that there is no way to efficiently compute the partial derivative matrices of $\mathbf{L}_\mathrm{a}$ and $\mathbf{L}_\mathrm{m}$. Therefore trying to solve

$$\mathbf{L}_\mathrm{a}(\mathbf{m}) = \mathbf{u}_\mathrm{a} \tag{15.7}$$

or

$$\mathbf{L}_\mathrm{t}(\mathbf{m}) - \mathbf{L}_\mathrm{m}(\mathbf{m}) = \mathbf{u}_\mathrm{a} \tag{15.8}$$

is not an option. On the other hand, solving

$$\mathbf{L}_\mathrm{t}(\mathbf{m}) = \mathbf{u}_\mathrm{a} + \mathbf{L}_\mathrm{m}(\mathbf{m}) \tag{15.9}$$

is possible because the partial derivative matrix of this last problem only depends on the sum of all the data, the missing and the available ones. Nevertheless, since the right-hand side of the last equation depends on $\mathbf{m}$, it requires an iterative scheme. This scheme is only interesting if we do not update the partial derivative matrix in each iteration, but since we can expect only a small number of missing data, this should not be a problem. The available solutions for $\mathbf{L}_\mathrm{m}(\mathbf{m})$ are

- $\mathbf{L}_\mathrm{m} = \mathbf{0}$, meaning that the missing data are ignored,
- synthetics in the spherically symmetric model ($\mathbf{L}_\mathrm{m}$ independent of $\mathbf{m}$) with normal mode summation,
- synthetics in $\mathbf{m}$ with normal mode summation first-order perturbation,
- synthetics in $\mathbf{m}$ with the spectral element method.

The first two solutions are numerically inexpensive but probably not very good, depending on the amount of missing data. The third one is probably a good compromise between numerical cost and precision and the last one is perfect but expensive. An equally good solution for a finalised model may be to use normal mode perturbation theory during the iterative process of the inversion and spectral-element synthetics in the last iteration.

To test the different strategies, we generate a data set for model SAW6 (Fig. 15.4), by computing synthetic seismograms from each of the 84 sources individually. We then randomly select the missing data among each source-receiver pair. The selected data are not used in the construction of the stacked data set. We adopt a rather extreme scenario with 35% of missing data, despite our experience with real cases where it was possible to have around 15–20% of missing data when working with 50 sources and 90 vertical-component receivers. We then perform three inversions. In the first one, the missing data are replaced with synthetic seismograms computed in the starting spherically symmetric model. This solution is numerically interesting because the scheme is still explicit: the data set completed with the synthetics of missing data does not depend on the inverted model. The drawback is that the

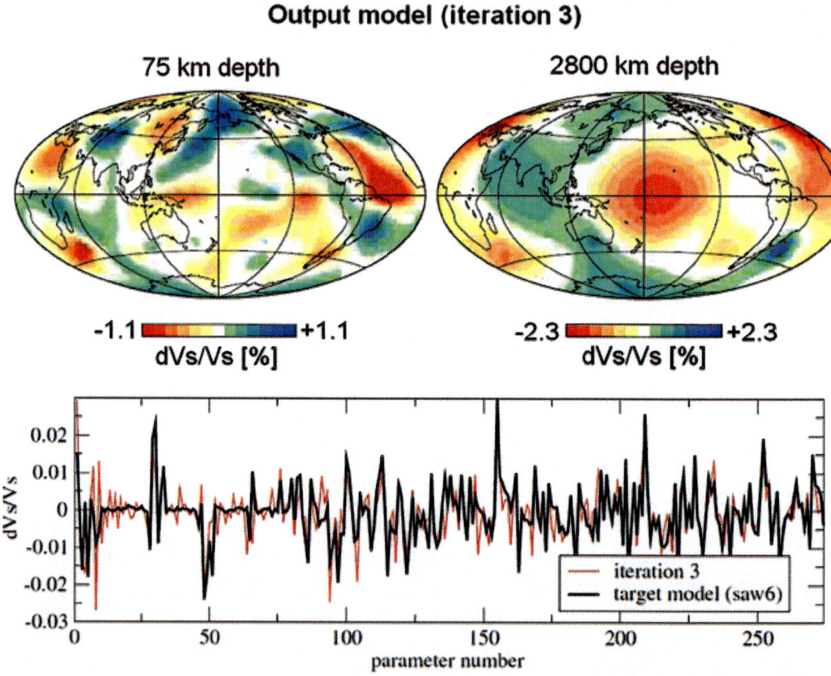

Fig. 15.13 Inversion results after three iterations for data generated with model SAW6 (Fig. 15.4, *bold line*) but with 35% of missing traces. The solution adopted to deal with the missing data in this test is to replace the missing data by synthetics computed in the starting model (1D). They are not updated during the iterative inversion. The result is noisy, as expected. The general features of the input model are nevertheless retrieved

final model cannot be accurate as the signature of the starting model will always be present and cannot be corrected.

The result of this inversion is presented in Fig. 15.13. As expected, the result is noisy, but the main features of the input model are still retrieved. In the second inversion, the missing data are replaced by synthetic seismograms computed in the current model ($\mathbf{m}_i$) with the Born approximation in the normal modes framework (Capdeville et al., 2000; Capdeville, 2005). This time the scheme becomes implicit in the sense that the synthetics of missing data depend on the current model.

The output model is shown in Fig. 15.14. Compared to the previous inversion, the result has significantly improved without being perfect. This is expected as the first-order Born approximation is not very accurate especially when the time series is long and when non-linear effects cannot be neglected.

We perform a final test where the missing data are replaced by synthetics computed in the current model with the spectral-element method. This solution is CPU time consuming as it requires to compute solutions for each source individually. To reduce the numerical cost we start from the model obtained in the last iteration of the previous test, and we perform only one extra iteration.

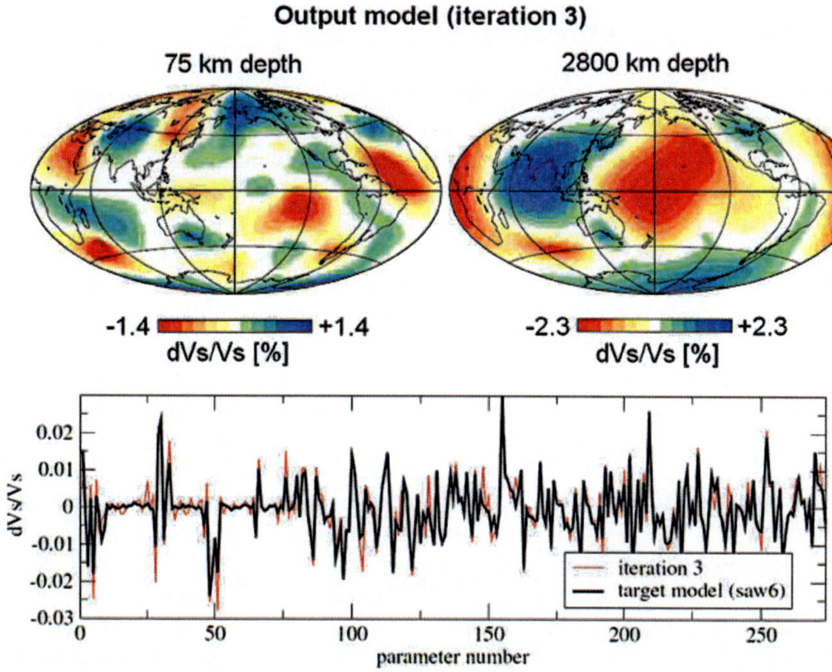

Fig. 15.14 The same as in Fig. 15.13, but the solution adopted to deal with the missing data is to replace them by synthetics computed in the current 3D model with the Born approximation. The 3D model is updated during the iterative inversion. The result is better as in the previous case where the missing data were replaced by 1D synthetics

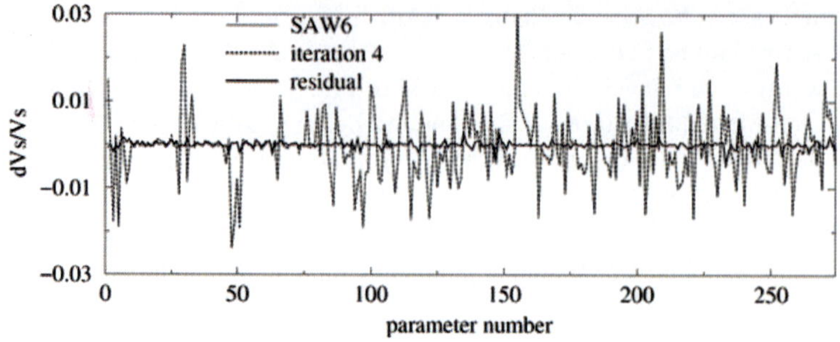

Fig. 15.15 The same as in Fig. 15.13, but with the missing data replaced by synthetics computed in the current model with spectral elements at iteration 4 starting from iteration 3 of the precedent test (Fig. 15.14). The result (*dotted line*) is in a good agreement with the target model

The result is presented in Fig. 15.15. Input and output model agree remarkably well, and only some more iterations would be required to obtain a precision similar to the one achieved in scenarios with complete synthetic data sets. Nevertheless, the result is already sufficiently accurate with respect to errors from noise or incomplete knowledge of the seismic source.

15.6 Discussion and Conclusions

In this chapter, we presented a method for non-linear full waveform inversion on the global scale, using the spectral-element method as a forward modelling tool and a data reduction scheme that drastically reduces the computational requirements. This scheme is based on the non-coherent trace stacking at a common receiver for a common source origin time.

The difference between the method presented here (see also Capdeville et al., 2005) and the one of Krebs et al. (2009) is the notion of source encoding. The idea is to convolve each individual shot by a different time function before stacking them. The expected result of the source encoding is to minimise possible drawbacks of the stacking process which could occur when sources are very close to each other, as is the case for typical configurations in exploration geophysics. Krebs et al. (2009) showed that a good source encoding is simply to randomly multiply each shot by $+1$ or -1 before stacking. The random sign of the source encoding is also changed through the iteration of the non-linear inversion process. Note that Capdeville et al. (2005) mentioned a similar idea by applying different time delays to each source before stacking. It appears that for distant sources, this source encoding process does not change much the converge rate and the result of the inversion.

The data reduction allows us to simulate the complete data set in one single spectral-element run and therefore to reduce the number of computations by a factor equal to the number of sources with respect to a classical approach. A series of tests shows very promising results. The main advantage of the approach is that it allows us to investigate global-scale full waveform inversion with comparatively moderate computational resources. There is also a data selection and processing advantage.

Phase identification, time picking or phase velocity measurements are not required, which saves time and minimises human error.

Clearly, there are also drawbacks to this approach. One of them is that some information is lost in the data reduction. Yet, all tomographic methods use some data reduction scheme. Travel time tomography, for instance, uses only a limited number of arrival times per trace, often only one or two. Here, we use traces of 12,000 s duration that contain a large amount of information, even at periods of 160 s and above. We therefore hope that long traces compensate the loss of information due to stacking.

If only one single stacked data set is used, there is a limit on the number of sources, in the sense that additional sources do not contribute any new constraints. This limit depends on the corner frequency and the length of the signal. When it is reached, the only way to obtain more information on the model is to use multiple stacked source data sets. This can be done by splitting the data set into two or more subsets for which the stacked seismograms are computed. Through the repetition of this splitting process, we will eventually converge towards the classical case where all sources are considered individually.

Another drawback of the source stacking is that it does not allow us to select specific time windows on traces in order to enhance some part of the signal with respect to others, for instance, to separate body waves from each other and from surface waves (e.g. Li & Romanowicz, 1996). Body waves have small amplitudes but contain information about the lower mantle whereas surface waves have a large amplitude but do not contain information about the lower mantle. As surface waves will dominate the stacked signal, there is little chance to recover the lower mantle before the upper mantle is very well explained. As we have seen, this was not a problem in our tests. We were, however, in a fortunate situation where we knew exactly what to invert for in order to explain the upper mantle. Thus, once the upper mantle was explained, the lower mantle was easy to retrieve. In a realistic case, it may be much more difficult, since we do not know for sure which elastic parameters are required to explain surface waves well enough and therefore to be able to access low-amplitude body waves and information about the lower mantle.

This last point leads to another difficulty that we will face in future work. What physical parameters (elastic, anelastic, density, etc.) do we need to invert for and at what resolution to explain our data set correctly? The resolution issue is not obvious: A too low resolution for a given frequency content will lead to aliasing. Too high resolution may lead to an unstable inversion scheme, as our data set may not have the information to constrain all the parameters. An equally difficult question concerns the number of physical parameters. Are v_S fluctuations enough to explain our data set? Probably not. Do we need v_S, v_P, density, anisotropy, 3D anelasticity and perturbations in source parameters? What is the relative sensitivity of our data set to those parameters? All these questions will need to be addressed in future work.

Finally, it is well known that the type of least-squares inversion can strongly constrain the possible models. The theoretical alternative consists in the random exploration of the model space. The numerical costs are, however, currently not affordable. Nevertheless, questions on error bars and on what we may be missing in the least-squares inversion will also need to be addressed in future work.

Appendix A
Mathematical Background for the Spectral-Element Method

This appendix gives a brief introduction to the mathematical foundations of the spectral-element method. It is far from being exhaustive but sufficient for most practical purposes. For more complete treatments the reader is referred to Quarteroni et al. (2000), Karniadakis & Sherwin (2005), Pozrikidis (2005) or Allaire (2007).

A.1 Orthogonal Polynomials

We consider a family of polynomials, p_n, defined on the interval $[a, b] \subset \mathbb{R}$ and where p_n is of degree $n \geq 0$. The polynomials are said to be *orthogonal* when any of their mutual projections satisfies the condition

$$\int_a^b w(x) p_n(x) p_m(x) \, \mathrm{d}x = A_n \delta_{nm} . \tag{A.1}$$

The symbol $w(x)$ denotes a positive weighting function, and A_n is a normalisation constant. Each integration weight together with particular integration limits uniquely determines a family of orthogonal polynomials. The two families that are most relevant in the context of the spectral-element method are the *Legendre polynomials* and the *Lobatto polynomials*.
Legendre polynomials, denoted by $L_n(x)$, are orthogonal with respect to the flat integration weight $w(x) = 1$ and the integration interval $[-1, 1]$. In symbols

$$\int_{-1}^{1} L_n(x) \, L_m(x) \, \mathrm{d}x = A_n \delta_{nm} . \tag{A.2}$$

The Legendre polynomials are explicitly given by

$$L_n(x) = \frac{1}{2^n n!} \frac{\mathrm{d}^n}{\mathrm{d}x^n} (x^2 - 1)^n , \tag{A.3}$$

A. Fichtner, *Full Seismic Waveform Modelling and Inversion*, Advances in Geophysical and Environmental Mechanics and Mathematics,
DOI 10.1007/978-3-642-15807-0,

and they can be shown to satisfy the *Legendre differential equation*

$$\frac{\mathrm{d}}{\mathrm{d}x}\left[(x^2-1)\,\frac{\mathrm{d}}{\mathrm{d}x}L_n(x)\right] = n(n+1)\,L_n(x)\,. \tag{A.4}$$

Lobatto polynomials, $Lo_n(x)$, are defined in terms of the Legendre polynomials:

$$Lo_n(x) := \frac{\mathrm{d}}{\mathrm{d}x}L_{n+1}(x). \tag{A.5}$$

Thus, the Lobatto polynomials satisfy the differential equation

$$\frac{\mathrm{d}}{\mathrm{d}x}[(x^2-1)Lo_{n-1}(x)] = n(n+1)\,L_n(x)\,. \tag{A.6}$$

The Lobatto polynomials are the family that is orthogonal with respect to the integration weight $w(x) = (1-x^2)$ and the integration interval $[-1, 1]$:

$$\int_{-1}^{1}(1-x^2)\,Lo_n(x)\,Lo_m(x)\,\mathrm{d}x = A_n\delta_{nm}\,. \tag{A.7}$$

In the following sections we will see that the roots of the Lobatto polynomials play an important role in polynomial interpolation and numerical quadrature.

A.2 Function Interpolation

A.2.1 Interpolating Polynomial

Finite-element methods in general and the spectral-element method in particular use interpolating functions for the representation of continuous functions that are known exactly only on a finite set of collocation points or grid points. The properties of a finite-element method depend strongly on the interpolation scheme.

Let $f(x)$ be a function that is known at $N+1$ data points x_i, where $i = 1, 2, \ldots, N+1$. We want to interpolate the function at an arbitrary point $x \in [x_1, x_{N+1}]$. For this, we replace $f(x)$ by an interpolating function $g(x)$ that satisfies the $N+1$ *interpolation or matching conditions*

$$g(x_i) = f(x_i)\,, \tag{A.8}$$

for $i = 1, 2, \ldots, N+1$. The properties of the interpolating function are chosen in accordance with the requirements of a particular application. The most straightforward choice of an interpolating function $g(x)$ is the polynomial of degree N,

$$P_N(x) = a_1 x^N + a_2 x^{N-1} + \cdots + a_N x + a_{N+1} \,. \tag{A.9}$$

Applying the interpolation condition (A.8) to the representation of the Nth-degree polynomial (A.9) gives a set of $N+1$ linear equations for the polynomial coefficients $a_1, a_2, \ldots, a_{N+1}$:

$$\begin{pmatrix} 1 & 1 & \ldots & 1 & 1 \\ x_1 & x_2 & \ldots & x_N & x_{N+1} \\ \vdots & \vdots & \vdots & \vdots & \vdots \\ x_1^{N-1} & x_2^{N-1} & \ldots & x_N^{N-1} & x_{N+1}^{N-1} \\ x_1^N & x_2^N & \ldots & x_N^N & x_{N+1}^N \end{pmatrix}^T \cdot \begin{pmatrix} a_{N+1} \\ a_N \\ \vdots \\ a_2 \\ a_1 \end{pmatrix} = \begin{pmatrix} f(x_1) \\ f(x_2) \\ \vdots \\ f(x_N) \\ f(x_{N+1}) \end{pmatrix}, \tag{A.10}$$

where the symbol ^{T} indicates matrix transposition. The $(N+1) \times (N+1)$ matrix on the left-hand side of Eq. (A.10) is the *Vandermonde matrix*, denoted by V. Computing the solution of the system (A.10) via Cramer's rule we find

$$a_i = \frac{\det V_i}{\det V} \,. \tag{A.11}$$

The symbol V_i denotes the Vandermonde matrix where the ith row has been replaced by the right-hand side of Eq. (A.10). By induction one can show that the *Vandermonde determinant* is explicitly given by

$$\det V = \prod_{i=1}^{N} \prod_{j=i+1}^{N} (x_i - x_j) \,. \tag{A.12}$$

Thus, when the collocation points x_i are mutually distinct, the Vandermonde matrix is non-singular, and the interpolation problem is well posed.

When the Vandermonde determinant is small, small variations in the right-hand side of Eq. (A.10) will result in large changes of the polynomial coefficients a_i and in large changes of the interpolated values between the collocation points. This implies that numerical errors, e.g. discretisation errors, will have a smaller effect when the Vandermonde determinant is large. The collocation points that maximise the Vandermonde matrix are called *Fekete points* (see Sect. A.2.4).

A.2.2 Lagrange Interpolation

Lagrange interpolation allows us to find an interpolating polynomial without explicitly computing the coefficients of the monomials in Eq. (A.9). For this we introduce the family of Nth-degree *Lagrange polynomials*

$$\ell_i^{(N)}(x) := \prod_{k=1,k\neq i}^{N+1} \frac{x - x_k}{x_i - x_k}, \qquad i = 1, 2, \ldots, N+1. \tag{A.13}$$

The Lagrange polynomials satisfy the *cardinal interpolation property*

$$\ell_i^{(N)}(x_j) = \delta_{ij}. \tag{A.14}$$

With the definition (A.13), the interpolating polynomial, $P_N(x)$, takes the form

$$P_N(x) = \sum_{i=1}^{N+1} f(x_i)\ell_i^{(N)}(x). \tag{A.15}$$

The Lagrange interpolation is, by construction, exact when $f(x)$ is a polynomial of degree $\leq N$. Choosing, as a special case, $f(x) = (x-a)^m$, where a is a constant and $m = 0, 1, \ldots, N$, gives

$$(x-a)^m = \sum_{i=1}^{N+1} (x_i - a)^m \ell_i^{(N)}(x). \tag{A.16}$$

For $m = 0$ we obtain the *first Cauchy relation*

$$1 = \sum_{i=1}^{N+1} \ell_i^{(N)}(x), \tag{A.17}$$

and for $a = x$ the *second Cauchy relation*

$$0 = \sum_{i=1}^{N+1} (x_i - x)^m \ell_i^{(N)}(x). \tag{A.18}$$

The Cauchy relations (A.17) and (A.18) will play an important role in the following paragraphs. As an alternative to the definition (A.13), we can represent the Lagrange polynomials in terms of the *generating polynomial*

$$\Phi_{N+1}(x) := \prod_{i=1}^{N+1} (x - x_i) = (x - x_1)(x - x_2) \cdot \cdots \cdot (x - x_{N+1}). \tag{A.19}$$

Differentiating (A.19) and substituting the result into (A.13) gives the identity

$$\ell_i^{(N)}(x) = \frac{1}{(x - x_i)} \frac{\Phi_{N+1}(x)}{\dot{\Phi}_{N+1}(x_i)}. \tag{A.20}$$

The dot in Eq. (A.19) denotes a differentiation with respect to the independent variable x. Equation (A.20) can be used to derive expressions for the derivatives of the Lagrange polynomials, $\dot{\ell}_i^{(N)}$, at the collocation points x_j:

$$\dot{\ell}_i^{(N)}(x_j) = \begin{cases} \frac{1}{(x_j - x_i)} \frac{\dot{\Phi}_{N+1}(x_j)}{\dot{\Phi}_{N+1}(x_i)} & \text{if} \quad x_i \neq x_j \\ \frac{\ddot{\Phi}_{N+1}(x_i)}{2\dot{\Phi}_{N+1}(x_i)} & \text{if} \quad x_i = x_j \, . \end{cases} \tag{A.21}$$

For our discussion of interpolation errors and Fekete points we will need a representation of the Lagrange polynomials in terms of the Vandermonde matrix. For this, we set $a = 0$ in Eq. (A.16) and obtain a set of $N + 1$ equations – one for each $m = 0, 1, \ldots, N$:

$$\begin{pmatrix} 1 & 1 & \ldots & 1 & 1 \\ x_1 & x_2 & \ldots & x_N & x_{N+1} \\ \vdots & \vdots & \vdots & \vdots & \vdots \\ x_1^{N-1} & x_2^{N-1} & \ldots & x_N^{N-1} & x_{N+1}^{N-1} \\ x_1^N & x_2^N & \ldots & x_N^N & x_{N+1}^N \end{pmatrix} \cdot \begin{pmatrix} \ell_1^{(N)}(x) \\ \ell_2^{(N)}(x) \\ \vdots \\ \ell_N^{(N)}(x) \\ \ell_{N+1}^{(N)}(x) \end{pmatrix} = \begin{pmatrix} 1 \\ x \\ \vdots \\ x^{N-1} \\ x^N \end{pmatrix} . \tag{A.22}$$

Applying Cramer's rule to the linear system (A.22) gives explicit expressions for the Lagrange polynomials:

$$\ell_i^{(N)}(x) = \frac{\det V(x_i = x)}{\det V} \, . \tag{A.23}$$

Having introduced the basic concepts of polynomial interpolation, we will now consider suitable choices of the collocation points x_i.

A.2.3 Lobatto Interpolation

In Sect. A.2.1 we already mentioned that it is desirable to use Fekete points as collocation points for polynomial interpolation. To systematically derive Fekete points, we first have to introduce Lobatto interpolation as a special case of Lagrange interpolation.

Without loss of generality, we restrict our attention to the interval $[-1, 1]$. For a given set of collocation points x_i, with $i = 1, 2, \ldots, N + 1$, the Lagrange polynomials satisfy the cardinal interpolation property $\ell_i^{(N)}(x_j) = \delta_{ij}$. In addition to this mandatory condition we shall now require that the collocation point distribution is such that the $N - 1$ optional conditions

$$\dot{\ell}_i^{(N)}(x_i) = 0 \tag{A.24}$$

are satisfied for $i = 2, 3, \ldots, N$, i.e. for the internal nodes. Property (A.24) ensures that the Lagrange polynomial $\ell_i^{(N)}(x)$ reaches a local maximum value of 1 at the internal collocation point x_i. Interestingly, this additional requirement uniquely specifies the internal node points as the roots of the Lobatto polynomial $Lo_{N-1}(x)$. To prove this statement, we first note that condition (A.24) together with Eq. (A.21) implies that the second derivative of the generating polynomial, $\Phi_{N+1}(x)$ vanishes at the node points x_i, with $i = 2, 3, \ldots, N$:

$$\ddot{\Phi}_{N+1}(x_i) = 0\,, \tag{A.25}$$

Since $\ddot{\Phi}(x)$ is a polynomial of degree $N-1$ that vanishes at the internal node points, just as $\Phi_{N+1}(x)$ itself, we must have

$$\ddot{\Phi}_{N+1}(x) = c\,\frac{\Phi_{N+1}(x)}{(x-1)(x+1)} = -c\,\frac{\Phi_{N+1}(x)}{(1-x^2)}\,, \tag{A.26}$$

where c is a constant. Equation (A.26) holds because a polynomial is uniquely determined by its zeros. The coefficient of the highest power monomial x^{N+1} of Φ_{N+1} is equal to 1. This implies $c = N(N+1)$, and therefore

$$(1-x^2)\ddot{\Phi}_{N+1}(x) + N(N+1)\Phi_{N+1}(x) = 0\,. \tag{A.27}$$

The generating polynomial satisfies the differential equation (A.27). From Eqs. (A.4) and (A.5) we infer that the solution of (A.27) is

$$\Phi_{N+1}(x) = a\,(x^2-1)\,\dot{L}_N(x) = a\,(x^2-1)\,Lo_{N-1}(x)\,, \tag{A.28}$$

where a is a normalisation constant. Thus, the collocation points that satisfy the optional condition (A.24) are the roots of the *completed Lobatto polynomial*

$$Lo^{\mathrm{c}}_{N-1}(x) := (1-x^2)\,Lo_{N-1}(x)\,. \tag{A.29}$$

They are commonly referred to as *Gauss–Lobatto–Legendre points* (GLL points). Figure A.1 shows the Lagrange polynomials of degree 3–6 with the GLL points as collocation points.

Using the GLL points for polynomial interpolation has an important consequence: The absolute values of the corresponding Lagrange polynomials are smaller or equal to 1; in symbols

$$|\ell_i^{(N)}(x)| \le 1\,, \qquad x \in [-1, 1]\,. \tag{A.30}$$

This property is interesting in the context of Runge's phenomenon. Runge's phenomenon, illustrated in Fig. A.2, consist in the overshooting of high-order interpolants near the edges of the interpolation interval when equidistant collocation points are used. This undesirable effect can be avoided with Lobatto interpolation.

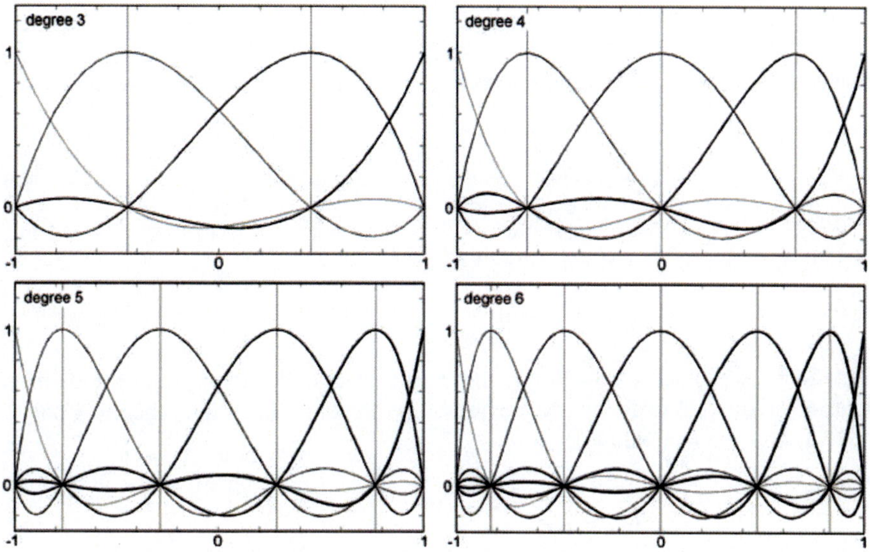

Fig. A.1 The Lagrange polynomials of degree 3–6. The collocation points, indicated by *vertical lines*, are the Gauss–Lobatto–Legendre (GLL) points

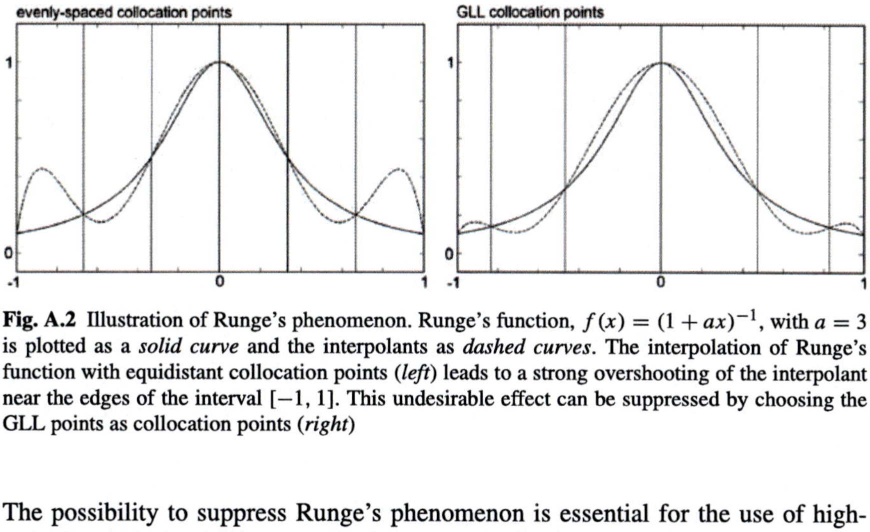

Fig. A.2 Illustration of Runge's phenomenon. Runge's function, $f(x) = (1 + ax)^{-1}$, with $a = 3$ is plotted as a *solid curve* and the interpolants as *dashed curves*. The interpolation of Runge's function with equidistant collocation points (*left*) leads to a strong overshooting of the interpolant near the edges of the interval $[-1, 1]$. This undesirable effect can be suppressed by choosing the GLL points as collocation points (*right*)

The possibility to suppress Runge's phenomenon is essential for the use of high-order interpolation in the spectral-element method.

To prove relation (A.30) we consider the $2N$-degree polynomial

$$Q_{2N}(x) := \left[\ell_1^{(N)}(x)\right]^2 + \left[\ell_2^{(N)}(x)\right]^2 + \cdots + \left[\ell_{N+1}^{(N)}(x)\right]^2 - 1\,. \tag{A.31}$$

The cardinal interpolation property ensures that

$$Q_{2N}(x_i) = 0\,, \qquad i = 1, 2, \ldots, N+1\,. \tag{A.32}$$

Evaluating the derivative of $Q_{2N}(x)$ at the internal collocation points gives

$$\dot{Q}_{2N}(x_j) = 2\sum_{i=1}^{N+1} \dot{\ell}_i^{(N)}(x_j)\ell_i^{(N)}(x_j)\,, \tag{A.33}$$

and in the light of the optional condition (A.24)

$$\dot{Q}_{2N}(x_j) = 0\,, \qquad j = 2, 3, \ldots, N\,. \tag{A.34}$$

Equations (A.32) and (A.34) imply that the internal collocation points are double roots of $Q_{2N}(x)$. We can therefore express $Q_{2N}(x)$ in terms of the generating polynomial, $\Phi_{N+1}(x)$ as follows:

$$Q_{2N}(x) = d\,\frac{\Phi_{N+1}^2(x)}{(x+1)(x-1)} = -d\,\frac{\Phi_{N+1}^2(x)}{(1-x^2)}\,, \tag{A.35}$$

where d is a constant. Substituting (A.28) into (A.35) yields

$$Q_{2N}(x) = -b\,(1-x^2)\,Lo_{N-1}^2(x)\,, \tag{A.36}$$

with a new constant b. To determine b, we evaluate $\dot{Q}_{2N}$ at the collocation point $x = 1$ with the help of Eq. (A.21):

$$\dot{Q}_{2N}(1) = 2\dot{\ell}_{N+1}^{(N)}(1) = \frac{\ddot{\Phi}_{N+1}(1)}{\dot{\Phi}_N(1)}\,. \tag{A.37}$$

Again substituting $a\,(x^2-1)\,Lo_{N-1}(x)$ for $\Phi_{N+1}(x)$ gives

$$\dot{Q}_{2N}(1) = \left.\frac{\partial_x^2[(x^2-1)\,Lo_{N-1}(x)]}{\partial_x[(x^2-1)\,Lo_{N-1}(x)]}\right|_{x=1}\,. \tag{A.38}$$

The differential equation (A.6) satisfied by the Legendre and Lobatto polynomials allows us to simplify (A.38):

$$\dot{Q}_{2N}(1) = \frac{Lo_{N-1}(1)}{L_N(1)}\,. \tag{A.39}$$

The Legendre polynomials are normalised such that $L_N(1) = 1$, and for the Lobatto polynomials evaluated at $x = 1$ we infer from (A.6)

$$\begin{aligned} N(N+1)\,L_N(1) = N(N+1) &= 2x\,Lo_{N-1}(x)|_{x=1} + (x^2-1)\,\partial_x Lo_{N-1}(x)|_{x=1} \\ &= 2Lo_{N-1}(1)\,. \end{aligned} \tag{A.40}$$

Thus, we have

$$\dot{Q}_{2N}(1) = \frac{1}{2} N(N+1)\,. \tag{A.41}$$

Directly differentiating (A.36) yields

$$\dot{Q}_{2N}(1) = 2b\, Lo^2_{N-1}(1) = \frac{1}{2} b N^2 (N+1)^2\,, \tag{A.42}$$

and therefore

$$b = \frac{1}{N(N+1)}\,. \tag{A.43}$$

Combining (A.36) and (A.43) gives the final expression of $Q_{2N}(x)$ in terms of the generating polynomial:

$$Q_{2N}(x) = -\frac{1-x^2}{N(N+1)}\, Lo^2_{N-1}(x)\,. \tag{A.44}$$

Equation (A.44) implies $Q_{2N}(x) \leq 0$ and therefore

$$\sum_{i=1}^{N+1} [\ell_i^{(N)}(x)]^2 \leq 1\,. \tag{A.45}$$

Thus, the relation $|\ell_i^{(N)}(x)| \leq 1$ holds for each individual $\ell_i^{(N)}(x)$.

A.2.4 Fekete Points

Fekete points are the collocation points that maximise the Vandermonde determinant (see Sect. A.2.1). Taking the Fekete points as collocation points ensures that small variations of the $f(x_i)$ – due for example to numerical inaccuracies or measurement errors – result in the smallest possible variations of the interpolated values between the grid points. In this paragraph we demonstrate that the GLL points are the Fekete points of the Vandermonde determinant. The grid points $x_1 = -1$ and $x_{N+1} = 1$ are assumed fixed.

The argument is very simple: Inside the interval $[-1, 1]$ the Lagrange polynomials are smaller than or equal to 1 (relation (A.30)), provided that the internal collocation points are the zeros of $Lo_{N-1}(x)$. The cardinal interpolation property ensures that $\ell_i^{(N)}(x_i) = 1$. Therefore, the Lagrange polynomial $\ell_i^{(N)}(x)$, with i between 2 and $N-1$, reaches a local maximum at $x = x_i$. Thus, we have $\dot{\ell}_i^{(N)}(x_i) = 0$. (This is just the optional condition (A.24).) Using the expression of the Lagrange polynomials in terms of the Vandermonde determinant (Eq. A.23) yields

$$\frac{\mathrm{d}}{\mathrm{d}x}\ell_i^{(N)}(x)|_{x=x_i} = \frac{\mathrm{d}}{\mathrm{d}x}\frac{\det V(x_i = x)}{\det V}\bigg|_{x=x_i} = \frac{1}{\det V}\frac{\mathrm{d}}{\mathrm{d}x_i}\det V = 0\,, \tag{A.46}$$

for $i = 2, 3, \ldots, N-1$, and therefore

$$\frac{\mathrm{d}}{\mathrm{d}x_i}V = 0\,, \qquad i = 2, 3, \ldots, N-1\,. \tag{A.47}$$

Since the internal collocation points are local maxima, we also have

$$0 > \frac{\mathrm{d}^2}{\mathrm{d}x^2}\ell_i^{(N)}(x)|_{x=x_i} = \frac{\mathrm{d}^2}{\mathrm{d}x^2}\frac{\det V(x_i = x)}{\det V}\bigg|_{x=x_i} = \frac{1}{\det V}\frac{\mathrm{d}^2}{\mathrm{d}x_i^2}\det V\,, \tag{A.48}$$

for $i = 2, 3, \ldots, N-1$, and

$$\frac{\mathrm{d}^2}{\mathrm{d}x_i^2}\det V < 0\,, \qquad i = 2, 3, \ldots, N-1\,. \tag{A.49}$$

Relations (A.47) and (A.49) imply that the zeros of $Lo_{N-1}(x)$ are indeed the Fekete points. The maximum of the Vandermonde matrix is global, but a proof of this statement is clearly beyond the scope of this brief overview (see Szegö, 1975).

A.2.5 Interpolation Error

We consider the max norm of the interpolation error $e(x)$:

$$||e(x)||_\infty := \max_{x\in[a,b]} |f(x) - P_N(f, x)|\,. \tag{A.50}$$

In Eq. (A.50) the dependence of the interpolating polynomial P_N on the function $f(x)$ is made explicit through the notation $P_N = P_N(f, x)$. Of all Nth-degree polynomials approximating the function $f(x)$, there is an optimal polynomial $P_N^{\text{opt}}(f, x)$ such that $||e(x)||_\infty$ is minimal. This minimal value of $||e(x)||_\infty$ is the *minimax error*, denoted by ε_N. The optimal polynomial $P_N^{\text{opt}}(f, x)$ is not necessarily an interpolating polynomial. The max norm $||e(x)||_\infty$ is related to the minimax error ε_N through the relation

$$\begin{aligned}||e(x)||_\infty &= ||P_N(f, x) - f(x)||_\infty = ||P_N(f, x) - P_N^{\text{opt}}(f, x) + P_N^{\text{opt}}(f, x) - f(x)||_\infty \\ &\le ||P_N(f, x) - P_N^{\text{opt}}(f, x)||_\infty + ||P_N^{\text{opt}}(f, x) - f(x)||_\infty \\ &= ||P_N(f, x) - P_N^{\text{opt}}(f, x)||_\infty + \varepsilon_N\,. \end{aligned} \tag{A.51}$$

The first summand can be transformed as follows:

$$
\begin{aligned}
||P_N(f,x) - P_N^{\mathrm{opt}}(f,x)||_\infty &= ||P_N(f,x) - P_N(P_N^{\mathrm{opt}},x)||_\infty \\
&\le ||P_N||_\infty ||f(x) - P_N^{\mathrm{opt}}(f,x)||_\infty = ||P_N||_\infty \varepsilon_N \,.
\end{aligned}
\tag{A.52}
$$

In Eq. (A.52) P_N is interpreted as an operator that acts on the function $f(x)$, and $||P_N||_\infty$ is the operator norm of P_N. For a fixed set of collocation points, P_N is linear in $f(x)$ because the polynomial coefficients are linear functions of the data values $f(x_i)$. Combining (A.51) and (A.52) gives

$$
||e(x)||_\infty \le (1 + ||P_N||_\infty)\,\varepsilon_N \,. \tag{A.53}
$$

The max norm of the interpolation operator, $||P_N||_\infty$, can be expressed in terms of the Lagrange polynomials:

$$
\begin{aligned}
||P_N||_\infty &= \sup_f \frac{||P_N(f,x)||_\infty}{||f(x)||_\infty} = \sup_f \frac{||\sum_{i=1}^{N+1} f(x_i)\ell_i^{(N)}(x)||_\infty}{||f(x)||_\infty} \\
&= \sup_f \frac{\max_{x\in[a,b]} |\sum_{i=1}^{N+1} f(x_i)\ell_i^{(N)}(x)|}{\max_{x\in[a,b]} |f(x)|} \le \max_{x\in[a,b]} \left|\sum_{i=1}^{N+1} \ell_i^{(N)}(x)\right| \\
&\le \max_{x\in[a,b]} \sum_{i=1}^{N+1} |\ell_i^{(N)}(x)| \,.
\end{aligned}
\tag{A.54}
$$

The function

$$
L_N(x) := \sum_{i=1}^{N+1} |\ell_i^{(N)}(x)| \tag{A.55}
$$

is the *Lebesgue function* and its max norm, denoted by Λ_N, is the *Lebesgue constant*. Using this terminology, Eq. (A.53) can be transformed to

$$
||e(x)||_\infty \le (1 + \Lambda_N)\,\varepsilon_N \,. \tag{A.56}
$$

Relation (A.56) reveals that there are two separate contributions to the interpolation error: (1) The minimax error ε_N depends only on the smoothness of the function $f(x)$. Smooth functions produce smaller minimax errors than rough functions. We will not further consider the influence of ε_N because we have no freedom of choice concerning the function that we need to interpolate. (2) The Lebesgue constant, Λ_N, depends only on the locations of the collocation points. It is desirable to choose the collocation points such that Λ_N is minimised.

The dependence of the Lebesgue constant on the polynomial order has received much attention in the literature. (See Hesthaven (1998) for a summary of important results.) When the collocation points are equidistant, the associated Lebesgue constant, $\Lambda_N^{\mathrm{equi}}$, can be shown to exhibit the asymptotic behaviour

$$\Lambda_N^{\text{equi}} \sim \frac{2^N}{N \log N}. \tag{A.57}$$

This renders high-order polynomial interpolation with equidistant nodes practically useless. That the Lebesgue constant cannot be made arbitrarily small is the content of Erdös' theorem. It states that

$$\Lambda_N > \frac{2}{\pi} \log(N+1) - c\,, \tag{A.58}$$

where c is a positive constant. This holds for any set of collocation points. A nearly logarithmic growth of Λ_N can be achieved when the roots of the Chebyshev polynomials are used as collocation points. For the GLL points we infer from Eq. (A.30) that

$$\left(L_N^{\text{GLL}}\right)^2 = \left(\sum_{i=1}^{N+1} |\ell_i^{(N)}(x)|\right)^2 \leq \sum_{i=1}^{N+1} |\ell_i^{(N)}(x)|^2 \leq N+1\,, \tag{A.59}$$

and therefore

$$\Lambda_N^{\text{GLL}} \leq \sqrt{N+1}\,. \tag{A.60}$$

The estimate (A.60) seems to be too pessimistic in practice. Based on numerical experiments, Hesthaven (1998) conjectures that Λ_N^{GLL} is bounded as

$$\Lambda_N^{\text{GLL}} \leq \frac{2}{\pi} \log(N+1) + 0.685\,. \tag{A.61}$$

Thus, the Lebesgue constant associated with the GLL points is nearly optimal.

A.3 Numerical Integration

The following paragraphs are concerned with the derivation of numerical quadrature formulas that are used in the context of the spectral-element method. The general strategy is to replace the function that we wish to integrate, $f(x)$, by an interpolating polynomial and to solve the resulting integral analytically.

A.3.1 Exact Numerical Integration and the Gauss Quadrature

Consider the weighted integral of a degree-$(2N+1)$ polynomial Q_{2N+1}

$$\int_a^b w(x) Q_{2N+1}(x)\, \mathrm{d}x\,, \tag{A.62}$$

where $w(x)$ is a positive weighting function. We can write Q_{2N+1} as the sum of a degree-N polynomial $P_N(x)$ that interpolates Q_{2N+1} at the collocation points $x_1, x_2, \ldots, x_{N+1}$ and another polynomial of degree $2N+1$:

$$Q_{2N+1}(x) = P_N(x) + R_N(x)\,(x-x_1)(x-x_2)\cdot\cdots\cdot(x-x_{N+1})\,. \tag{A.63}$$

The polynomial $R_N(x)$ is of degree N. We are interested in the integration error, ε, incurred by replacing the degree-$(2N+1)$ polynomial Q_{2N+1} by the degree-N polynomial $P_N(x)$:

$$\begin{aligned}\varepsilon &= \int_a^b w(x)Q_{2N+1}(x)\,\mathrm{d}x - \int_a^b w(x)P_N(x)\,\mathrm{d}x = \\ &= \int_a^b w(x)R_N(x)\,(x-x_1)(x-x_2)\cdot\cdots\cdot(x-x_{N+1})\,\mathrm{d}x\,. \end{aligned} \tag{A.64}$$

The integration error is equal to zero when the $N+1$ collocation points x_i are the roots of the degree-$(N+1)$ orthogonal polynomial $p_{N+1}(x)$ that corresponds to the integration weight $w(x)$. To prove this assertion, we note that $p_{N+1}(x)$ is proportional to $(x-x_1)(x-x_2)\cdot\cdots\cdot(x-x_{N+1})$ when each x_i is a root of $p_{N+1}(x)$. Therefore, we have

$$(x-x_1)(x-x_2)\cdot\cdots\cdot(x-x_{N+1}) = c\,p_{N+1}(x)\,, \tag{A.65}$$

where c is a constant. The degree-N polynomial $R_N(x)$ can be expressed through the orthogonal polynomials up to degree N:

$$R_N(x) = \sum_{i=1}^{N} c_i\,p_i(x)\,. \tag{A.66}$$

The numbers c_i, with $i = 1, 2, \ldots, N$, are the expansion coefficients. Combining Eqs. (A.64), (A.65) and (A.66) and using the orthogonality of the polynomials $p_i(x)$ yields

$$\varepsilon = \sum_{i=1}^{N} c \int_a^b c_i\,p_i(x)p_{N+1}(x)\,\mathrm{d}x = 0\,. \tag{A.67}$$

Thus, we can integrate a degree-$(2N+1)$ polynomial exactly with only $N+1$ collocation points, given that the collocation points are the roots of the degree-$(N+1)$ orthogonal polynomial that corresponds to the integration weight $w(x)$.

In the case where $f(x)$ is any function, not necessarily a polynomial, we can construct working formulas that approximate the integral. For this, we replace $f(x)$ by its interpolating polynomial

$$P_N(x) = \sum_{i=1}^{N+1} f(x_i)\ell_i^{(N)}(x) \tag{A.68}$$

and introduce this approximation into the weighted integral:

$$\int_a^b w(x) f(x)\,\mathrm{d}x \approx \int_a^b w(x) P_N(x)\,\mathrm{d}x = \sum_{i=1}^{N+1} w_i f(x_i)\,. \tag{A.69}$$

The *integration weights*, w_i, are independent of $f(x)$:

$$w_i = \int_a^b w(x)\ell_i^{(N)}(x)\,\mathrm{d}x\,. \tag{A.70}$$

Even when $f(x)$ is not a polynomial, the collocation points should be the roots of an orthogonal polynomial, simply because $f(x)$ is closer to a degree-$(2N+1)$ polynomial than to a degree-N polynomial. Equation (A.69) is known as *Gauss quadrature rule.*

A.3.2 Gauss–Legendre–Lobatto Quadrature

A disadvantage of the Gauss quadrature – especially in the context of the spectral-element method – is that the roots of orthogonal polynomials are generally located inside the integration interval $[a, b]$ but never directly on its boundaries. Explicitly imposing the requirement that two collocation points coincide with the boundaries leads to Gauss–Lobatto–Legendre (GLL) quadrature formulas.

In the interest of simplicity we consider the integration interval $[-1, 1]$ and the flat weighting function $w(x) = 1$. The integral over a degree-$(2N-1)$ polynomial Q_{2N-1} is then

$$\int_{-1}^{1} Q_{2N-1}(x)\,\mathrm{d}x\,. \tag{A.71}$$

Following the developments of the previous section, we decompose Q_{2N-1} into an interpolating polynomial of degree N and a polynomial of degree $2N-1$:

$$Q_{2N-1}(x) = P_N(x) + R_{N-2}(x)\,(x+1)(x-x_2)\cdot\ldots\cdot(x-x_N)(x-1)\,. \tag{A.72}$$

The collocation points -1 and 1 are now imposed explicitly. For the integration error we find

$$\varepsilon = -\int_{-1}^{1} (1 - x^2)\, R_{N-2}(x)\, (x - x_2)(x - x_3) \cdot \dots \cdot (x - x_N)\, \mathrm{d}x\,. \tag{A.73}$$

To make ε vanish, we need to choose the $N - 1$ internal collocation points $x_2, x_3, \dots, x_N$ such that they are the roots of the degree-$(N-1)$ orthogonal polynomial that corresponds to the integration weight $(1 - x^2)$, i.e. to the Lobatto polynomial Lo_{N-1}. Thus, a polynomial of degree $2N - 1$ can be integrated exactly when replaced by the degree-N interpolating polynomial $P_N(x)$. The collocation points are -1, 1 and the $N - 1$ roots of the Lobatto polynomial Lo_{N-1}.

In the case of an arbitrary function $f(x)$, we approximate the integral over $f(x)$ with the integral over the interpolating polynomial $P_N(x)$:

$$\int_{-1}^{1} f(x)\, \mathrm{d}x \approx \int_{-1}^{1} P_N(x)\, \mathrm{d}x = \sum_{i=1}^{N+1} w_i\, f(x_i)\,, \tag{A.74}$$

The integration weights, w_i, are

$$w_i = \int_{-1}^{1} \ell_i^{(N)}(x)\, \mathrm{d}x\,. \tag{A.75}$$

Equation (A.74) is referred to as *Gauss–Lobatto–Legendre quadrature rule.* Table A.1 summarises the collocation points and integration weights for the GLL quadrature and the polynomial degrees that are most frequently used in spectral-element simulations: 4–7.

Table A.1 Collocation points and integration weights for the GLL quadrature and the polynomial degrees 4–7

Polynomial degree	Collocation points	Integration weights
4	0	0.7111111111
	±0.6546536707	0.5444444444
	±1	0.1
5	±0.2852315164	0.5548583770
	±0.7650553239	0.3784749562
	±1	0.0666666666
6	0	0.4876190476
	±0.4688487934	0.4317453812
	±0.8302238962	0.2768260473
	±1	0.0476190476
7	±0.2092992179	0.4124587946
	±0.5917001814	0.3411226924
	±0.8717401485	0.2107042271
	±1	0.0357142857

Appendix B
Time–Frequency Transformations

We briefly review some of the principal definitions and results from time–frequency analysis. The emphasis will be on the definition of time–frequency transforms and the derivation of their inverses. For an excellent overview of time–frequency analysis, the reader is referred to Strang & Nguyen (1996). Throughout the text we work with the following definition of the Fourier transform of a function f:

$$\tilde{f}(\omega) := F[f](\omega) := \frac{1}{\sqrt{2\pi}} \int_{\mathbb{R}} f(t) \mathrm{e}^{-\mathrm{i}\omega t} \, \mathrm{d}t. \tag{B.1}$$

where $\mathbf{i} = \sqrt{-1}$ denotes the imaginary unit. The inverse Fourier transform corresponding to (B.1) is

$$f(t) = F^{-1}[\tilde{f}](t) = \frac{1}{\sqrt{2\pi}} \int_{\mathbb{R}} \tilde{f}(\omega) \mathrm{e}^{\mathrm{i}\omega t} \, \mathrm{d}\omega. \tag{B.2}$$

For any two Fourier-integrable functions f and g we obtain *Parseval's relation* by combining Eqs. (B.1) and (B.2):

$$\begin{aligned}
(\tilde{f}, \tilde{g}) &= \int_{\mathbb{R}} \tilde{f}(\omega) \tilde{g}^*(\omega) \, \mathrm{d}\omega = \frac{1}{2\pi} \int_{\mathbb{R}} \left[\int_{\mathbb{R}} f(t) \mathrm{e}^{-\mathrm{i}\omega t} \, \mathrm{d}t \right] \left[\int_{\mathbb{R}} g^*(\tau) \mathrm{e}^{\mathrm{i}\omega\tau} \, \mathrm{d}\tau \right] \mathrm{d}\omega \\
&= \frac{1}{2\pi} \int_{\mathbb{R}} \int_{\mathbb{R}} f(t) g^*(\tau) \int_{\mathbb{R}} \mathrm{e}^{\mathrm{i}\omega(\tau - t)} \, \mathrm{d}\omega \, \mathrm{d}t \, \mathrm{d}\tau \\
&= \int_{\mathbb{R}} \int_{\mathbb{R}} f(t) g^*(\tau) \, \delta(\tau - t) \, \mathrm{d}\tau \, \mathrm{d}t = \int_{\mathbb{R}} f(t) g^*(t) \, \mathrm{d}t = (f, g) \, .
\end{aligned} \tag{B.3}$$

The symbol g^* denotes the complex conjugate of g. Parseval's relation states that the scalar product of f and g in the time domain is equal to the scalar product of $\tilde{f}$ and $\tilde{g}$ in the frequency domain. *Plancherel's formula* follows immediately by setting $f = g$:

$$||\tilde{f}||_2 = ||f||_2 \, . \tag{B.4}$$

The Fourier transform has zero time resolution. This means, in a seismological context, that the spectrum $\tilde{f}(\omega)$ of a seismogram $f(t)$ does not contain any information about the time localisation of different waveforms (e.g. body and surface waveforms) and their respective frequency content. To overcome this deficiency we define the *windowed Fourier transform* of a function f in terms of the regular Fourier transform of $f(\tau)h^*(\tau - t)$, where h is a sliding window:

$$\tilde{f}_h(t,\omega) := F_h[f](t,\omega) := \frac{1}{\sqrt{2\pi}\,||h||_2}\int_{\mathbb{R}} f(\tau)h^*(\tau - t)\mathrm{e}^{-\mathrm{i}\omega\tau}\,\mathrm{d}\tau. \qquad \text{(B.5)}$$

The energy of the window h,

$$||h||_2 = \sqrt{\int_{\mathbb{R}} h(t)\,h^*(t)\,\mathrm{d}t}, \qquad \text{(B.6)}$$

is assumed to be finite. We note that the use of h^* instead of h in the definition (B.5) is a common convention in time frequency analysis. The windowed Fourier transform has non-zero time resolution because it provides the spectrum of individual waveforms that have been isolated by the multiplication of the seismogram f with the sliding window function h. By defining a time shift operator T_{t} through $T_{\mathrm{t}}h(\tau) = h(\tau - t)$, we can express F_h in terms of F:

$$F_h[f](t,\omega) = \frac{1}{||h||_2}\,F[fT_{\mathrm{t}}h^*](\omega)\,. \qquad \text{(B.7)}$$

Making use of Parseval's relation for the Fourier transform (B.3) we can then derive a similar result for the windowed Fourier transform:

$$\begin{aligned}(\tilde{f}_h,\tilde{g}_h) &= \int_{\mathbb{R}^2} F_h[f](t,\omega)F_h^*[g](t,\omega)\,\mathrm{d}t\,\mathrm{d}\omega \\ &= \frac{1}{||h||_2^2}\int_{\mathbb{R}^2} F[fT_{\mathrm{t}}h^*](\omega)F^*[gT_{\mathrm{t}}h^*](\omega)\,\mathrm{d}t\,\mathrm{d}\omega \\ &= \frac{1}{||h||_2^2}\int_{\mathbb{R}^2} f(\tau)h^*(\tau - t)g^*(\tau)h(\tau - t)\,\mathrm{d}\tau,\,\mathrm{d}t \\ &= \frac{1}{||h||_2^2}\int_{\mathbb{R}^2} f(\tau)g^*(\tau)|h(\tau - t)|^2\,\mathrm{d}t\,\mathrm{d}\tau = \int_{\mathbb{R}} f(\tau)g^*(\tau)\,\mathrm{d}\tau = (f,g)\,.\end{aligned} \qquad \text{(B.8)}$$

Setting $f = g$ gives an analogue of Plancherel's formula: $||\tilde{f}_h||_2 = ||f||_2$. From (B.8) we can derive an expression for the inverse of the windowed Fourier transform. For this, we write

$$\int_{\mathbb{R}} f(t)g^*(t)\,\mathrm{d}t = \int_{\mathbb{R}^2} F_h[f](\tau,\omega)F_h^*[g](\tau,\omega)\,\mathrm{d}\tau\,\mathrm{d}\omega$$
$$= \frac{1}{\sqrt{2\pi}\,||h||_2}\int_{\mathbb{R}^2} F_h[f](\tau,\omega)\int_{\mathbb{R}} g^*(t)h(t-\tau)\mathrm{e}^{\mathrm{i}\omega t}\,\mathrm{d}t\,\mathrm{d}\tau\,\mathrm{d}\omega$$
$$= \frac{1}{\sqrt{2\pi}\,||h||_2}\int_{\mathbb{R}}\left[\int_{\mathbb{R}^2} F_h[f](\tau,\omega)h(t-\tau)\mathrm{e}^{\mathrm{i}\omega t}\,\mathrm{d}\tau\,\mathrm{d}\omega\right]g^*(t)\,\mathrm{d}t\,. \tag{B.9}$$

Since g can be any function, we deduce that the inverse F_h^{-1} of F_h is given by

$$f(t) = F_h^{-1}[\tilde{f}_h](t) = \frac{1}{\sqrt{2\pi}\,||h||_2}\int_{\mathbb{R}^2} F_h[f](\tau,\omega)h(t-\tau)\mathrm{e}^{\mathrm{i}\omega t}\,\mathrm{d}\omega\,\mathrm{d}\tau\,. \tag{B.10}$$

Computing the inverse time–frequency transform therefore requires an integration over both time and frequency. Another important result can be derived by invoking Parseval's relation for the regular Fourier transform: Defining $g_{\omega,t}(\tau) := h(\tau - t)\,\mathrm{e}^{\mathrm{i}\omega\tau}$ gives

$$F_h[f](t,\omega) = \frac{1}{\sqrt{2\pi}\,||h||_2}\int_{\mathbb{R}} f(\tau)g^*_{\omega,t}(\tau)\,\mathrm{d}\tau = \frac{1}{\sqrt{2\pi}\,||h||_2}\int_{\mathbb{R}} \tilde{f}(\nu)\tilde{g}^*_{\omega,t}(\nu)\,\mathrm{d}\nu. \tag{B.11}$$

To interpret Eq. (B.11) we consider a fixed point (ω, t) in the time–frequency space. The time–frequency representation $F_h[f]$ can be computed in two complementary ways: First, by integrating over the time representation $f(\tau)$ multiplied by the time window $g^*_{\omega,t}(\tau)$ shifted by t or, second, by integrating over the frequency representation $\tilde{f}(\nu)$ multiplied by the frequency window $\tilde{g}^*_{\omega,t}(\nu)$ shifted by ω. A narrow time window, i.e. a high time resolution, generally leads to a broad frequency window and therefore to a low frequency resolution. Analogously, a low time resolution results in a high frequency resolution. This trade-off depends strongly on the choice of the sliding time window h in the definition of F_h (B.5), and it can be quantified with the well-known *uncertainty principle* that we state here without proof:

$$\Delta_{\tilde{g}}\Delta_g \geq \frac{1}{2}||g_{\omega,t}||_2 = \frac{1}{2}||h||_2\,, \tag{B.12}$$

The symbols $\Delta_{\tilde{g}}$ and Δ_g denote the variances of the frequency window $\tilde{g}_{\omega,t}$ and the time window $g_{\omega,t}$, respectively. Figure B.1 illustrates the trade-off between time and frequency resolution. An effective window function h should allow us to increase the time resolution (reduce Δ_g) while reducing the frequency resolution (increase $\Delta_{\tilde{g}}$) as little as possible. The optimal choice for h is the Gaussian

$$h_\sigma(t) = (\pi\sigma^2)^{-1/4}\mathrm{e}^{-x^2/2\sigma^2}, \tag{B.13}$$

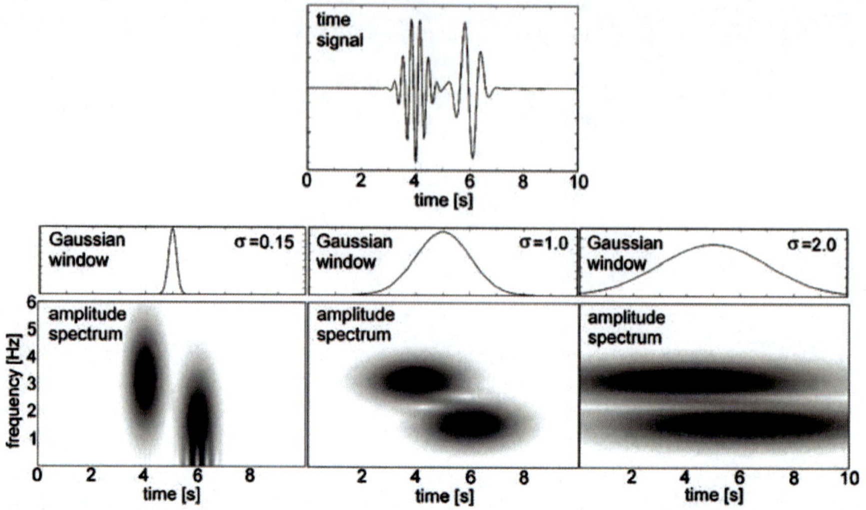

Fig. B.1 Time–frequency analysis of the time-domain signal shown in the *top panel*. The time and frequency resolution depends on the width σ of the Gaussian window. Narrow windows (*left*) generate a high time resolution at the expense of a small frequency resolution. Wide windows improve the frequency resolution at the expense of the time resolution (*right*)

which generates an equal sign in the uncertainty principle (B.12). The L_2 norm of h_σ is 1, i.e. $||h_\sigma||_2 = 1$ for all $\sigma > 0$. We omit the lengthy but straightforward derivations of these results that may be found in Strang & Nguyen (1996). For $h = h_\sigma$ the windowed Fourier transform is termed the *Gabor transform*. In order to simplify the notation, we introduce the symbolisms $F_{h_\sigma} = G$ and $G[f](t,\omega) = \tilde{f}(t,\omega)$. The dependence of G and $\tilde{f}(t,\omega)$ on σ is implicit.

We close this short review with the expressions for the Gabor transform pair because of its outstanding importance for the analysis in Sect. 11.5:

$$\tilde{f}(t,\omega) = G[f](t,\omega) = \frac{1}{\sqrt{2\pi}} \int_{\mathbb{R}} f(\tau) h_\sigma^*(\tau - t) \mathrm{e}^{-\mathrm{i}\omega\tau} \, \mathrm{d}\tau \,, \tag{B.14}$$

$$f(t) = G^{-1}[\tilde{f}](t) = \frac{1}{\sqrt{2\pi}} \int_{\mathbb{R}^2} \tilde{f}(\tau,\omega) h_\sigma(t - \tau) \mathrm{e}^{\mathrm{i}\omega t} \, \mathrm{d}\omega \, \mathrm{d}\tau \,. \tag{B.15}$$

References

Abarbanel, S., Gottlieb, D.: A mathematical analysis of the PML method. J. Comput. Phys. **134**, 357–363 (1997)

Abdulah, A.: Seismic body wave attenuation tomography beneath the Australasian region. PhD Thesis, The Australian National University (2007)

Aki, K., Christoffersson, A., Husebye, E. S.: Determination of the three-dimensional seismic structure of the lithosphere. J. Geophys. Res. **82**, 277–296 (1977)

Aki, K., Richards, P. G.: Quantitative Seismology, 2nd edn. University Science Books, Herndon, VA (2002)

Alford, R. M., Kelly, K. R., Boore, D. M.: Accuracy of finite difference modeling of the acoustic wave equation. Geophysics **39**, 834–842 (1974)

Allaire, G.: Numerical analysis and optimization. Oxford University Press, Oxford (2007)

Alterman, Z., Karal, F. C.: Propagation of elastic waves in layered media by finite-difference methods. Bull. Seism. Soc. Am. **58**, 367–398 (1968)

Anderson, D. L., Archambeau, C.: Anelasticity of Earth. J. Geophys. Res. **69**, 2071–2084 (1964)

Askan, A., Akcelik, V., Bielak, J., Ghattas, O.: Full waveform inversion for seismic velocity and anelastic losses in heterogeneous media. Bull. Seism. Soc. Am. **97**, 1990–2008 (2007)

Askan, A., Bielak, J.: Full anelastic waveform tomography including model uncertainty. Bull. Seism. Soc. Am. **98**, 2975–2989 (2008)

Babuska, V., Cara, M.: Seismic anisotropy in the Earth. Kluwer Academic Publishers, Dordrecht (1991)

Backus, G. E.: Long-wave elastic anisotropy produced by horizontal layering. J. Geophys. Res. **67**(11), 4427–4440 (1962)

Baig, A. M., Dahlen, F. A., Hung, S.-H.: Traveltimes of waves in three-dimensional random media. Geophys. J. Int. **153**, 467–482 (2003)

Bamberger, A., Chavent, G., Lailly, P.: Une application de la théorie du contrôle à un problème inverse sismique. Ann. Geophys. **33**, 183–200 (1977)

Bamberger, A., Chavent, G., Lailly, P.: About the stability of the inverse problem in 1-D wave equations – application to the interpretation of seismic profiles. Appl. Math. Optm. **5**, 1–47 (1979)

Bamberger, A., Chavent, G., Hemons, Ch., Lailly, P.: Inversion of normal incidence seismograms. Geophysics **47**(5), 757–770 (1982)

Bao, H., Bielak, J., Ghattas, O., Kallivokas, L. F., O'Hallaron, D. R., Shewchuk, J. R., Xu, J.: Large-scale simulation of elastic wave propagation in heterogeneous media on parallel computers. Comput. Methods Appl. Mech. Eng. **152**, 85–102 (1998)

Bassin, C., Laske, G., Masters, G.: The current limits of resolution for surface wave tomography in North America. EOS Trans. AGU **81**, F897 (2000)

Båth, M.: Introduction to seismology, 2nd edn. Birkhäuser, Basel, Boston, MA, Stuttgart (1979)

Bécache, E., Fauqueux, S., Joly, P.: Stability of perfectly matched layers, group velocities and anisotropic waves. J. Comput. Phys. **188**, 399–433 (2003)

Ben Hadj Ali, H., Operto, S., Virieux, J., Sourbier, F.: Efficient 3D frequency-domain full waveform inversion with phase encodings. In: 71st Conference and Technical Exhibition, EAGE, Extended Abstracts, 5812 (2009a)
Ben Hadj Ali, H., Operto, S., Virieux, J., Sourbier, F.: Three-dimensional frequency-domain full waveform inversion with phase encoding. In 79th SEG Meeting, Extended Abstracts, 2288–2292 (2009b)
Bérenger, J.-P.: A perfectly matched layer for the absorption of electromagnetic waves. J. Comput. Phys. **114**, 185–200 (1994)
Blanch, J. O., Robertsson, J. O. A., Symes, W. W.: Modelling of a constant Q: methodology and algorithm for an efficient and optimally inexpensive viscoelastic technique. Geophysics **60**(1), 176–184 (1995)
Bleibinhaus, F., Hole, J. A., Ryberg, T., Fuis, G. S.: Structure of the California Coast Ranges and San Andreas Fault at SAFOD from seismic waveform inversion and reflection imaging. J. Geophys. Res. **112**, doi:10.1029/2006JB004611 (2007)
Bleibinhaus, F., Lester, R. W., Hole, J. A.: Applying waveform inversion to wide-angle seismic surveys. Tectonophysics **472**, 238–248 (2009)
Bleibinhaus, F., Rondenay, S.: Effects of surface scattering in full waveform inversion. Geophysics **74**, WCC87–WCC95 (2009)
Bohlen, T.: Parallel 3-D viscoelastic finite difference modelling. Comput. Geosci. **28**, 887–899 (2002)
Bolt, B. A.: Velocity of the seismic waves Lg and Rg across Australia. Nature **180**, 495 (1958)
Bolt, B. A., Doyle, H. A., Sutton, D. J.: Seismic observations from the 1956 atomic explosions in Australia. Geophys. J. R. Astron. Soc. **1** (1958)
Bolt, B. A., Niazi, M.: Dispersion of Rayleigh waves across Australia. Geophys. J. R. Astron. Soc. **9**, 21–35 (1964)
Boore, D. M.: Love waves in nonuniform waveguides: finite difference calculations. J. Geophys. Res. **75**, 1512–1527 (1970)
Boore, D. M.: Finite-difference methods for seismic wave propagation in heterogeneous materials. In: Methods in Computational Physics, vol. **11**, Academic Press, New York, NY (1972)
Boschi, L.: Global multiresolution models of surface wave propagation: comparing equivalently regularized Born and ray theoretical solutions. Geophys. J. Int. **167**, 238–252 (2006)
Bozdağ, E., Trampert, J.: On crustal corrections in surface wave tomography. Geophys. J. Int. **172**, 1066–1082 (2008)
Bozdağ, E., Trampert, J.: Misfit functions for full waveform inversion based on instantaneous phase and envelope measurements. Geophys. J. Int., in press (2010)
Brenders, A. J., Pratt, R. G.: Full waveform tomography for lithospheric imaging: results from a blind test in a realistic crustal model. Geophys. J. Int. **168**, 133–151 (2007)
Brenders, A. J., Pratt, R. G.: Efficient waveform tomography for lithospheric imaging: implications for realistic, two-dimensional acquisition geometries and low-frequency data. Geophys. J. Int. **168**, 152–170 (2007)
Brossier, R., Operto, S., Virieux, J.: Seismic imaging of complex onshore structures by 2D elastic frequency-domain full waveform inversion. Geophysics **74**, WCC105–WCC118 (2009)
Brossier, R., Operto, S., Virieux, J.: Robust elastic frequency-domain full waveform inversion using the l-1 norm. Geophys. Res. Lett. **36**, L20310 (2009)
Brossier, R., Operto, S., Virieux, J.: Which data residual norm for robust elastic frequency-domain full waveform inversion? Geophysics **75**, R37–R46 (2010)
Bullen, K. E., Bolt, B.: An introduction to the theory of seismology, 4th edn. Cambridge University Press, Cambridge (1985)
Bunge, H.-P., Davies, J. H.: Tomographic images of a mantle circulation model. Geophys. Res. Lett. **28**, 77–80 (2001)
Bunks, C., Saleck, F. M., Zaleski, S., Chavent, G.: Multiscale seismic waveform inversion. Geophysics **60**, 1457–1473 (1995)

Burstedde, C., Ghattas, O.: Algorithmic strategies for full waveform inversion: 1D experiments. Geophysics **74**, WCC37–WCC46 (2009)
Capdeville, Y., Chaljub, E., Vilotte, J. P., Montagner, J.-P.: Coupling the spectral-element method with a modal solution for elastic wave propagation in global earth models. Geophys. J. Int. **152**, 34–67 (2003a)
Capdeville, Y., Romanowicz, B., To, A.: Coupling spectral elements and modes in a spherical Earth: an extension to the "sandwich" case. Geophys. J. Int. **154**, 44–57 (2003b)
Capdeville, Y., Gung, Y., Romanowicz, B.: Towards global earth tomography using the spectral element method: a technique based on source stacking. Geophys. J. Int. **162**, 541–554 (2005)
Capdeville, Y., Marigo, J.-J.: Second-order homogenisation of the elastic wave equation for non-periodic layered media. Geophys. J. Int. **170**, 823–838 (2007)
Capdeville, Y., Marigo, J. J.: Shallow layer correction for spectral element like methods. Geophys. J. Int. **172**, 1135–1150 (2008)
Capdeville, Y.: Contributions aux problès direct et inverse en sismologie. Thesis, Université Paris 7 (2010)
Capdeville, Y., Guillot, L., Marigo, J.-J.: 1-D non-periodic homogenisation for the seismic wave equation. Geophys. J. Int. **181**, 897–910 (2010a)
Capdeville, Y., Guillot, L., Marigo, J.-J.: 2-D non-periodic homogenisation to upscale elastic media for P-SV waves. Geophys. J. Int. **182**, 903–922 (2010b)
Cara, M.: Filtering of dispersed wave trains. Geophys. J. R. astron. Soc. **33**, 65–80 (1973)
Cara, M., Lévêque, J. J.: Waveform inversion using secondary observables. Geophys. Res. Lett. **14**(10), 1046–1049 (1987)
Cerjan, C., Kosloff, D., Kosloff, R., Reshef, M.: A nonreflecting boundary condition for discrete acoustic and elastic wave equations. Geophysics **50**, 705–708 (1985)
Chaljub, E., Capdeville, Y., Vilotte, J. P.: Solving elastodynamics in a fluid-solid heterogeneous sphere: a parallel spectral-element approximation on non-conforming grids. J. Comput. Phys. **187**, 457–491 (2003)
Chaljub, E., Valette, B.: Spectral element modelling of three-dimensional wave propagation in a self-gravitating Earth with an arbitrarily stratified outer core. Geophys. J. Int. **158**, 131–141 (2004)
Chaljub, E., Komatitsch, D., Capdeville, Y., Vilotte, J. P., Valette, B., Festa, G.: Spectral-element analysis in seismology. In: Advances in Geophysics, vol. **48**, 365–419. Editors: Wu, R.-S., Maupin, V. Elsevier, Amsterdam (2007)
Charney, J., Frjørtoft, R., von Neumann, J.: Numerical integration of the barotropic vorticity equation. Tellus **2**, 237–254 (1950)
Charpentier, I.: Checkpointing schemes for adjoint codes: application to the meteorological model Meso-NH. SIAM J. Sci. Comput. **22**(6), 2135–2151 (2001)
Chen, P., Zhao, L., Jordan, T. H.: Full 3D tomography for the crustal structure of the Los Angeles region. Bull. Seism. Soc. Am. **97**(4), 1094–1120 (2007)
Cheng, H.-X., Kennett, B. L. N.: Frequency dependence of seismic wave attenuation in the upper mantle beneath the Australian region. Geophys. J. Int. **150**, 45–57 (2002)
Choi, Y., Min, D.-J., Shin, C.: Two-dimensional waveform inversion of multi-component data in acoustic coupled media. Geophys. Prosp. **56**, 863–881 (2008)
Červený, V.: Seismic Ray Theory. Cambridge University Press, Cambridge (2001)
Chevrot, S., van der Hilst, R. D.: The Poisson ratio of the Australian crust: geological and geophysical implications. Earth Planet. Sci. Lett. **183**, 121–132 (2000)
Claerbout, J. F., Johnson, A. G.: Extrapolation of time dependent waveforms along their path of propagation. Geophys. J. **26**, 285–295 (1971)
Claerbout, J. F.: Fundamentals of geophysical data processing. McGraw-Hill, New York, NY (1976)
Clayton, R. W., Engquist, B.: Absorbing boundary conditions for acoustic and elastic wave equations. Bull. Seism. Soc. Am. **67**, 1529–1540 (1977)

Clayton, R. W., Engquist, B.: Absorbing boundary conditions for wave-equation migration. Geophysics **45**, 895–904 (1980)

Cleary, J. R.: P times to Australian stations from nuclear explosions. Bull. Seism. Soc. Am. **57**, 773–781 (1967)

Cleary, J. R., Simpson, D. W., Muirhead, K. J.: Variations in the Australian upper mantle structure, from observations of the Cannikin explosion. Nature **236**, 111–112 (1972)

Clitheroe, G., Gudmundsson, O., Kennett, B. L. N.: The crustal thickness of Australia. J. Geophys. Res. **105**, 13697–13713 (2000)

Cole, J. B.: A nearly exact second-order finite-difference time-domain wave propagation algorithm on a coarse grid. Comput. Phys. **8**, 730–734 (1994)

Cole, J. B.: High accuracy solution of Maxwell's equations using nonstandard finite differences. Comput. Phys. **11**, 287–292 (1997)

Collino, F.: High-order absorbing boundary conditions for wave propagation models. Straight line boundary and corner cases. In: Proceedings of the Second International Conference on Mathematical and Numerical Aspects of Wave Propagation, SIAM, Delaware, 161–171 (1993)

Collino, F., Tsogka, C.: Application of the perfectly matched absorbing layer model to the linear elastodynamic problem in anisotropic heterogeneous media. Geophysics **66**(1), 294–307 (2001)

Cormack, A. M.: Representation of a function by its line integrals, with some radiological applications. J. Appl. Phys. **34**(9), 2722–2727 (1963)

Courant, R., Friedrichs, K., Lewy, H.: Über die partiellen Differenzengleichungen der mathematischen Physik. Mathematische Annalen **100**, 32–74 (1928)

Crank, J., Nicolson, P.: A practical method for numerical evaluation of solutions of partial differential equations of heat conduction type. Proc. Camb. Philos. Soc. **43**, 50–67 (1947)

Crase, E., Pica, A., Noble, M., McDonald, J., Tarantola, A.: Robust elastic nonlinear waveform inversion – Application to real data. Geophysics **55**(5), 527–538 (1990)

Cummins, P., Geller, R., Hatori, T., Takeuchi, N.: DSM synthetic seismograms: SH, spherically symmetric case. Geophys. Res. Lett. **21**, 533–536 (1994a)

Cummins, P., Geller, R., Takeuchi, N.: DSM complete synthetic seismograms: P-SV, spherically symmetric case. Geophys. Res. Lett. **21**, 1663–1666 (1994b)

Dablain, M. A.: The application of high-order differencing to the scalar wave equation. Geophysics **51**, 54–66 (1986)

Dahlen, F. A., Hung, S.-H., Nolet, G.: Fréchet kernels for finite-frequency traveltimes – I. Theory. Geophys. J. Int. **141**, 157–174 (2000)

Dahlen, F. A., Tromp, J.: Theoretical global seismology. Princeton University Press, Princeton, NJ (2001)

Dahlen, F. A., Baig, A. M.: Fréchet kernels for body-wave amplitudes. Geophys. J. Int. **150**, 440–466 (2002)

Dai, N. X., Vafidis, A., Kanasewich, E. R.: Seismic migration and absorbing boundaries with a one-way wave system for heterogeneous media. Geophys. Prosp. **44**, 719–739 (1996)

De Basabe, J. D., Sen, M. K.: Grid dispersion and stability criteria of some common finite-element methods for acoustic and elastic wave equations. Geophysics **72**, T81–T95 (2007)

Debayle, E., Kennett, B. L. N.: The Australian continental upper mantle: structure and deformation inferred from surface waves. J. Geophys. Res. **105**, 25423–25450 (2000)

Debayle, E., Sambridge, M.: Inversion of massive surface wave data sets: model construction and resolution assessment. J. Geophys. Res. **109**, B02316 (2004)

de Hoop, A. T.: Representation theorems for the displacement in an elastic solid and their applications to elastodynamic diffraction theory. D.Sc Thesis, Technische Hogeschool Delft (1958)

de Jersey, N. J.: Seismological evidence bearing on the crustal thickness in the South-West Pacific. University of Queensland, Brisbane, Paper v. 3, no. 2 (1946)

de la Puente, J., Käser, M., Dumbser, M., Igel, H.: An arbitrary high-order discontinuous Galerkin method for elastic waves on unstructured meshes – IV. Anisotropy. Geophys. J. Int. **169**, 1210–1228 (2007)

de la Puente, J., Dumbser, M., Käser, M., Igel, H.: Discontinuous Galerkin methods for wave propagation in poroelastic media. Geophysics **73**, T77–T97 (2008)
de la Puente, J., Ampuero, J.-P., Käser, M.: Dynamic rupture modeling on unstructured meshes using a discontinuous Galerkin method. J. Geophys. Res. **114**, doi:10.1029/2008JB006271 (2009)
Deng, H. L.: A complexity analysis of generic optimization problems: characterizing the topography of high-dimensional functions. Ph.D. Thesis, Center for Wave Phenomena, Colorado School of Mines (1997)
Dessa, J. X., Pascal, G.: Combined traveltime and frequency-domain seismic waveform inversions: a case of study on multioffset ultrasonic data. Geophys. J. Int. **154**, 117–133 (2003)
Dessa, J. X., Operto, S., Kodaira, S., Nakanishi, A., Pascal, G., Virieux, J., Kaneda, Y.: Multiscale seismic imaging of the eastern Nankai trough by full waveform inversion. Geophys. Res. Lett. **31**, doi:10.1029/2004GL020453 (2004)
Devaney, A. J.: Geophysical diffraction tomography. IEEE Trans. Geosci. Remote Sens. **22**(1), 3–13 (1984)
Djikpéssé, H., Tarantola, A.: Multiparameter ℓ_1-norm waveform fitting: interpretation of Gulf of Mexico reflection seismograms. Geophysics **64**, 1023–1035 (1999)
Dumbser, M., Käser, M.: An arbitrary high-order discontinuous Galerkin method for elastic waves on unstructured meshes – II. The three-dimensional isotropic case. Geophys. J. Int. **167**, 319–336 (2006)
Dumbser, M., Käser, M., de la Puente, J.: Arbitrary high-order finite volume schemes for seismic wave propagation on unstructured meshes in 2D and 3D. Geophys. J. Int. **171**(2), 665–694 (2007)
Dziewonski, A. M., Mills, J, Bloch, S.: Residual dispersion measurement – a new method of surface wave analysis. Bull. Seism. Soc. Am. **62**, 129–139 (1972)
Dziewonski, A. M., Hales, A. L., Lapwood, E. R.: Parametrically simple Earth models consistent with geophysical data. Phys. Earth Planet. Int. **10**, 12–48 (1975)
Dziewonski, A. M., Hager, B. H., O'Connell, R. J.: Large-scale heterogeneities in the lower mantle. J. Geophys. Res. **82**, 239–255 (1977)
Dziewonski, A. M., Anderson, D. L.: Preliminary reference Earth model. Phys. Earth Planet. Int. **25**, 297–356 (1981)
Emerman, S. H., Stephen, R. A.: Comment on 'Absorbing boundary conditions for acoustic and elastic wave equations' by R. Clayton and B. Engquist. Bull. Seism. Soc. Am. **73**, 661–665 (1983)
Engquist, B., Majda, A.: Absorbing boundary conditions for the numerical simulation of waves. Math. Comput. **31**, 629–651 (1977)
Epanomeritakis, I., Akcelik, V., Ghattas, O., Bielak, J.: A Newton-CG method for large-scale three-dimensional elastic full waveform seismic inversion. Inverse Probl. **24**, doi:10.1088/0266-5611/24/3/034015 (2008)
Faccioli, E., Maggio, F., Paolucci, R., Quarteroni, A.: 2D and 3D elastic wave propagation by a spseudo-spectral domain decomposition method. J. Seismol. **1**, 237–251 (1997)
Festa, G., Nielsen, S.: PML absorbing boundaries. Bull. Seism. Soc. Am. **93**, 891–903 (2003)
Festa, G., Vilotte, J.-P.: The Newmark scheme as velocity-stress time-staggering: an efficient PML implementation for spectral element simulations for elastodynamics. Geophys. J. Int. **161**, 789–812 (2005)
Festa, G., Delavaud, E., Vilotte, J.-P.: Interaction between surface waves and absorbing boundaries for wave propagation in geological basins: 2D numerical simulations. Geophys. Res. Lett. **32**, doi:10.1029/2005GL024091 (2005)
Flanagan, M. P., Wiens, D. A.: Attenuation of broadband P and S waves in Tonga: observation of frequency-dependent Q. Pure Appl. Geophys. **153**, 345–375 (1998)
Fichtner, A., Bunge, H.-P., Igel, H.: The adjoint method in seismology – I. Theory. Phys. Earth Planet. Int. **157**, 86–104 (2006a)

Fichtner, A., Bunge, H.-P., Igel, H.: The adjoint method in seismology – II. Applications: traveltimes and sensitivity functionals. Phys. Earth Planet. Int. **157**, 105–123 (2006b)

Fichtner, A., Igel, H.: Efficient numerical surface wave propagation through the optimization of discrete crustal models – a technique based on non-linear dispersion curve matching (DCM). Geophys. J. Int. **173**, 519–533 (2008)

Fichtner, A., Kennett, B. L. N., Igel, H., Bunge, H.-P.: Theoretical background for continental- and global-scale full-waveform inversion in the time-frequency domain. Geophys. J. Int. **175**, 665–685 (2008)

Fichtner, A., Kennett, B. L. N., Igel, H., Bunge, H.-P.: Spectral-element simulation and inversion of seismic waves in a spherical section of the Earth. J. Numer. Anal. Ind. Appl. Math. **4**, 11–22 (2009)

Fichtner, A., Kennett, B. L. N., Igel, H., Bunge, H.-P.: Full seismic waveform tomography for upper-mantle structure in the Australasian region using adjoint methods. Geophys. J. Int. **179**, 1703–1725 (2009)

Fichtner, A., Kennett, B. L. N., Igel, H., Bunge, H.-P.: Full waveform tomography for radially anisotropic structure: new insights into present and past states of the Australasian upper mantle. Earth Planet. Sci. Lett. **290**, 270–280 (2010)

Fink, M.: Time-reversed acoustics. Sci. Am. **281**, 91–97 (1999)

Fishwick, S., Kennett, B. L. N., Reading, A. M.: Contrasts in lithospheric structure within the Australian craton – insights from surface wave tomography. Earth Planet. Sci. Lett. **231**, 163–176 (2005)

Fishwick, S., Reading, A. M.: Anomalous lithosphere beneath the Proterozoic of western and central Australia: a record of continental collision and intraplate deformation? Precambrian Research **166**, 111–121 (2008)

Fishwick, S., Heintz M., Kennett, B. L. N., Reading, A. M., Yoshizawa, K.: Steps in lithospheric thickness within eastern Australia, evidence from surface wave tomography. Tectonics **27**, TC0049, doi:10.129/2007TC002116 (2008)

Flanagan, M. P., Wiens, D. A.: Attenuation of broadband P and S waves in Tonga: observation of frequency-dependent Q. Pure Appl. Geophys. **153**, 345–375 (1998)

Fletcher, R., Reeves, C. M.: Function minimization by conjugate gradients. Comput. J. **7**, 149–154 (1964)

Fletcher, R.: Practical methods of optimization. Wiley, New York, NY (1987)

Frankel, A., Clayton, R. W.: A finite-difference simulation of wave propagation in two-dimensional random media. Bull. Seism. Soc. Am. **74**, 2167–2186 (1984)

Frankel, A., Clayton, R. W.: Finite-difference simulations of seismic scattering: implications for the propagation of short-period seismic waves in the crust and models of crustal heterogeneity. J. Geophys. Res. **91**, 6465–6489 (1986)

Frankel, A., Vidale, J.: A 3-dimensional simulation of seismic waves in the Santa Clara valley, California, from a Loma Prieta aftershock. Bull. Seism. Soc. Am. **82**, 2045–2074 (1992)

Friederich, W., Dalkolmo, J.: Complete synthetic seismograms for a spherically symmetric Earth by a numerical computation of the Green's function in the frequency domain. Geophys. J. Int. **122**, 537–550 (1995)

Friederich, W.: Propagation of seismic shear and surface waves in a laterally heterogeneous mantle by multiple forward scattering. Geophys. J. Int. **136**, 180–204 (1999)

Friederich, W.: The S-velocity structure of the East Asian mantle from inversion of shear and surface waveforms. Geophys. J. Int. **153**, 88–102 (2003)

Furumura, T., Kennett, B. L. N., Furumura, M.: Seismic wavefield calculation for laterally heterogeneous whole Earth models using the pseudospectral method. Geophys. J. Int. **135**, 845–860 (1998)

Furumura, T., Chen, L.: Large-scale parallel simulation and visualisation of 3D seismic wave field using the Earth Simulator. Comput. Model. Eng. Sci. **6**, 153–168 (2004)

Furumura, T., Kennett, B. L. N.: Subduction zone guided waves and the heterogeneity structure of the subducted plate: intensity anomalies in northern Japan. J. Geophys. Res. **110**, doi:10.1029/2004JB003486 (2005)

Gao, F., Levander, A., Pratt, R. G., Zelt, C. A., Fradelizio, G. L.: Waveform tomography at a groundwater contamination site: surface reflection data. Geophysics **72**, G45–G55 (2007)

Gauthier, O., Virieux, J., Tarantola, A.: Two-dimensional nonlinear inversion of seismic waveforms: numerical results. Geophysics **51**(7), 1387–1403 (1986)

Gazdag, J.: Wave equation migration with the phase shift method. Geophysics **43**, 1342–1351.

Gee, L. S., Jordan, T. H.: Generalized seismological data functionals. Geophys. J. Int. **111**, 363–390 (1992)

Geller, R. J., Ohminato, T.: Computation of synthetic seismograms and their partial derivatives for heterogeneous media with arbitrary natural boundary conditions using the direct solution method. Geophys. J. Int. **116**, 421–446 (1994)

Geller, R. J., Takeuchi, N.: A new method for computing highly accurate DSM synthetic seismograms. Geophys. J. Int. **123**, 449–470 (1995)

Geller, R. J., Takeuchi, N.: Optimally accurate second-order time-domain finite difference schemes for the elastic equation of motion: one-dimensional case. Geophys. J. Int. **135**, 48–62 (1998)

Gill, P. E., Murray, W., Wright, M. H.: Practical optimization. Academic Press, London (1981)

Givoli, D.: High-order local non-reflecting boundary conditions: a review. Wave Motion **39**, 319–326 (2003)

Goncz, J. H., Cleary, J. R.: Variations in the structure of the upper mantle beneath Australia from Rayleigh wave observations. Geophys. J. R. Astron. Soc. **44**, 507–516 (1976)

Gottschämmer, E., Olsen, K. B.: Accuracy of the explicit planar free-surface boundary condition implemented in a fourth-order staggered-grid velocity-stress finite-difference scheme. Bull. Seism. Soc. Am. **91**, 617–623 (2001)

Graves, R. W.: Simulating seismic wave propagation in 3D elastic media using staggered-grid finite differences. Bull. Seism. Soc. Am. **86**, 1091–1106 (1996)

Griewank, A., Walther, A.: Revolve: an implementation of checkpointing for the reverse or adjoint mode of computational differentiation. Trans. Math. Softw. **26**, 19–45 (2000)

Gudmundsson, O., Kennett, B. L. N., Goody, A.: Broadband observations of upper-mantle seismic phases in northern Australia and the attenuation structure in the upper mantle. Phys. Earth Planet. Int. **84**, 207–226 (1994)

Guillot, L., Capdeville, Y., Marigo, J.-J.: 2-D non-periodic homogenisation of the elastic wave equation – SH case. Geophys. J. Int., **182**, 1438–1454 (2010)

Gutenberg, B.: Über die Konstitution des Erdinnern, erschlossen aus Erdbebenbeobachtungen. Physikalische Zeitschrift **14**, 1217–1218 (1913)

Gutenberg, B.: Untersuchungen zur Frage bis zu welcher Tiefe die Erde kristallin ist. Zeitschrift für Geophysik, **2**, 24–29 (1926)

Gutenberg, B.: Seismological evidence for roots of mountains. Bull. Geol. Soc. Am. **54**, 473–498 (1943)

Hestenes, M. R., Stiefel, E. L.: Methods of conjugate gradients for solving linear systems. J. Res. Natl. Bur. Stand. **49**, 409–436 (1952)

Hesthaven, J. S.: From electrostatics to almost optimal nodal sets for polynomial interpolation in a simplex. SIAM J. Numer. Anal. **35**(2), 655–676 (1998)

Hestholm, S., Ketcham, S., Greenfield, R., Moran, M., McMechan, G.: Quick and accurate parameterization in viscoelastic wave modeling. Geophysics **71**, 147–150 (2006)

Hicks, G. J., Pratt, R. G.: Reflection waveform inversion using local descent methods: estimating attenuation and velocity over a gas-sand deposit. Geophysics **66**, 598–612.

Hicks, G. J.: Arbitrary source and receiver positioning in finite-difference schemes using Kaiser windowed sinc functions. Geophysics **67**, 156–166 (2002)

Higdon, R. L.: Radiation boundary conditions for elastic wave propagation. SIAM J. Numer. Anal. **27**, 831–870 (1990)

Higdon, R. L.: Absorbing boundary conditions for elastic waves. Geophysics **56**, 231–241 (1991)

Holberg, O.: Computational aspects of the choice of operator and sampling interval for numerical differentiation in large-scale simulation of wave phenomena. Geophys. Prosp. **35**, 629–655 (1987)

Hole, J. A., Ryberg, T., Fuis, G. S., Bleibinhaus, F., Sharma, A. K.: Structure of the San Andreas fault zone at SAFOD from a seismic refraction survey. Geophys. Res. Lett. **33**, L07312 (2006)
Hu, F. Q.: On absorbing boundary conditions for linearized Euler equations by a perfectly matched layer. J. Comput. Phys. **129**, 201–219 (1996)
Hung, S. H., Dahlen, F. A., Nolet, G.: Fréchet kernels for finite-frequency traveltimes – II. Examples. Geophys. J. Int. **141**, 175–203 (2000)
Igel, H., Mora, P., Riollet, B.: Anisotropic wave propagation through finite-difference grids. Geophysics **60**, 1203–1216 (1995)
Igel, H., Djikpéssé, H., Tarantola, A.: Waveform inversion of marine reflection seismograms for P impedance and Poisson's ratio. Geophys. J. Int. **124**(2), 363–371 (1996)
Igel, H.: Wave propagation in three-dimensional spherical sections by the Chebyshev spectral method. Geophys. J. Int. **136**, 559–566 (1999)
Igel, H., Nissen-Meyer, T., Jahnke, G.: Wave propagation in 3D spherical sections. Effects of subduction zones. Phys. Earth Planet. Int. **132**, 219–234 (2002)
Ikelle, L., Diet, J. P., Tarantola, A.: Linearized inversion of multi-offset seismic reflection data in the $\omega - k$ domain: depth-dependent reference medium. Geophysics **53**, 50–64 (1988)
Jackson, D. D.: Attenuation of seismic waves by grain boundary relaxation. Proc. Natl Acad. Sci. **68**, 1577–1579 (1971)
Jackson, I.: Laboratory measurements of seismic wave dispersion and attenuation: recent progress. In: Earth's Deep Interior: Mineral Physics and Tomography from the Atomic to the Global Scale, Geophys. Monogr. **117**, 265–289 (2000)
Jaiswal, P., Zelt, C. A., Dasgupta, R., Nath, K. K.: Seismic imaging of the Naga thrust using multiscale waveform inversion. Geophysics **74**, WCC129–WCC140 (2009)
Jeffreys, H.: The rigidity of the Earth's central core. Mon. Not. R. Astron. Soc. Geophys. Suppl. **1**, 371–383 (1926)
Jeffreys, H., Bullen, K.: Seismological Tables. British Association for the Advancement of Science, London (1940)
Jo, C.-H., Shin, C., Suh, J. H.: An optimal 9-point, finite-difference, frequency-space 2-D scalar wave extrapolator. Geophysics **61**, 529–537 (1996)
Jordan, T. H.: Composition and development of the continental tectosphere. Nature **274**, 544–548 (1978)
Kaiho, Y., Kennett, B. L. N.: Three-dimensional structure beneath the Australasian region from refracted wave observations. Geophys. J. Int. **142**, 651–668 (2000)
Kane, C., Marsden, J. Ortiz, M., West, M.: Variational integrators and the Newmark algorithm for conservative and dissipative mechanical systems. Int. J. Numer. Methods Eng. **49**, 1295–1325 (2003)
Karniadakis, G. M., Sherwin, S.: Spectral/hp Element Methods for Computational Fluid Dynamics, 2nd edn. Oxford University Press, Oxford (2005)
Käser, M., Dumbser, M., de la Puente, J., Igel, H.: An arbitrary high-order discontinuous Galerkin method for elastic waves on unstructured meshes – III. Viscoelastic attenuation. Geophys. J. Int. **168**, 224–242 (2007)
Kawai, K., Takeuchi, N., Geller, R. J.: Complete synthetic seismograms up to 2 Hz for transversely isotropic spherically symmetric media. Geophys. J. Int. **164**, 411–424 (2006)
Kawai, K., Geller, R. J.: Waveform inversion for localised seismic structure and an application to D" structure beneath the Pacific. J. Geophys. Res. **115**, doi:10.1029/2009JB006503 (2010)
Kawakatsu, H., Montagner, J.-P.: Time-reversal seismic-source imaging and moment tensor inversion. Geophys. J. Int. **175**, 686–688 (2008)
Keilis-Borok, V. I., Yanovskaya, T. B.: Inverse problems in seismology. Geophys. J. **13**, 223–234 (1967)
Kelley, C. T.: Iterative methods for optimization. SIAM, Philadelphia, PA (1999)
Kelly, K. R., Ward, R. W., Treitel, S., Alford, R. M.: Synthetic seismograms: a finite-difference approach. Geophysics **41**, 2–27 (1976)
Kennett, B. L. N.: Seismic waves in a stratified half space. Geophys. J. R. Astron. Soc. **57**, 557–583 (1979)

Kennett, B. L. N.: Seismic waves in a stratified half space II. Theoretical seismograms. Geophys. J. R. Astron. Soc. **61**, 1–10 (1980)
Kennett, B. L. N.: Elastic wave propagation is stratified media. Adv. Mech. **21**, 79–167, Academic Press, New York, NY (1981)
Kennett, B. L. N., Engdahl, E. R., Buland, R.: Constraints on seismic velocities in the Earth from traveltimes. Geophys. J. Int. **122**, 108–124 (1995)
Kennett, B. L. N.: On the density distribution within the Earth. Geophys. J. Int. **132**(2), 374–382 (1998)
Kennett, B. L. N.: The seismic wavefield I. Introduction and theoretical development. Cambridge University Press, Cambridge (2001)
Kennett, B. L. N., Furumura, T.: Stochastic waveguide in the lithosphere: Indonesian subduction zone to Australian craton. Geophys. J. Int. **172**, 363–382 (2008)
Keys, R. G.: Absorbing boundary conditions for acoustic media. Geophysics **50**, 892–902.
Kirkpatrick, S., Gelatt, C. D., Vecchi, M. P.: Optimization by simulated annealing. Science **220**, 671–680 (1983)
Klingelhoefer, F., Lafoy, Y., Collot, J., Cosquer, E., Géli, L., Nouzé, H., Vially, R.: Crustal structure of the basin and ridge system west of New Caledonia (southwest Pacific) from wide-angle and reflection seismic data. J. Geophys. Res. **112**, doi:10.1029/2007JB005093 (2007)
Komatitsch, D.: Méthodes spectrales et éléments spectraux pour l'équation de l'élastodynamique 2D et 3D en milieu hétérogène. Ph.D. thesis, Institut de Physique du Globe de Paris (1997)
Komatitsch, D., Vilotte, J. P.: The spectral-element method: an efficient tool to simulate the seismic response of 2D and 3D geological structures. Bull. Seism. Soc. Am. **88**, 368–392 (1998)
Komatitsch, D., Barnes, C., Tromp, J.: Wave propagation near a fluid-solid interface: a spectral-element approach. Geophysics **65**, 623–631 (2000)
Komatitsch, D., Martin, R., Tromp, J., Taylor, M. A., Wingate, B. A.: Wave propagation in 2D elastic media using a spectral element method with triangles and quadrangles. J. Comput. Acoust. **9**, 703–718 (2001)
Komatisch, D., Tromp, J.: Spectral-element simulations of global seismic wave propagation – I. Validation. Geophys. J. Int. **149**(2), 390–412 (2002)
Komatitsch, D., Tromp, J.: Spectral-element simulations of global seismic wave propagation-II. 3-D models, oceans, rotation and self-gravitation. Geophys. J. Int. **150**, 303–318 (2002)
Komatitsch, D., Tromp, J.: A perfectly matched layer absorbing boundary condition for the second-order seismic wave equation. Geophys. J. Int. **154**, 146–153 (2003)
Komatitsch, D., Liu, Q., Tromp, J., Süss, P., Stidham, C., Shaw, J. H.: Simulation of ground motion in the Los Angeles basin based upon the spectral-element method. Bull. Seism. Soc. Am. **94**, 187–206 (2004)
Komatitsch, D., Tsuboi, S., Tromp, J.: The spectral-element method in seismology. In: Seismic Earth: Array Analysis of Broadband Seismograms, Eds. Levander, A. Nolet, G., AGU Monogr. **157**, 205–228 (2005)
Konishi, K., Kawai, K., Geller, R. J., Fuji, N.: MORB in the lowermost mantle beneath the western Pacific: evidence from waveform inversion. Earth Planet. Sci. Lett. **278**, 219–225 (2009)
Korenaga, J., Holbrook, W., Singh, S., Minshull, T.: Natural gas hydrates on the Southeast U.S. margin: constraints from full waveform and travel time inversions of wide-angle seismic data. J. Geophys. Res. **102**, 15345–15365 (1997)
Kosloff, D. D., Baysal, E.: Forward modelling by a Fourier method. Geophysics **47**, 1402–1412 (1982)
Kosloff, D. D., Kessler, D., Filho, A. Q., Tessmer, E., Behle, A., Strahilevitz, R.: Solution of the equations of dynamic elasticity by a Chebyshev spectral method. Geophysics **55**, 734–748 (1990)
Krebs, J., Anderson, J., Hinkley, D., Neelamani, R., Baumstein, A., Lacasse, M. D., Lee, S.: Fast full-wavefield seismic inversion using encoded sources. Geophysics **74**, WCC177, doi:10.1190/1.3230502 (2009)
Kristek, J., Moczo, P., Archuleta, R. J.: Efficient methods to simulate planar free surface in the 3D 4th-order staggered-grid finite-difference schemes. Stud. Geophys. Geod. **46**, 355–381 (2002)

Kristek, J. Moczo, P.: On the accuracy of the finite-difference schemes: the 1D elastic problem. Bull. Seism. Soc. Am. **96**, 2398–2414 (2006)

Kristek, J., Moczo, P., Galis, M.: A brief summary of some PML formulations and discretizations for the velocity-stress equation of seismic motion. Stud. Geophys. Geod. **53**, 459–474 (2009)

Kristekova, M., Kristek, J., Moczo, P., Day, S. M.: Misfit criteria for quantitative comparison of seismograms. Bull. Seism. Soc. Am. **96**, 1836–1850 (2006)

Kristekova, M., Kristek, J., Moczo, P.: Time-frequency misfit and goodness-of-fit criteria for quantitative comparison of time signals. Geophys. J. Int. **178**, 813–825 (2009)

Lamb, H.: On the propagation of tremors over the surface of an elastic solid. Philos. Trans. R. Soc. Lond. **A203**, 1–42 (1904)

Lambeck, K., Burgess, G., Shaw, R. D.: Teleseismic travel-time anomalies and deep crustal structure in central Australia. Geophys. J. Int. **94**, 105–124 (1988)

Larmat, C., Montagner, J.-P., Fink, M., Capdeville, Y., Tourin, A., Clévédé, E.: Time-reversal imaging of seismic sources and application to the great Sumatra earthquake. Geophys. Res. Lett. **33**, doi:10.1029/2006GL026336 (2006)

Larmat, C., Guyer, R. A., Johnson, P. A.: Tremor source location using time reversal: selecting the appropriate imaging field. Geophys. Res. Lett. **36**, doi:10.1029/2009GL040099 (2009)

Larner, K., Beasley, C.: Cascaded migration: improving the accuracy of finite-difference migration. Geophysics **52**, 618–643 (1987)

Lay, T., Wallace, T. C.: Modern global seismology. Academic Press, San Diego, CA (1995)

Lerner-Lam, A. L., Jordan, T. H.: Earth structure from fundamental and higher-mode waveform analysis. Geophys. J. R. Astron. Soc. **75**, 759–797 (1983)

Levander, A. R.: Fourth-order finite-difference P-SV seismograms. Geophysics **53**, 1425–1436 (1988)

Levenberg, K.: A method for the solution of certain non-linear problems in least-squares. Q. Appl. Math. **2**, 164–168 (1944)

Lévêque, J. J., Rivera, L., Wittlinger, G.: On the use of the checkerboard test to assess the resolution of tomographic inversions. Geophys. J. Int. **115**, 313–318 (1993)

Li, X., Romanowicz, B.: Global mantle shear velocity model developed using nonlinear asymptotic coupling theory. J. Geophys. Res. **101**, 11245–11271 (1996)

Li, Y. G., Vidale, J. E., Cochran, E. S.: Low-velocity structure of the San Andreas fault at Parkfield from fault zone trapped waves. Geophys. Res. Lett. **31**, L12S06 (2004)

Liu, H.-P., Anderson, D. L., Kanamori, H.: Velocity dispersion due to anelasticity; implications for seismology and mantle composition. Geophys. J. R. Astron. Soc. **47** 41–58 (1976)

Liu, Q., Tromp, J.: Finite-frequency kernels based on adjoint methods. Bull. Seism. Soc. Am. **96**, 2383–2397 (2006)

Liu, Q., Tromp, J.: Finite-frequency sensitivity kernels for global seismic wave propagation based upon adjoint methods. Geophys. J. Int. **174**, 265–286 (2008)

Lokmer, I., O'Brien, G. S., Stich, D., Bean, C. J.: Time reversal imaging of synthetic volcanic tremor sources. Geophys. Res. Lett. **36**, doi:10.1029/2009GL038178 (2009)

Luo, Y., Schuster, G. T.: Wave-equation traveltime inversion. Geophysics **56**, 645–653 (1991)

Lysmer, J., Drake, L. A.: A finite-element method for seismology. Methods Comput. Phys. **11**, Academic Press, New York, NY (1972)

Madariaga, R.: Dynamics of an expanding circular fault. Bull. Seism. Soc. Am. **67**, 163–182 (1976)

Maday, Y., Patera, A. T.: Spectral element methods for the incompressible Navier-Stokes equations. In: Noor, A. K., Oden, J. T. (eds.) State of the Art Survey in Computational Mechanics, pp. 71–143.ASME, New York, NY (1989)

Mahrer, K. D.: An empirical study of instability and improvement of absorbing boundary conditions for the elastic wave equation. Geophysics **51**, 1499–1501 (1986)

Mahrer, K. D.: Numerical time step instability and Stacey's and Clayton-Engquist's absorbing boundary conditions. Bull. Seism. Soc. Am. **80**, 213–217 (1990)

Malin, P., Shalev, E., Balven, H., Lewis-Kenedi, C.: Structure of the San Andreas fault at SAFOD from P wave tomography and fault-guided wave mapping. Geophys. Res. Lett. **33**, L13314 (2006)

Malinowski, M., Operto, S.: Quantitative imaging of the permo-mesozoic complex and its basement by frequency-domain waveform tomography of wide-aperture seismic data from the Polish basin. Geophys. Prosp. **56**, 805–825 (2008)

Marquardt, D.: An algorithm for least-squares estimation of nonlinear parameters. SIAM J. Appl. Math. **11**, 431–441 (1963)

Marquering, H., Nolet, G., Dahlen, F. A.: Three-dimensional waveform sensitivity kernels. Geophys. J. Int. **132**, 521–534 (1998)

Marquering, H., Dahlen, F. A., Nolet, G.: Three-dimensional sensitivity kernels for finite-frequency traveltime: the banana-doughnut paradox. Geophys. J. Int. **137**, 805–815 (1999)

Maupin, V.: A multiple-scattering scheme for modelling surface wave propagation in isotropic and anisotropic three-dimensional structures. Geophys. J. Int. **146**, 332–348 (2001)

McDougall, I., Duncan, R. A.: Age progressive volcanism in the Tasmantid Seamounts. Earth Planet. Sci. Lett. **89**, 207–220 (1988)

Mégnin, C., Romanowicz, B.: The 3D shear velocity structure of the mantle from the inversion of body, surface and higher-mode waveforms. Geophys. J. Int. **143**, 709–728 (2000)

Meier, U., Curtis, A., Trampert, J.: Global crustal thickness from neural network inversion of surface wave data. Geophys. J. Int. **169**, 706–722 (2007a)

Meier, U., Curtis, A., Trampert, J.: Fully nonlinear inversion of fundamental mode surface waves for a global crustal model. Geophys. Res. Lett. **34**, doi:10.1029/2007GL030989 (2007b)

Meju, M. A.: Regularised extremal bounds analysis (REBA): an approach to quantifying uncertainty in nonlinear geophysical inverse problems. Geophys. Res. Lett. **36**, doi:10:1029/2008GL036407 (2009)

Mercerat, E. D., Vilotte, J. P., Sánchez-Sesma, F. J.: Triangular spectral-element simulation of two-dimensional elastic wave propagation using unstructured triangular grids. Geophys. J. Int. **166**, 679–698 (2006)

Meza-Fajardo, K. C., Papageorgiou, A. S.: A nonconvolutional, split-field, perfectly matched layer for wave propagation in isotropic and anisotropic elastic media: stability analysis. Bull. Seism. Soc. Am. **98**(4), 1811–1836 (2008)

Michell, J.: Conjectures concerning the cause, and observations upon the phenomena of earthquakes; particularly of that great earthquake of the first of November, 1755, which proved so fatal to the city of Lisbon, and whose effects were felt as far as Africa, and more or less throughout almost all Europe. Philos. Trans. R. Soc. Lond. **51**, 566–634 (1760)

Mizutani, H. Geller, R. J., Takeuchi, N.: Comparison of accuracy and efficiency of time-domain schemes for calculating synthetic seismograms. Phys. Earth Planet. Int. **119**, 75–97 (2000)

Moczo, P., Bystrický, Kristek, J., Carcione, J. M., Bouchon, M.: Hybrid modeling of P-SV seismic motion at inhomogeneous viscoelastic topographic structures. Bull. Seism. Soc. Am. **87**, 1305–1323 (1997)

Moczo, P., Kristek, J., Halada, L.: 3D fourth-order staggered-grid finite-difference schemes: stability and grid dispersion. Bull. Seism. Soc. Am. **90**, 587–603 (2000)

Moczo, P., Kristek, J., Vavrycuk, V., Archuleta, R. J., Halada, J.: 3D heterogeneous staggered-grid finite-difference modeling of seismic motion with volume harmonic and arithmetic averaging of elastic moduli. Bull. Seism. Soc. Am. **92**, 3042–3066 (2002)

Moczo, P., Kristek, J., Gális, M.: Simulation of the planar free surface with near-surface lateral discontinuities in the finite-difference modeling of seismic motion. Bull. Seism. Soc. Am. **94**, 760–768 (2004)

Moczo, P., Kristek, J.: On the rheological methods used for the time-domain methods of seismic wave propagation. Geophys. Res. Lett. **32**(1), Art. No. L01306 (2005)

Moczo, P., Kristek, J., Galis, M., Pazak, P., Balazovjech, M.: The finite-difference and finite-element modeling of seismic wave propagation and earthquake motion. Acta Phys. Slovaca **57**(2), 177–406 (2007)

Moczo, P., Kristek, J., Galis, M., Pazak, P.: On accuracy of the finite-difference and finite-element schemes with respect to P-wave to S-wave speed ratio. Geophys. J. Int. **182**, 493–510 (2010)

Mohorovičić, A.: Das Beben vom 8. Oktober 1909. Jahrbuch des meteorologischen Observatoriums in Zagreb für 1909, **4**(1), 1–67 (1910) (translated title)

Montagner, J.-P.: Seismic anisotropy of the Pacific Ocean inferred from long-period surface waves dispersion. Phys. Earth Planet. Int. **38**, 28–50 (1985)

Montagner, J.-P.: Upper mantle low anisotropy channels below the Pacific Plate. Earth Planet. Sci. Lett. **202**, 263–274 (2002)

Montelli, R., Nolet, G., Dahlen, F. A., Masters, G., Engdahl, E. R., Hung, S. H.: Finite-frequency tomography reveals a variety of plumes in the mantle. Science **303**, 338–343 (2004)

Mora, P.: Elastic finite differences with convolutional operators. Stanford Explor. Proj. Rep. **48**, 277–289 (1986)

Mora, P.: Nonlinear two-dimensional elastic inversion of multioffset seismic data. Geophysics **52**, 1211–1228 (1987)

Mora, P.: Elastic wave-field inversion of reflection and transmission data. Geophysics **53**, 750–759 (1988)

Mora, P.: Inversion=migration+tomography. Geophysics **54**, 1575–1586 (1989)

Morelli, A., Dziewonski, A. M.: Body wave traveltimes and a spherically symmetric P- and S-wave velocity model. Geophys. J. Int. **122**, 178–194 (1993)

Muir, F., Dellinger, J., Etgen, J., Nichols, D.: Modeling elastic fields across irregular boundaries. Geophysics **57**, 1189–1193 (1992)

Myers, J. S., Shaw, R. D., Tyler, I. M.: Tectonic evolution of Proterozoic Australia. Tectonics **15**, 1431–1446 (1996)

Nataf, H. C., Ricard, Y.: 3SMAC: an a priori tomographic model of the upper mantle based on geophysical modeling. Phys. Earth Planet. Int. **95**, 101–122 (1995)

Nehrbass, J. W., Lee, J. F., Lee, R., 1996. Stability analysis for perfectly matched layered absorbers. Electromagnetics, **16**(4), 385–397.

Nettles, M., Dziewonski, A. M.: Radially anisotropic shear velocity structure of the upper mantle globally and beneath North America. J. Geophys. Res. **113**, B02303, doi:10.1029/2006JB004819 (2008)

Newmark, N. M.: A method of computation for structural dynamics. ASCE J. Eng. Mech. Div. **85**, 67–94 (1959)

Nissen-Meyer, T.: Numerical simulation of 3D seismic wave propagation through subduction zones. Diploma Thesis, Ludwig-Maximilians-Universität München (2001)

Nissen-Meyer, T., Fournier, A., Dahlen, F. A.: A two-dimensional spectral-element method for computing spherical-Earth seismograms – I. Moment-tensor source. Geophys. J. Int. **168**, 1067–1092 (2007a)

Nissen-Meyer, T., Dahlen, F. A., Fournier, A.: Spherical Earth Fréchet sensitivity kernels. Geophys. J. Int. **168**, 1051–1066 (2007b)

Nissen-Meyer, T., Fournier, A., Dahlen, F. A.: A two-dimensional spectral-element method for computing spherical-Earth seismograms – II. Waves in solid-fluid media. Geophys. J. Int. **174**, 873–888 (2008)

Nolet, G., Montelli, R.: Optimal parametrization of tomographic models. Geophys. J. Int. **161**, 1–8 (2005)

Nolet, G.: A Breviary of Seismic Tomography. Cambridge University Press (2008)

Nyman, D. C., Landisman, M.: The display-equalized filter for frequency-time analysis. Bull. Seism. Soc. Am. **67**, 393–404 (1977)

Ohminato, T., Chouet, B.: A free-surface boundary condition for including 3D topography in the finite-difference method. Bull. Seism. Soc. Am. **87**, 494–515 (1997)

Oldham, R. D.: On the propagation of earthquake motion to great distances. Philos. Trans. R. Soc. Lond. **194**, 135–174 (1900)

Oldham, R. D.: The constitution of the interior of the Earth as revealed by earthquakes. Q. J. Geol. Soc. Lond. **62**, 456–475 (1906)

Operto, S., Virieux, J., Dessa, J. X., Pascal, G.: Crustal seismic imaging from multifold ocean bottom seismometer data by frequency domain full waveform tomography: application to the eastern Nankai trough. J. Geophys. Res. **111**, doi:10.1029/2005JB003835 (2006)

Pan, G. S., Phinney, R. A., Odom, R. I.: Full-waveform inversion of plane-wave seismograms in stratified acoustic media: theory and feasibility. Geophysics **53**, 21–31 (1988)

Patera, A. T.: A spectral element method for fluid dynamics: laminar flow in a channel expansion. J. Comput. Phys. **54**, 468–488 (1984)

Pica, A., Diet, J. P., Tarantola, A.: Non-linear inversion of seismic reflection data in a laterally invariant medium. Geophysics **55**, 284–292 (1990)

Polak, E., Ribière, G.: Note sur la convergence de la méthode de directions conjuguées. Revue Fr. Inf. Rech. Oper. **16-R1**, 35–43 (1969)

Polak, E.: Optimization. Springer, New York, NY (1997)

Pozrikidis, C.: Finite and spectral element methods using Matlab. Chapman & Hall/CRC, Boca Raton, FL (2005)

Pratt, R. G., Worthington, M. H.: Inverse theory applied to multi-source cross-hole tomography. Part I: acoustic wave equation method. Geophys. Prosp. **38**, 287–310 (1990)

Pratt, R. G., Song, Z. M., Williamson, P., Warner, M.: Two-dimensional velocity models from wide-angle seismic data by wavefield inversion. Geophys. J. Int. **124**, 323–340 (1996)

Pratt, R. G., Shin, C., Hicks, G. J.: Gauss-Newton and full Newton methods in frequency-space seismic waveform inversion. Geophys. J. Int. **133**, 341–362 (1998)

Pratt, R. G.: Seismic waveform inversion in the frequency domain, Part 1: Theory and verification in a physical scale model. Geophysics **64**, 888–901 (1999)

Pratt, R. G., Shipp, R. M.: Seismic waveform inversion in the frequency domain, Part 2: fault delineation in sediments using crosshole data. Geophysics **64**(3), 902–914 (1999)

Press, F.: Earth models obtained by Monte Carlo Inversion. J. Geophys. Res. **73**, 5223–5234 (1968)

Press, W. H., Teukolsky, S. A., Vetterling, W. T., Flannery, B. P.: Numerical Recipes, 3rd edition. Cambridge University Press, Cambridge (2007)

Priolo, E., Carcione, J. M., Seriani, G.: Numerical simulation of interface waves by high-order spectral modeling techniques. J. Acoust. Soc. Am. **95**, 681–693 (1994)

Quarteroni, A., Sacco, R., Saleri, F.: Numerical Mathematics. Springer, New York, NY (2000)

Ravaut, C., Operto, S., Improta, L., Virieux, J., Herrero, A., Dell'Aversana, P.: Multiscale imaging of complex structures from multifold wide-aperture seismic data by frequency-domain full-waveform tomography: application to a thrust belt. Geophys. J. Int. **159**, 1032–1056 (2004)

Ritsema, J., van Heijst, H. J., Woodhouse, J. H.: Complex shear velocity structure imaged beneath Africa and Iceland. Science **286**, 1925–1928 (1999)

Ritsema, J., van Heijst, H. J.: Seismic imaging of structural heterogeneity in Earth's mantle: Evidence for large-scale mantle flow. Sci. Progr. **83**, 243–259 (2000)

Robertsson, J. O. A., Blanch, J. O., Symes, W. W.: Viscoelastic finite-difference modeling. Geophysics **59**(9), 1444–1456 (1994)

Robertsson, J. O. A.: A numerical free-surface condition for elastic/visco-elastic finite-difference modeling in the presence of topography. Geophysics **61**, 1921–1934 (1996)

Romanowicz, B.: The upper mantle degree 2: Constraints and inferences from global mantle wave attenuation measurements. J. Geophys. Res. **95**, 11051–11071 (1990)

Ronchi, C., Ianoco, R., Paolucci, P. S.: The 'Cubed Sphere': a new method for the solution of partial differential equations in spherical geometry. J. Comput. Phys. **124**, 93–114 (1996)

Sadourny, R.: Conservative finite-difference approximations of the primitive equation on quasi-uniform spherical grids. Mon. Weather Rev. **100**, 136–144 (1972)

Sambridge, M., Mosegaard, K.: Monte Carlo methods in geophysical inverse problems. Rev. Geophys. **40**, doi:10.1029/2000RG000089 (2002)

Santosa, F., Symes, W. W.: Computation of the Hessian for least-squares solutions of inverse problems of reflection seismology. Inverse Probl. **4**, 211–233.

Schoenberg, M., Muir, F.: A calculus for finely layered anisotropic media. Geophysics **54**(5), 581–589 (1989)

Schuberth, B. S. A., Bunge, H.-P., Steinle-Neumann, G., Moder, C., Oeser, J.: Thermal versus elastic heterogeneity in high-resolution mantle circulation models with pyrolite composition: High plume excess temperatures in the lowermost mantle. Geochem. Geophys. Geosys. **10**, Q01W01, doi:10.1029/2008GC002235 (2009a)

Schuberth, B. S. A., Bunge, H.-P., Ritsema, J.: Tomographic filtering of high-resolution mantle circulation models: Can seismic heterogeneity be explained by temperature alone? Geochem. Geophys. Geosys. **10**, Q05W03, doi:10.1029/2009GC002401 (2009b)

Seriani, G., Priolo, E., Pregarz, A.: Modelling waves in anisotropic media by a spectral element method. Proceedings of the 3rd International Conference on Mathematical and Numerical Aspects of Wave Propagation, 289–298 (1995).

Seriani, G.: 3-D large-scale wave propagation modeling by a spectral element method on a Cray T3E multiprocessor. Comput. Methods Appl. Mech. Eng. **164**, 235–247 (1998)

Seriani, G., Oliveira, S. P.: Dispersion analysis of spectral element methods for elastic wave propagation. Wave Motion **45**, 729–744 (2008)

Sheng, J. M., Leeds, A., Buddensiek, M., Schuster, G. T.: Early arrival waveform tomography on near-surface refraction data. Geophysics **71**, U47–U57 (2006)

Shibutani, T., Sambridge, M., Kennett, B. L. N.: Genetic algorithm inversion for receiver functions with application to crust and uppermost mantle structure beneath Eastern Australia: Geophys. Res. Lett. **23**, 1829–1832 (1996)

Shin, C., Min, D. J.: Waveform inversion using a logarithmic wave field. Geophysics **71**, R31–R42 (2006)

Shin, C., Cha, Y. H.: Waveform inversion in the Laplace domain. Geophys. J. Int. **173**, 922–931 (2008)

Shito, A., Karato, S.-I., Park, J.: Frequency dependence of Q in Earth's upper mantle inferred from continuous spectra of body waves. Geophys. Res. Lett. **31**, doi:10.1029/2004GL019582 (2004)

Sieminski, A., Lévêque, J.-J., Debayle, E.: Can finite-frequency effects be accounted for in ray theory surface wave tomography? Geophys. Res. Lett. **31**(24), L24614 (2004)

Sieminski, A., Liu, Q., Trampert, J., Tromp, J.: Finite-frequency sensitivity of surface waves to anisotropy based upon adjoint methods. Geophys. J. Int. **168**(3), 1153–1174 (2007a)

Sieminski, A., Liu, Q., Trampert, J., Tromp, J.: Finite-frequency sensitivity of body waves to anisotropy based upon adjoint methods. Geophys. J. Int. **171**(1), 368–389 (2007b)

Sigloch, K., Nolet, G.: Measuring finite-frequency body-wave amplitudes and traveltimes. Geophys. J. Int. **167**, 271–287 (2006)

Sigloch, K., McQuarrie, N., Nolet, G.: Two-stage subduction history under North America inferred from multiple-frequency tomography. Nat. Geosci. **1**, 458–462 (2008)

Simone, A., Hestholm, S.: Instabilities in applying absorbing boundary conditions to high-order seismic modelling algorithms. Geophysics **63**, 1017–1023 (1998)

Simons, F. J., Zielhuis, A., van der Hilst, R. D.: The deep structure of the Australian continent from surface wave tomography. Lithos **48**, 17–43 (1999)

Simons, F. J., van der Hilst, R. D., Montagner, J.-P., Zielhuis, A.: Multimode Rayleigh wave inversion for heterogeneity and azimuthal anisotropy of the Australian upper mantle. Geophys. J. Int. **151**, 738–754 (2002)

Song, Z. M., Williamson, P. R., Pratt, R. G.: Frequency-domain acoustic wave modeling and inversion of crosshole data: Part 2 – inversion method, synthetic experiments and real-data results. Geophysics **60**, 796–809 (1995)

Spetzler, J., Trampert, J., Snieder, R.: Are we exceeding the limits of the great circle approximation in global surface wave tomography? Geophys. Res. Lett. **28**(12), 2341–2344 (2001)

Štekl, I., Pratt, R. G.: Accurate viscoelastic modeling by frequency-domain finite differences using rotated operators. Geophysics **63**, 1779–1794 (1998)

Stich, D., Danecek, P., Morelli, A., Tromp, J.: Imaging lateral heterogeneity in the northern Apennines from time reversal of reflected surface waves. Geophys. J. Int. **177**, 543–554 (2009)

Stolt, R. H.: Migration by Fourier transform. Geophysics **43**, 23–48 (1978)

Strang, G., Nguyen, T.: Wavelets and Filter Banks. Wellesley-Cambridge Press, Cambridge (1996)

Stuart, A., Pavliotis, G.: Multiscale Methods: Averaging and Homogenization. Springer, New York (2007)

Stupazzini, M.: 3D ground motion simulation of the Grenoble valley by GeoELSE. Proceedings of the 3rd International Symposium on the Effects of Surface Geology on Seismic Motion (ESG), Grenoble, France (2006).

Stupazzini, M., Paolucci, R., Igel, H.: Near-fault earthquake ground-motion simulation in the Grenoble valley by a high-performance spectral-element code. Bull. Seism. Soc. Am. **99**, 286–301 (2009).

Szegö, G.: Orthogonal Polynomials, 5th edn. American Mathematical Society, Providence, RI (1975)

Takeuchi, H., Saito, M.: Seismic surface waves. In: Methods in Computational Physics, vol. **11**, pp. 217–295, Academic Press, London (1972)

Takeuchi, N., Geller, R. J.: Optimally accurate second-order time-domain finite difference scheme for computing synthetic seismograms in 2-D and 3-D media. Phys. Earth Planet. Int. **119**, 99–131 (2000)

Takeuchi, N., Geller, R. J., Cummins, P. R.: Complete synthetic seismograms for 3-D heterogeneous Earth models computed using modified DSM operators and their applicability to inversion for Earth structure. Phys. Earth Planet. Int. **119**, 25–36 (2000)

Tape, C., Liu, Q. Y., Maggi, A., Tromp, J.: Adjoint tomography of the southern California crust. Science **325**, 988–992 (2009)

Tape, C., Liu, Q. Y., Maggi, A., Tromp, J.: Seismic tomography of the southern California crust based upon spectral-element and adjoint methods. Geophys. J. Int. **180**, 433–462 (2010)

Tarantola, A., Valette, B.: Generalised nonlinear inverse problems solved using the least squares criterion. Rev. Geophys. **20**, 219–232 (1982)

Tarantola, A.: Inversion of seismic reflection data in the acoustic approximation. Geophysics **49**, 1259–1266 (1984)

Tarantola, A.: Theoretical background for the inversion of seismic waveforms, including elasticity and attenuation. Pure Appl. Geophys. **128**, 365–399 (1988)

Tarantola, A.: Inverse problem theory and methods for model parameter estimation, 2nd edn. Society for Industrial and Applied Mathematics, Philadelphia, PA (2005)

Teixeira, F. L., Chew, W. C.: Systematic derivation of anisotropic PML absorbing media in cylindrical and spherical coordinates. IEEE Microw. Guided Wave Lett. **7**, 371–373 (1997)

Tessmer, E., Kosloff, D. D.: 3D elastic modelling with surface topography by a Chebyshev spectral method. Geophysics **59**, 464–473 (1994)

Thomson, D. J.: Spectrum estimation and harmonic analysis. Proc. IEEE **70**, 1055–1096 (1982)

Thurber, C., Roecker, S., Roberts, K., Gold, M., Powell, L., Rittger, K.: Earthquake locations and three-dimensional fault zone structure along the creeping section of the San Andreas fault near Parkfield, CA: Preparing for SAFOD. Geophys. Res. Lett. **30**, doi:10.1029/2002GL016004 (2003)

Tibuleac, I. M., Nolet, G., Michaelson, C., Koulakov, I.: P wave amplitudes in a 3-D Earth. Geophys. J. Int. **155**, 1–10 (2003)

Tikhonov, A. N.: Solution of incorrectly formulated problems and the regularisation method. Dokl. Akad. Nauk. SSSR **151**, 501–504 (1963)

Toshinawa, T., Ohmachi, T.: Love-wave propagation in a three-dimensional sedimentary basin. Bull. Seism. Soc. Am. **82**, 1661–1677 (1992)

Trampert, J., Spetzler, J.: Surface wave tomography: finite-frequency effects lost in the null space. Geophys. J. Int. **164**(2), 394–400 (2006)

Tromp, J., Tape, C., Lui, Q.: Seismic tomography, adjoint methods, time reversal and banana-doughnut kernels. Geophys. J. Int. **160**, 195–216 (2005)

Udías, A.: Principles of Seismology. Cambridge University Press, Cambridge (1999)

van der Hilst, R. D., Widiyantoro, S., Engdahl, E. R.: Evidence for deep mantle circulation from global tomography. Nature **386**, 578–584 (1997)

van der Hilst, R. D., Kárason, H.: Compositional heterogeneity in the bottom 1000 kilometers of Earth's mantle: Toward a hybrid convection model. Science **283**, 1885–1888 (1999)

van der Hilst, R. D., de Hoop, M. V.: Banana-doughnut kernels and mantle tomography. Geophys. J. Int. **163**(3), 956–961 (2005)

van Leeuwen, T., Mulder, W. A.: A correlation-based misfit criterion for wave-equation traveltime tomography. Geophys. J. Int. **182**, 1383–1394 (2010)

Virieux, J.: SH-wave propagation in heterogeneous media: Velocity-stress finite-difference method. Geophysics **49**, 1933–1957 (1984)

Virieux, J.: P-SV wave propagation in heterogeneous media: Velocity-stress finite-difference method. Geophysics **51**, 889–901 (1986)

Virieux, J., Operto, S.: An overview of full waveform inversion in exploration geophysics. Geophysics **74**, WCC1–WCC26 (2009)

Vogel, C. R., Oman, M. E.: Iterative methods for total variation denoising. SIAM J. Sci. Comput. **17**, 227–238 (1996)

von Rebeur-Paschwitz, E. L. A.: The earthquake of Tokyo 18. April 1889. Nature **40**, 32–32 (1889)

Wang, Y. H., Rao, Y.: Crosshole seismic waveform tomography I.: Strategy for real data application. Geophys. J. Int. **166**, 1224–1236 (2006)

Wang, H. J., Igel, H., Gallovic, F., Cochard, A., Ewald, M.: Source-related variations of ground motions in 3D media: application to the Newport-Inglewood fault, Los Angeles basin. Geophys. J. Int. **175**, 202–214 (2008)

Williamson, P. R.: A guide to the limits of resolution imposed by scattering in ray tomography. Geophysics **56**(2), 202–207 (1991)

Williamson, P. R., Worthington, M. H.: Resolution limits in ray tomography due to wave behavior: Numerical experiments. Geophysics **58**(5), 727–735 (1993)

Woodhouse, J. H., Dahlen, F. A.: Effect of a general aspherical perturbation on free oscillations of the Earth. Geophys. J. R. Astron Soc. **53**, 335–354 (1978)

Woodhouse, J. H.: The joint inversion of seismic waveforms for lateral variations in Earth structure and earthquake source parameters. In: Physics of the Earth's Interior, vol. **85**, pp. 366–397. North Holland, Amsterdam (1983)

Woodhouse, J. H.: The calculation of eigenfrequencies and eigenfunctions of the free oscillations of the earth and the sun. In: Seismological Algorithms, pp. 321–370, Academic Press, London (1988)

Woodward, M. J.: Wave-equation tomography. Geophysics **57**, 15–26 (1992)

Wu, R.-S., Toksöz, M. N.: Diffraction tomography and multisource holography applied to seismic imaging. Geophysics **52**(1), 11–25 (1987)

Yomogida, K.: Fresnel zone inversion for lateral heterogeneities in the Earth. Pure Appl. Geophys. **138**, 391–406 (1992)

Yoshizawa, K., Kennett, B. L. N.: Multimode surface wave tomography for the Australian region using a three-stage approach incorporating finite frequency effects. J. Geophys. Res. **109**, doi:10.1029/2002JB002254 (2004)

Yoshizawa, K., Kennett, B. L. N.: Sensitivity kernels for finite-frequency surface waves. Geophys. J. Int. **162**, 910–926 (2005)

Zahradník, J., Urban, L.: Effect of a simple mountain range on underground seismic motion. Geophys. J. R. Astron. Soc. **79**, 167–183 (1984)

Zahradník, J., Moczo, P., Hron, F.: Testing four elastic finite difference schemes for behaviour at discontinuities. Bull. Seism. Soc. Am. **83**, 107–129 (1993)

Zhang, G. Q.: High-order approximation of one-way wave equations. J. Comput. Math. **3**, 90–97 (1985)

Zhao, L., Jordan, T. H., Chapman, C. H.: Three-dimensional Fréchet differential kernels for seismic delay times. Geophys. J. Int. **141**, 558–576 (2000)

Zhao, L., Jordan, T. H., Olsen, K. B., Chen, P.: Fréchet kernels for imaging regional Earth structure based on three-dimensional reference models. Bull. Seism. Soc. Am. **95**, 2066–2080 (2005)

Zheng, Y., Huang, X.: Anisotropic perfectly matched layers for elastic waves in cartesian and curvilinear coordinates. MIT Earth Resources Laboratory, Consortium Report (2002)

Zhou, C., Cai, W., Luo, Y., Schuster, G., Hassanzadeh, S.: Acoustic wave-equation traveltime and waveform inversion of crosshole seismic data. Geophysics **60**, 765–773 (1995)

Zhou, Y., Dahlen, F. A., Nolet, G.: Three-dimensional sensitivity kernels for surface wave observables. Geophys. J. Int. **158**, 142–168 (2004)

Zhou, Y., Dahlen, F. A., Nolet, G., Laske, G.: Finite-frequency effects in global surface wave tomography. Geophys. J. Int. **163**, 1087–1111 (2005)
Zhou, Y.: Multimode surface wave sensitivity kernels in radially anisotropic Earth media. Geophys. J. Int. **176**, 865–888 (2009)
Zielhuis, A., Nolet, G.: Shear-wave velocity variations in the upper mantle beneath central Europe. Geophys. J. Int. **117**, 695–715 (1994)
Zielhuis, A., van der Hilst, R. D.: Upper-mantle shear velocity beneath eastern Australia from inversion of waveforms from SKIPPY portable arrays. Geophys. J. Int. **127**, 1–16 (1996)
Zoeppritz, K. B.: Über Erdbebenwellen II. Laufzeitkurven. Nachrichten der Königlichen Gesellschaft der Wissenschaften zu Göttingen, Mathematisch-Physikalische Klasse, pp. 529–549 (1907)
Zoeppritz, K. B., Geiger, L.: Über Erdbebenwellen III. Berechnung von Weg und Geschwindigkeit der Vorläufer. Die Poissonsche Konstante im Erdinnern. Nachrichten der Königlichen Gesellschaft der Wissenschaften zu Göttingen, Mathematisch-Physikalische Klasse, pp. 400–428 (1909)
Zoeppritz, K. B., Geiger, L., Gutenberg, B.: Über Erdbebenwellen V. Konstitution des Erdinnern, erschlossen aus dem Bodenverrückungsverhalten der einmal reflektierten zu den direkten longitudinalen Erdbebenwellen, und einige andere Beobachtungen über Erdbebenwellen. Nachrichten der Königlichen Gesellschaft der Wissenschaften zu Göttingen, Mathematisch-Physikalische Klasse, pp. 121–206 (1912)

Index